Research Series on the Chinese Dream and China's Development Path

Drawing on a large body of empirical studies done over the last two decades, this Series provides its readers with in-depth analyses of the past and present and forecasts for the future course of China's development. It contains the latest research results made by members of the Chinese Academy of Social Sciences. This series is an invaluable companion to every researcher who is trying to gain a deeper understanding of the development model, path and experience unique to China. Thanks to the adoption of Socialism with Chinese characteristics, and the implementation of comprehensive reform and opening-up, China has made tremendous achievements in areas such as political reform, economic development, and social construction, and is making great strides towards the realization of the Chinese dream of national rejuvenation. In addition to presenting a detailed account of many of these achievements, the authors also discuss what lessons other countries can learn from China's experience.

Xiaoyan Lin

High-Speed Railways and New Structure of Socio-economic Development in China

With Contributions by Wei Bu, Tianshan Liu, Sichen Liu, Juan Chen, Xiaojun Chen, Yuhui Zhou, Lihua Guo, Xinmei Han, Haoran Pan

SSAP

Springer

Xiaoyan Lin
School of Economics and Management
Beijing Jiaotong University
Beijing, China

Translated by
Yanqing Chen
Yunnan University of Finance
and Economics
Kunming, China

Yuntao Zhu
Yunnan University
Kunming, China

Sponsored by the Chinese Fund for the Humanities and Social Sciences.
国家社科基金重大项目资助 (Grant Number: 17ZDA084)

ISSN 2363-6866 ISSN 2363-6874 (electronic)
Research Series on the Chinese Dream and China's Development Path
ISBN 978-981-19-6386-5 ISBN 978-981-19-6387-2 (eBook)
https://doi.org/10.1007/978-981-19-6387-2

Jointly published with Social Sciences Academic Press
The print edition is not for sale in China (Mainland). Customers from China (Mainland) please order the print book from: Social Sciences Academic Press.

This Springer imprint is published by the registered company Springer Nature Singapore Pte Ltd.
The registered company address is: 152 Beach Road, #21-01/04 Gateway East, Singapore 189721, Singapore

Series Preface

Since China's reform and opening in 1978, the country has come a long way on the path of Socialism with Chinese characteristics, under the leadership of the Communist Party of China. Over 30 years of reform, efforts and sustained spectacular economic growth have turned China into the world's second largest economy and brought many profound changes in the Chinese society. These historically significant developments have been garnering increasing attention from scholars, governments, and the general public alike around the world since the 1990s, when the newest wave of China studies began to gather steam. Some of the hottest topics have included the so-called China miracle, Chinese phenomenon, Chinese experience, Chinese path, and the Chinese model. Homegrown researchers have soon followed suit. Already hugely productive, this vibrant field is putting out a large number of books each year, with Social Sciences Academic Press alone having published hundreds of titles on a wide range of subjects.

Because most of these books have been written and published in Chinese, however, readership has been limited in China—even among many who study China—for whom English is still the lingua franca. This language barrier has been an impediment to efforts by academia, business communities, and policymakers in other countries to form a thorough understanding of contemporary China. What is distinct about China's past and present may mean not only for her future but also for the future of the world. The need to remove such an impediment is both real and urgent, and the *Research Series on the Chinese Dream and China's Development Path* is my answer to the call.

This series features some of the most notable achievements from the last 20 years by scholars in China in a variety of research topics related to reform and opening. They include both theoretical explorations and empirical studies and cover economy, society, politics, law, culture, and ecology, the six areas in which reform and opening policies have had the deepest impact and farthest-reaching consequences for the country. Authors for the series have also tried to articulate their visions of the "Chinese Dream" and how the country can realize it in these fields and beyond.

All of the editors and authors for the *Research Series on the Chinese Dream and China's Development Path* are both longtime students of reform and opening and recognized authorities in their respective academic fields. Their credentials and expertise lend credibility to these books, each of which having been subject to a rigorous peer-review process for inclusion in the series. As part of the Reform and Development Program under the State Administration of Press, Publication, Radio, Film, and Television of the People's Republic of China, the series is published by Springer, a Germany-based academic publisher of international repute, and distributed overseas. I am confident that it will help fill a lacuna in studies of China in the era of reform and opening.

Shouguang Xie

Preface

By the end of December 2012, the total length of high-speed rail lines in operation and under construction world wide was 45,297 km, with more than 20 countries and regions running high-speed railways. At the Second Session of the 12th National People's Congress on March 5, 2014, Premier Li Keqiang pointed out that China's high-speed railway mileage reached 11,000 kilometers, ranking first in the world. As of September 16, 2014, China had 34 high-speed railways in operational service with a total operating mileage of 11,683 km. There are 34 high-speed railways under construction with a total mileage of 14,806 km, and 10 high-speed railway projects under planning, with a total mileage of 3296 km. By 2020, the total length of China's high-speed railway network will be close to 30,000 km. With the increasing number of high-speed railways, the pattern and structure of transportation in China have been undergoing constant change. In the first half of 2014, the number of passengers sent by the railway was 1.11 billion, an annual increase of 9.7%. The number of direct passengers reached 380 million, an increase of 6.2% year-on-year, indicating that the high-speed rails and Multiple Units (EMUs) played an important role in driving passenger growth. According to the statistics, the passenger volume of high-speed rail and EMU increased 31% and 33.3%, respectively, from January to May 2014.

Looking back on the development of China's high-speed railways, after more than 20 years of preparation, construction, and operation in the last ten years, a new network is coming into being, structuring Chinese transportation and shaping Chinese people's travel culture. For such a huge investment in the national development strategy, the significance of high-speed railways to China goes beyond the improvement of transportation capacity, bringing huge impact on both regional and national economic development. From a broader research perspective, the development of high-speed railways has made for tremendous changes in urban and rural structures, the process of urbanization, regional social progress, energy conservation, and ecological environmental protection. Wherefore, the development of high-speed railways has been given more abundant scientific development content. At the same time, in view of the fact that the development of high-speed railways is in line with the government's economic intervention policy in response to the financial crisis in 2008, high-speed railways have played very strategicand political roles in the sustainable growth of Chinese economy, the transformation of the mode of economic development, the realization of inclusive regional economic growth, social harmony, and stability.

This book is an outcome of our enduring effort in careful planning, research, data collection, drafting, proof-reading, and revising. For a long time, the research on the role of transport infrastructure in a country or a region was mainly based on externality theory and some ideas of welfare economics. Quantitative evaluation has been the focus and innovation of a large number of studies. Specifically, their theoretical contexts include: the study of the relationship between transportation and economy, supported by economic theory; the relationship between transportation and society, supported by sociological theory; and the relationship between transportation, resources, and the environment, supported by environmental resources theory with an emphasis on the progress and innovation of econometric models. A prominent feature of existing literature is that they take project evaluation or feasibility study as its main platform and carrier, and seek to apply the above theoretical ideas and measurement models to a certain traffic investment project, calculating and evaluating its economic feasibility. But this book is different in that: (1) it resorts to different theoretical bases. Regional economics and transport economics are the main theoretical foundations. It is one of the characteristics of this book to systematically and comprehensively explore the micro and macro issues in the relationship between high-speed railways and regional economic and social development. (2) It seeks different target heights. This book focuses on the level of development strategy. It is another feature of this book to explore theories and methods without confining the discussion to specific projects, but take high-speed railways as an integral part of national and regional development strategies. (3) It tries to incorporate different overall concepts. Viewed as a relatively independent system, the contribution of high-speed rails to regional socio-economic development is analyzed and evaluated in a more objective and reasonable way. (4) The analysis of the mechanism precedes the measurement results. As the top priority of this book, mechanism analysis will be elaborated more fully to highlight the place and value of high-speed rails in the harmonious development of regional economy.

In writing up this book, we move from something very general to more specific. While depicting what high-speed rails mean to Chinese socio-economic development, we include some detailed case studies both in theory and in practice. Overall, what makes our writing unique lies in the following two aspects. First, theoretically speaking, this book highlights the microscopic theory of the impact of high-speed railways on regional economy and society, that is, the economic factors brought by the high-speed railway and the changes in social development. In addition to promoting the flow of economic elements in its region, the development of high-speed railways is also extensive and far-reaching in its role of radiating and driving social and cultural development. At the macro level, the development of high-speed railways directly boosts the development of the regional economy and promotes the smooth implementation of the national strategic transformation. This will enable the regional trade linkage to form a unified domestic market and break the dual structure of urban and rural areas, so as to truly implement the Scientific Outlook on Development (proposed by the Chinese formal president Hu Jintao) and realize a comprehensive, coordinated, and sustainable development. Second, methodologically speaking, this book does not rigidly adhere to the study of the role of high-speed railways in regional economic and social development. In fact, the relationship between high-speed railways and the regional economy is put in a larger logical framework, which not only shows the logic between the high-speed railway and the development of the regional economy, but also proves the grand logic between the two. The two complement each other and together constitute the method of argumentation in this book.

The full implementation of China's high-speed railway development strategy was born on the occasion of the global financial crisis. It is an important part of China's stable economy and its strategic response to the crisis and greatly benefits strong and sustainable growth of Chinese economy. Practice has initially proved that this major move is scientific, correct, and forward-looking. Moreover, under the current slowdown of world economy and the increasing pressure of Chinese economy, Chinese high-speed rail development serves as not only a sound alternative approach to deal with the complex social economic context in China and abroad, but also provides needed policy support for the Chinese government. Therefore, the construction and operation of high-speed railway and the improvement of the road network may be both impetus for economic development and for the overall betterment of Chinese technology, social wellbeings, people's life standards, and for the protection of natural resources and environment as well.

In this book the role of Chinese high-speed rails in regional socio-economic development is discussed from the following three aspects. The first level is the technical and economic advantages and the optimization of transport structure. On the basis of combining the technical and economic advantages of high-speed railways, it focuses on analyzing the market competitive advantage of high-speed railways and its optimization effect on the regional transportation system. The second level is regional economic development. This book discusses the positive effect and basic role of high-speed railways in regional economies, focusing on high-speed railways driving regional economic growth, optimizing regional economic structures; forming the high-speed railway economic belt, and so on. The third level is regional social progress. This book analyses the huge external benefits (such as land resource utilization, time and space competitiveness, travel habits and transportation service quality, energy conservation, and environmental protection) generated by high-speed railway construction and operation and their significance in promoting regional social progress. Its benign development will be an important part of regulating the harmonious coexistence between human beings and the regional social environment. At the same time, if the corresponding policy support can be given, the comprehensive external benefits of high-speed railways in regional social development can be more effectively brought into play. The book consists of four parts:

Part I: An Overview of the Birth and Development of High-Speed Railways

Chapter 1 Introduction. This chapter explains the source and research background of this book, including the history and current status of high-speed railways in China and around the world with detailed statistics. At the same time, this chapter introduces the research results and the latest research trends regarding of railways' impact on regional economic and social development, clarifying the research scope and logical framework, identifying research questions, and our approach to look at the basic role of the construction and development of high-speed railways in regional socio-economic development. It is evident that the construction of China's high-speed railways has shown an unprecedented climax since 2008. The high-speed railways has not only continued to be viewed as a high-quality basic system to support high-level economic development, but also a means for the government to regulate and control the economy during the economic downturn.

Part II: The Regional Fundamental Role of High-Speed Railways: The Rationale for the Formation and Its Reality

Part II consists of four chapters. Chapter 2 is Mechanism of the Fundamental Role of High-Speed Railway on Regional Development, Chap. 3 is Technical and Economic Characteristics of High-Speed Railway. Chapter 4 is Time Benefits of High-Speed Railways (HSR) and Calculation of Its Time and Space Competitiveness (TSC), and Chap. 5 is High-Speed Railway and Regional Traffic Structure. The above four chapters focus on the role that high-speed railways play in reshaping space and time due to its unique technical and economic properties. Chapters 2 and 3 are mainly concerned with theories of economy, while Chaps. 4 and 5 more on calculations and methods.

Chapter 2 Mechanism of the Fundamental Role of High-Speed Railway on Regional Development. This chapter provides the theoretical basis for the following chapters. This section identifies the theoretical foundations used in the study, looks at the construction and operation of high-speed railways in the context of regional economic and social development, and sorts out the intrinsic links between high-speed railways and regional socio-economic development. The high-speed railways will be singled out from the transportation economy and regional economic theory, and then be reconstructed into the theoretical system of regional economic and social development from another perspective. It is found that the relationship between the transport system and economic development exists in three levels: transport-pulling, economic-push and mutual interaction. The relationship between high-speed railways and the economy is more of an interaction and interdependence, especially when both the economy and the transport system are maturing.

Chapter 3 Technical and Economic Characteristics of High-Speed Railway. This section focuses on the characteristics of high-speed rail transport relative to other modes of transport from its inception, analysing the techno-economic characteristics of high-speed rail, i.e. the elasticity of speed, the complexity of costs, the rigidity of environmental impacts, and the derivative and cross-border nature of demand. The comparative advantages of high-speed railways are also presented on the basis of comparison with other modes of transport, i.e. the capacity of high-speed railways, the efficiency of land resource use in the regions along the route, energy saving, emission reduction, travel time, safety and comfort, etc. One of the key areas of focus is the cross-bordering or inter-regional characteristics of high-speed rail.

Chapter 4 Time Benefits of High-Speed Railways (HSR) and Calculation of Its Time and Space Competitiveness (TSC). Because of its remodeling of time and space, high-speed railways can rapidly improve regional accessibility, achieve considerable time benefits, and enhance regional space-time competitiveness. This is the fundamental mechanism for high-speed railways to play a fundamental role in promoting regional economic and social development. The construction and operation of the high-speed railways can fundamentally shorten the travel time between the regional node cities and improve the accessibility of the area. This chapter will focus on establishing a set of indicators for measuring the time and space competitiveness of high-speed railways, which will be used to evaluate the effect of high-speed railways on the time-space competitiveness of areas along the line. Evidence has it that when the high-speed railway is opened, it not only reduces the travel time between cities along the route, but also increases the accessibility of cities along the route to different degrees, thus improving the space-time competitiveness of the node cities along the route.

Chapter 5 High-Speed Railway and Regional Traffic Structure. As a new mode of transportation, high-speed railways will inevitably bring about great influence when it is introduced into the original regional transportation system. Of course, soft factors such as market impact, the completeness of the original transportation system, and the embedded system of the high-speed rail will exert their influence. In order to verify these effects, this chapter puts forward the influence mechanism of high-speed railways in the regional transportation system by analyzing some cases such as the Shinkansen in Japan, high-speed railways in Taiwan (China), and also those in mainland Hainan Province: that is, competition before coordination. No matter when the high-speed railways enters the regional transportation system, it will help to optimize the original traffic pattern in the long term, but its evolution path and influence mechanism may be varied in different transportation market structures.

Part III: Performance of the Regional Basic Functions of High-Speed Railways

The basic role of high-speed railways in modern regional economic and social development will be discussed in Chaps. 6–9, namely, the High-Speed Railway and Regional Economic Aggregate (Chap. 6), High-Speed Rail Economic Belt and the Improvement of Regional Economic Structure (Chap. 7), High-Speed Railway and Regional Land Development (Chap. 8), and High Speed Railways and Regional Service Quality (Chap. 9).

Chapter 6 High-Speed Railway and Regional Economic Aggregate. The investment itself is economic growth. High-speed railway construction is a transportation infrastructure investment, which may directly affect economic growth through demand pull and capital accumulation in the short term, and also have long-term effects. This chapter takes high-speed railways as the axis of economic growth in the region and quantitatively measures its impact on the total economic growth along the line by analyzing its guiding role in the investment, labor, and other factors of production in the regions. It is found that high-speed railways, as an important transportation infrastructure, have a fundamental role in the construction and operation of regional economic aggregates along the line. Ever since the planning stage, they have been deeply affecting the development activities around the site.

Chapter 7 High-Speed Rail Economic Belt and the Improvement of Regional Economic Structure. The theory of economic belt has made progress because of the unique technical and economic characteristics of high-speed railways, especially its role in restructuring time and space. Based on the analysis of the impact of high-speed railways on regional economic growth and structural adjustment, this chapter puts forward the formation and development mechanism of the high-speed railway economic belt. Through the analysis of the Guangxi high-speed railways and other classical cases of different regions, it summarizes the impact of high-speed railway economic belt construction on regional economic development. It is crucial that the various entities concerned should grasp the characteristics of the high-speed railway economic belt and create good transportation conditions and material basis for the transformation of regional economic development. The construction and operation of high-speed railways not only drives the growth of regional economic aggregate along the line, stimulate the upgrading of relevant industrial structure, and produce obvious industrial agglomeration effect, providing rich elements or resources for further adjustment and optimization of regional economic structures.

Chapter 8 High-Speed Railway and Regional Land Development. This chapter considers that the corridor effects of rail transit on land quantity and land type are equally applicable in the field of high-speed railways. The high-speed railway affects the nature of the land and the inherent law of optimal allocation of land resources. In addition, the high-speed railway construction has a far-reaching impact on the nature of land use, the intensity of land development, and the spatial form of regional land use, furthering the high-density development of land along the track. Such a land development and utilisation model makes for a higher land use value, forming a win-win situation for different sectors concerned, and is in line with the national policy of conserving, intensifying and improving land use.

Chapter 9 High Speed Railways and Regional Service Quality. Transportation service quality is an important part of regional service quality. High-speed railways can rapidly improve the quality of transportation service in the region. On the basis of an extensive literature review, this chapter mainly discusses the high-speed railway passenger service quality evaluation system and its control technology, and puts forward a set of high-speed railway passenger service quality evaluation systems and a model of service quality control selection area. It is hoped that we can study the service quality of high-speed railways from the comprehensive perspective of evaluation and control, and provide useful references for improving the service quality management of high-speed railways. It is found that the choice of the standard of high-speed rail service quality often depends on the economic quality of the region through which it passes. Therefore, the control of high-speed rail service quality should be based on the principle of coordinating with regional economic quality.

Part IV: Strategic Options for More Emphasis on the Significant Role of High-Speed Railway in Regional Development

High-speed railways is an important part of transportation. Transportation is the subsystem of the region, and the region is the macro subsystem. The role of high-speed railways in regional development, regardless of their size or sustainability, is subject to two dimensions. First, on the micro level, the high-speed railways itself has a good organizational management system and a correct development model. Second, on the macro level, high-speed railways play an important part and unique role in the national macro strategy. Only by combining the two can the coordinated development of high-speed railways with the main bodies of transportation, industry, city, region, and country be formed. Chapter 10 (The Development Models and Policies of High-Speed Railway), and Chap. 11 (High-Speed Railway and National Macro-strategy) correspond to the above two dimensions, respectively.

Chapter 10 The Development Models and Policies of High-Speed Railway. As the backbone transportation mode in the transportation system, high-speed railways must provide high-quality transportation services to the society under the dual constraints of existing technology and funds. It is necessary to carefully select its development model to meet the requirements of economic and social development for people and goods, and realize the "balance between supply and demand" between the economy and the social system. This chapter positions the future development model of China's high-speed railways from the dual perspective of industry and enterprise development. That is to say, from the perspective of industrial development, it is necessary to follow the development model of circular economy and pay more attention to achieving sustainable development from a strategic perspective. At the same time, it is also necessary to attach importance to the branding of high-speed railways as transportation enterprises to enhance their market competitiveness, and to regard them as the basic guarantee for China's high-speed railway enterprises that will occupy a place in the future international and domestic market competition.

Chapter 11 High-Speed Railway and National Macro-strategy. The development of the world's high-speed railways has its own unique and profound economic and social background. Countries undergoing economic growth have regarded high-speed railways as the engines for economic development, since they are mostly located in areas that are economically developed and where there is a relatively large population, and strong demand for travel. In countries with economic maturity, high-speed railways is an important means of realizing the transformation of technology and economic growth mode that focuses on technology investment. At the same time, the high-speed railways has become the supporting force for the balanced development of the economy, which vigorously turns to the exportation of technology and the high-speed rail strategy based on land development. When environmental issues become a solid constraint on global sustainability, high-speed railways tends to be constructed as an important starting point for energy conservation and emission reduction as well as for building green eco-cities. In this chapter, the development and role of high-speed railways is discussed in the context of the national strategies of high-end urbanisation, technology transfer and energy saving and emission reduction.

In particular, it is important to note that this book examines the high-speed railway transport system as a matter of fact from a more physical and strategic perspective, maintaining a neutral and impartial research position. Similarly, as a mega state-owned enterprise, or state-owned industry, and an important link in the industrial chain due to its own public interest, the relationship between high-speed railways and the state, especially its investment and financing dilemma, the debt risk, which has accumulated over the years due to its role as an important instrument of macroeconomic regulation by the state and the characteristics of the financial system itself, and the possible development of the system of the China Railway Corporation after the separation of the government and enterprises, are also important, but they are is beyond the scope of this book.

It should also be noted that although the book is the result of the post-funding project granted by the National Social Science Fund, the contents of the book contain a number of research results of the writers in recent years. Many of these achievements have been supported by research projects of various government departments and enterprises such as the former Ministry of Railways, the Guangxi Railway Construction Office, the Beijing-Shanghai High-Speed Railway Co., Ltd., and the Beijing Planning Office of Philosophy and Social Sciences. In the course of the above projects, we have received many experts' and scholars' guidance and very good suggestions. In addition, a number of graduate students participated in the research and investigation of relevant chapters, including: Wang Yingping, Chu Shan, Liu Xiuying, Luo Shen, Wang Huiyun, Wu Di, Xu Jianping, Guo Xiaoxiao, Liu Jianchuan, Kuang Zhensheng, Hao Yaping, Zhou Xi, Chen Xiaobin and so on. We would like to give our heartfelt thanks to all of them.

Of course, the high-speed railways has only 50 years of operation history in the world, and it is a new endeavor in China. Such a short practice is undoubtedly far from enough for the establishment of abundant mechanisms and theories. Also, due to the great differences in historical trajectories and backgrounds among countries, whether the experience of those high-speed rail pioneer countries can be used for general theoretical support, remains to be tested by continuous follow-up studies. Therefore, as an explorer of the regional economic theory of high-speed railway, this book is welcomed to be discussed and criticized by experts and scholars from all walks of life. We thank the authors of many references. Finally, we would like to express our sincere gratitude to the authors of all the references. Above all, the views in this book are the responsibility of the authors.

Beijing, China
November 2014

Xiaoyan Lin

Contents

Part I
An Overview of the Birth and Development of High-Speed Railways

Chapter 1
Introduction

1.1 An Overview of the History and the Status Quo of High-Speed Railway Development

As an international and contemporary concept, high-speed railway generally refers to the railway with a maximum speed of more than 200 km per hour (as defined by the International Union of Railways).[1] The initial research and development of high-speed railway can generally be traced back to the beginning of the twentieth century. In order to improve the speed of trains and make the railway adapt to social development, Germany, France, Japan and other countries carried out a large number of theoretical research and experimental work on high-speed trains. On October 27, 1903, Germany pioneered the test record of a speed of 210 km/h, using electric vehicles. On March 28, 1955, France succeeded in driving three passenger cars pulled by two electric locomotives; the test speed reached 331 km/h, setting a new world record for the speed of high-speed railway. Making full use of the high-speed train test experience of Germany, France and other countries, and relying on its own technical force, Japan built the first high-speed railway in the world in 1964–the Hokkaido Shinkansen (from Tokyo to Osaka, 515.4 km in length, 210 km/h) and put it into operation. The Hokkaido Shinkansen won the support and welcome of the government and the public for its advantages of safety, speed, punctuality, comfort, large transport capacity, light environmental pollution, energy and land resources saving. The Shinkansen went into operation in 1964, became profitable in 1966 and recovered all its investment in 1972. See Table 1.1 for the operating mileage of high-speed railways around the world (Qing et al. 2014).

[1] In addition to the International Union of Railways, in May 1985, the United Nations Economic Commission for Europe set the maximum speed of high-speed trains to 300 km/h for passenger trains and 250 km/h for passenger and freight mixed trains. In 1996, the latest definition of high-speed railway by Council Directive 96/48/EC of 23 July 1996 on the interoperability of the trans-European high-speed rail system was any railway running at least 250 km/h on a new high-speed dedicated line.

X. Lin, *High-Speed Railways and New Structure of Socio-economic Development in China*, Research Series on the Chinese Dream and China's Development Path,
https://doi.org/10.1007/978-981-19-6387-2_1

Table 1.1 Operation mileage statistics of high-speed rails (HSR) around the world

No	Country	Operation mileage of HSR (km)	Mileage of HSR under construction (km)	Total mileage of HSR (km)
1	China	9356	12,729	22,085
2	Japan	2388	776	3164
3	France	1872	730	2602
4	Germany	1032	378	1410
5	Spain	2665	1781	4446
6	Italy	1342	92	1434
7	Russia	780	400	1180
8	Taiwan, China	345	–	345
9	Belgium	209	–	209
10	Portugal	–	1057	1057
11	Turkey	447	758	1205
12	South Korea	412	302	714
13	United Kingdom	1574	–	1574
14	Netherlands	120	–	120
15	Switzerland	35	72	107
16	Norway	–	63	63
17	Uzbekistan	344	–	344
18	Austria	322	221	543
19	Poland	224	–	224
20	Bulgaria	81	380	461
21	Greece	10	570	580
22	Denmark	–	60	60
23	Saudi Arabia	–	440	440
24	Sweden	782	82	864
25	Algeria	–	66	66
	Total	24,340	20,957	45,297

Note Statistics in this table are as of December 2012

1.1.1 The Development and Present Situation of High-Speed Rails Around the World

With the operation of the Japanese Shinkansen in 1964, the railway industry that was once considered a "sunset industry" was revived and showed great vitality. This revitalization of the railway industry marks the advent of the second era of railway development.

According to the technical characteristics, scale and scope of high-speed railway, its development can be divided into four stages (see Table 1.2 for details).

Table 1.2 Construction and development of high-speed railway around the world

Phases	Time	Major countries involved
The initial development	1964–1990	Japan, France, Italy, Germany
The expansion	Late 1980s to mid-1990s	Major EU member states
The rapid development	1990's to 2004	Japan, major EU member states, Russia, China's mainland and Taiwan region, South Korea, etc
Intensive development	2004–	Mainland China and more…

The initial stage of high-speed railway construction started from the late 1960s and lasted until the late 1980s. The high-speed railways built and put into operation during this period included Tokaido, Sanyo, Tohoku and Joetsu Shinkansen in Japan, the southeast TGV line in France, the TGV line in the Atlantic, the Rome-Florence line in Italy, and the Hannover-Würzburg high-speed railway in Germany, with a high-speed mileage of 3,198 km. During this time, Japan built the main structure of Shinkansen network. Apart from North America, the most economically and technologically advanced countries such as Japan, France, Italy and Germany have been the main stakeholders of the first high-speed railway construction climax. Due to adoption of new technologies during this period of high-speed railway construction, the competitiveness of the railway has been enhanced; the share of railway passenger transport in the market has rebounded; and the economic benefits have begun to improve. In addition, the emergence of HSR has helped to create greater transportation capacity, balancing regional economic development and other related industries along the line, and save energy and reduce environmental pollution.

The period of expanded development of high-speed railway construction lasted from the late 1980s to the mid-1990s. As the achievements of high-speed rail construction in Japan and France influenced many countries, the late 1980s saw a second surge in national interest and research in high-speed rail construction. For example, in 1991, Sweden launched the X2000 tilting train. In 1992, Spain introduced French and German technology to build the 471 km long Madrid-Seville high-speed railway. In 1994, the Channel Tunnel[2] connected France with the United Kingdom, creating the first international high-speed rail link. The construction of high-speed railways during this period has the following characteristics: (1) The countries that have built high-speed railways have entered the era of planning and construction of high-speed road networks. (2) The construction of high-speed railways does not only

[2] The Channel Tunnel, also called the Chunnel or the Euro Tunnel, is a railway tunnel that lies underneath the water of the English Channel and connects the island of Great Britain with mainland France. The Channel Tunnel, completed in 1994 and officially opened on May 6 of that year, is considered one of the most amazing engineering feats of the twentieth century. For further details, please go to Jennifer Rosenberg's article "How the Channel Tunnel Was Built and Designed" (https://www.sogou.com/link?url=hedJjaC291P8Yl3GwImTpfrZ3-yWJCqUTFrplhcbK1DrzZVEalrclwd3YCn8DT3NE7OLAc8Mtik).

respond to the needs of the railway sector, but also the political needs of interconnected regions. And (3) the call for energy-saving and environmental protection has spurred the development of pollution-free high-speed railways, including high-speed railway networks within and across borders.

The rapid development of high-speed railway construction lasted from the 1990s to the beginning of the twenty-first century (or the year 2004). The development of HSR in this period spread throughout Asia, North America, Oceania and Europe, forming a revival of the railway in the transport sector. Since 1992, Russia, Korea, Taiwan (China), Australia, the United Kingdom, the Netherlands and other countries and regions have successively started constructing new high-speed rail lines. According to some statistics, the Czech Republic, Hungary, Poland, Austria, Greece, Romania and other countries in eastern and central Europe have begun to carry out trunk railway transformation so that they could keep pace with the high-speed railway network planning and construction in other places in Europe. The countries in the third HSR construction climax differed from the other periods in the following ways. One, most countries have drawn up national plans for the construction of high-speed railway at the initial stage of designing a new high-speed rail line. Two, although the construction of high-speed railway requires huge funds, it is recognized by many governments that high-speed railways are of great benefit to the whole country in terms of social benefits, energy conservation and environmental pollution control. Three, high-speed railways have benefited regional exchanges and balanced development. European countries have listed the construction of high-speed railway as a political task, and have called for breaking the shackles of borders in the construction of high-speed railway. Four, the financing of high-speed railways projects has shifted from national public welfare investment to a mix of various financing methods, shaping the new trend of raising construction funds. And five, the technological innovation of high-speed railway has been extended to many other related fields.

The concentrated development period of high-speed railway construction began in 2004, and the development of China's high-speed railway became the focus of this period. At the beginning of the twenty-first century, China had mastered the R&D and production system of independent innovation technology for high-speed railways. Wherefore, China has made rapid progress in high-speed railway development, not only making high-speed railways rapidly networked in China, but also making China play a more important role in the international high-speed railway market through technology export.

Due to the many features and advantages of high-speed railways, traditional rail transport has been revitalised and is flourishing around the world, thus accelerating the pace of high-speed railway modernisation and laying a good foundation for the formation and development of high-speed railway networks around the world.

1.1.1.1 Shinkansen in Japan

The Japanese high-speed railway, represented by the Tokaido Shinkansen, started early. After the Second World War, the Japanese economy recovered rapidly and the government realized the important role of railway transportation in economic and social development. In 1957, the "Japanese National Railway Trunk Line Investigation Committee", composed of experts and scholars, was established to specialize in the construction of new Shinkansen and the development of the railway transportation economy. In 1958, the construction plan of the Tokaido Shinkansen proposed by the committee was approved by the Cabinet. The president of the Japanese National Railways, Shinji Sogō, and the chief engineer, Hideo Ohshima, made a plan to establish a new standard line. In April 1959, the Tokaido Shinkansen was officially built. Prior to this, the "Power Disperse Mode" train designed by the research team under the leadership of the chief engineer in the Shinkansen project reached a speed of 145 km per hour on the narrow track of Tokaido, breaking the world record of running speed on the narrow track; in 1959, the speed increased to 163 km. This provides strong technical support for the Shinkansen to use "Power Disperse Mode" trains. Moreover, the Japanese National Railways first developed an AC power supply technology for use in high-speed rail, which was more efficient than the DC technology widely used in the world at that time.

Shortly before the opening of the 1964 Summer Olympic Games, in Tokyo, the Tokaido Shinkansen in Japan was officially opened to traffic. Its technology and operating speed were the highest in the world at the time, and Japan became the country that succeeded in building the world's first high-speed railway line. With the opening of the Tokaido Shinkansen, the distance between city clusters such as Keihin, Zhongjing[3] and the Hanshin has been greatly shortened. The formation of a 4-h economic circle greatly facilitated the flow of people, capital, technology and information between cities, which has led to rapid economic growth along the route. In the half century since the Tokaido Shinkansen was opened, high-speed railway technology has fully matured, with high operational stability and safety, generating huge social and economic benefits, and to a certain extent contributing to the transformation of the life pattern of the Japanese people and urban development.

With the opening of the Tokaido Shinkansen in 1964, traveling between the two major cities Tokyo and Osaka changed from a two-day trip to several hours. The speed and convenience of high-speed rail has greatly changed Japanese people's business thinking and leisure habits. The wave of high-speed rail construction has started to rise in Japan. Three years after the opening of the Tokaido Shinkansen, in 1967, the Sanyo Shinkansen a westward route began construction. In 1972, Okayama had high-speed rails in operation, and Hakata in 1975. While Japan was witnessing a booming construction of high-speed railway, the government issued the "National Shinkansen Railway Expansion Law." This led to more HSR lines. For example, in 1971, the

---

[3] Zhongjing, (nicknamed Nagoya), is known as "Zhongjing", because it is located between the capital cities of Tokyo and Kyoto).

Tohoku Shinkansen and the Joetsu Shinkansen started to operate in the northeast. In 1974, the Narita Shinkansen[4] was put into construction. And in February 2012, the Hayabusa bullet train was put into operation, and its speed reached the highest speed of the Japanese Shinkansen. The Japanese high-speed railway train technology has also undergone several technological innovations and upgrading, and is one of the leaders in the world. Shinkansen also has a great influence in the world, and it has become a business card of Japan. In the first few years of operation, foreign tourists to Japan, including former Chinese President Deng Xiaoping, former British Premier Margaret Thatcher and other national leaders, would go and spend some time on Shinkansen in Japan. The Japanese Shinkansen has demonstrated and promoted the development of high-speed railways in other countries.

From a general research perspective, the reasons for the great success of high-speed railway development in Japan can be summarised in the following three points: firstly, the sound planning and layout. Japan's first high-speed railway, the Tokaido Shinkansen, was built to solve the problem of linking the two economic centres of Japan at the time, Tokyo and Osaka. Before the advent of high speed rail, the travel time between the two areas was 6.5 h, whereas after the Tokaido Shinkansen was in operation the travel time was reduced to 3 h, achieving immediate time benefits as well as economic benefits. Following the success of the first railway, five more high-speed railways were built and opened in the same way, creating a high-speed railway network centred on large cities such as Tokyo, Osaka and Fukuda, with a radius of 500 kilometres, forming Japan's 'high-speed railway economic belt'.

Second, Japan's high-speed rail continued to maintain technological innovation. While expanding its scale, the technological level of Japan's Shinkansen has also been increasing day by day, and has been a leader in the world. As the Tokaido Shinkansen adopted electric locomotives, standard double track railway, it reached 210 km per hour on opening to the public. However, Japan's high-speed rail technology has not stopped there, but, with each passing day, keeps innovating and leading the world. Japan's train control system was a good case in point. With it, Shinkansen keeps its time accuracy error control within one minute. Moreover, its magnetic levitation technology is also a leader in the world.

Third, the Japanese government has attached great importance to the development of related supporting policies. The revolutionary role of the policy in the development of Japan's high-speed rail can be traced back to the very beginning of the development of the high-speed rail project. At that time, Japan originally had three different plans to optimize the railway system. On the basis of strategic considerations, the Japanese government eventually adopted the third high-speed rail program, which were not favored at that time and were even more risky, and required continued funding and policies for the program. During the first few years after the high-speed railway project was completed, it suffered great losses because the railway operation was managed by Japanese state-owned companies. Therefore, the Japanese government

[4] The Narita Shinkansen is an unfinished Shinkansen from the heart of Tokyo to Narita Airport. Due to the strong opposition to the construction by the local government along the line, and the oil crisis of the 1970s and other factors, construction was stopped in the process of purchasing land.

reformed the management by splitting the state-owned high-speed railway company into seven private-owned companies and introducing private capital to the running of the HSR project, which cleared the way for the development of Japan's high-speed railway.

1.1.1.2 TGV in France

In 1955, France, one of the leading countries in the world to study how to speed up trains, created a world record of 331 km/h by using an electric locomotive. In order to solve the problem of rail transport saturation and pursue economic benefits, in 1976 France began building the famous Southeast High-speed Railway of TGV (short for train à grande vitesse).[5] Since then, the rapid development of the TGV high-speed railway has made major technological breakthroughs and won great social and economic benefits. TGV has thus become a good option for travelers to travel more quickly, conveniently, safely and comfortably.

Since the opening of the TGV, the French high-speed railways have been at the forefront of the development of high-speed railways in the world, particularly in the study of speed targets, with the TGV high-speed trains running at 380 km/h on part of the South-East line in 1981, much faster than conventional railways. In 1990, the TGV set a new record of 515.3 km/h on the Atlantic line. In the same year, the Mediterranean high-speed railway was inaugurated, operating at 350 km/h, and a high-speed double-decker train with a speed of 300 km/h was developed and put into service. France then went on to develop a fourth generation of power-dispersed high-speed trains at 350 km/h.

After more than 40 years of development, France has nearly 2,000 km of high-speed rail in operation. Moreover, France's high-speed railways and existing railways operate in network, rapidly building a system of high-speed railways throughout the country, usually operating at a speed of 300 km/h. According to the order of completion, the French TGV high-speed railway network mainly includes seven basic routes: the Southeast Line (1981), the Atlantic Line (1989), the Northern Line (1993), the Southeast Extension Line (Rhone-Alpes Line) (1994), the Paris Area Tie Line (1994), the Mediterranean Line (2001) and the Eastern Line (2007). In May 1994, the Northern Line, the Southeast Line and the Atlantic Line were linked into a high-speed railway network with the opening up of the outer ring road in the Greater Paris Area. In the meantime, the high-speed rails connecting France to other European countries were completed and in operation. In November 1994, the Eurostar started running between the capitals of France, Britain and Belgium. Three years

[5] TGV, or "train à grande vitesse", is the French high-speed railway system. It is developed by Alstom and the state-owned company SNCF, and is operated by SNCF. TGV trains travel to and from neighboring cities of Paris, including Belgium, Germany, and Switzerland. Railway companies in some countries have purchased TGV trains or technology from France, including the Netherlands, South Korea, Spain, the United Kingdom and the United States. The TGV train is produced by Alstom.

later, the TGV-PBKA (named after the four cities Paris, Brussels, Cologne (Köln in German), and Amsterdam), which connects the French capital Paris, Brussels, Belgium, Cologne, Germany, and Amsterdam, was put into operation. The French high-speed rail has developed rapidly and even surpassed Japan in certain technical and economic measures.

1.1.1.3 ICE in Germany

Germany's high-speed rail, or the intercity high-speed train, is often referred to as ICE (Inter-city Express). The first generation of ICE was built in 1991. The ICE runs at 280 km per hour across Germany and connects Switzerland and Austria. In 1998, Germany made the electric swing ICT by using the tilting technology, which increased the speed and stability of the train curve. Since then, the technology has been continuously upgraded, with ICE-3 reaching 300 km per hour and ICE-5 adopting the world's leading magnetic levitation technology. In addition to the advanced technology, second only to Japan, Germany's HSR development merits reasonable strategic planning. These include (1) the through trains are dispatched every hour; (2) on large platforms where trains will pause, passengers can make transfers within the station; and (3) the "21 Road Network" plan has been formulated and subdivided into long-term strategic planning, short- and medium-term plans for three years and five years respectfully. Whereby not only the road network system has been rationally arranged, but also the rapid development of high-speed railway has been made possible.

Later, Germany's high-speed magnetic levitation technology attracted the attention of the world due to its cooperation with China. On January 5, 2000, the former State Development Planning Commission of the People's Republic of China (SDPC) submitted to the State Council "the Request for Instructions of the SDPC on Approving the Proposal for the Construction of a Demonstration Line of Magnetic Levitation Train from Shanghai Pudong International Airport to Lujiazui" (No. 1102 document of NDPC in 2000), and was approved. The demonstration line starts at the Station of Longyang Road on Metro Line 2 in the west and goes to Pudong International Airport in the east. It passes through villages, towns and units such as Huamu, Kangqiao, Sunqiao, Huangqiao, Huanglou, Chuansha and Airport along the route. The main line is 30 km in length (double lines), and the access section is about 5 km long. Two stations (Longyang Road and Pudong Airport) are scheduled on the whole line. At the northern end of Pudong Airport, there will be a comprehensive maintenance base, an operations control center with corresponding vehicles, line structure, driving power supply system, operation control system, maintenance facilities and operation infrastructure.

According to the reform plan of the Shanghai rail transit investment and financing, and the requirements of the project-based legal person system, in August 2000, the Shanghai Maglev Transportation Development Co., Ltd. was established by the joint effort of seven companies. These included the Shanghai Shentong Group Co., Ltd., the Shenergy (Group) Co., Ltd., the Shanghai International Group Co., Ltd., the Shanghai Baosteel Group Corporation, the Shanghai Automotive Industry (Group), the Shanghai Electric (Group) Corporation, and the Shanghai Pudong Development (Group) Co., Ltd. They jointly invested three billion yuan to set up the new company, which was to implement the construction and operation of the Shanghai Maglev project in an enterprise-oriented and market-oriented manner. Shanghai Maglev Transportation Development Co., Ltd. is mainly engaged in the investment, construction, operation and management of the Shanghai maglev transportation line projects, as well as the comprehensive development along the line and station area. The company is also engaged in various businesses such as locomotive and rolling stock, property leasing and management, commodity management, parking lot services, technical consulting services and the tourism catering services. After 22 months of hard work, on December 31, 2002 the Shanghai Maglev demonstration operation line officially passed the single-line operation test. After the line was put into operation, a special operating company under Shanghai Maglev Transportation Development Co., Ltd. was set up and responsible for the operation so as to meet the reform requirements of "Four Separates" of investment, construction, operation and supervision.

1.2 An Overview on the Development and Prospect of High-Speed Railway in China

By the end of the twentieth century, with the advantages of high-speed railway gradually recognized by the world, experts believe that the construction period of the global high-speed railway network has arrived. Since the first railroad[6] (from Wusong to Shanghai) in China was put into use in 1876, more than 120 years has passed. However, it is a pity that over the past century, compared with other countries, China's railways are far behind in terms of its operating mileage, transportation efficiency, technical standards and equipment quality. Moreover, since the implementation of the policy of reform and opening up in 1978, the rapid development of the national economy has failed to meet the tremendous demand for railway transportation. This has not only made the railway a long-term "bottleneck" industry, but also seriously hindered the development of the national economy. Therefore, since the 1990s, the Chinese government has assembled relevant experts and officially to carry out high-speed railway technology and application feasibility studies. Looking back, the government's decision at that time not only echoed China's national needs, but also the need and choice to rejuvenate China's railway.

[6] In 1874, a number of foreign firms, such as Jardine Bank in Britain, set up Woosung Railway Co. ltd. Approved by Shen Bingcheng (who is the late grandfather of famous late calligrapher Shen Mishi), Shanghai Taotai(similar to the vice governor of a province) at the time, the Songhu Railway project (from Shanghai to Wusong) was started in December of the same year. On July 3, 1876, the section from Shanghai to Shanghai Jiangwan opened, 6 times a day.

Compared with developed countries, the planning and construction of China's high-speed railways started very late, but its development was overwhelming. On October 12, 2003, the T60 train bound for Beijing from Changchun pulled out of Shenyang North Railway Station and started its debut show on the Qinhuangdao-Shenyang High-speed Railway, the first high-speed passenger railway constructed in China. With its operation, some industry insiders predicted at the time that this would be a milestone of China's entry into the era of high-speed railway. In 2004, China introduced the technology of high-speed train with a speed of 200 km/h. On this basis, the "China Star" high-speed train running at a test speed of 250 km/h was designed and manufactured. This is an important step in the development of high-speed railway. By the end of 2008, the Beijing-Tianjin Inter-city Railway, with independent intellectual property rights, officially operated at a speed of 350 km per hour, indicating that China has already mastered the core technologies for high-speed railways with a running speed of more than 300 km per hour. In 2010, the 380 km/h CRH380A was born, a symbol that China has the potential to lead the world in the field of high-speed rail technology.

As of September 2014, 34 high-speed railways were put into operation in China, with a total operating mileage of 11,683 km (Table 1.3), and 26 high-speed railways were under construction, with a total mileage of 14,806 km (Table 1.4).[7]

With the continuous development of China's high-speed railway technology, the rapid increase of commercial operation speed, the saving of travel time, the improvement of travel conditions, the reduction of travel expenses, and the enhancement of the global environmental awareness of the people's survival on the earth, the high-speed railway has shown a strong momentum of vigorous development in China. In addition, China is a typical continental country with a vast territory, with a span of 5,200 km from north to south and 5,400 km from east to west. The average distance between large and medium-sized cities and provincial capitals is between 300 and 1000 km, which means a huge demand for passengers and freight in China's medium and long distance. As a convenient and fast transportation mode with large traffic volume, it has great potential. According to China's Medium-and Long-Term Railway Network Plan (2008 Adjustment) (see Table 1.5 for details), there are 10 high-speed railway projects being scheduled, with a total mileage of 3,296 km. By 2020, the total length of China's railway high-speed lines will be close to 30,000 km. With scientific and well-designed construction and operational organization, HSRs will bring huge business opportunities for regional economic development along the tracks and make for the steady growth of China's national economy. As a late arrival to HSR, it is vital for China to make the best use of the infrastructure characteristics of the high-speed railway to drive its economic growth. However, it is crucial that China vigorously insist on innovating its high-speed railway technology, taking the full advantage of the specific national conditions in the organization and management of HSR projects placing emphasis on new requirement for urbanization and the harmonious development between Chinese railway modernization and economy.

[7] The English names of the lines in Table1.3 and Table 1.4 are mainly from *the full wiki* on line. Here is the link for High-speed rail in China: Wikis http://www.thefullwiki.org/High-speed_rail_in_China.

Table 1.3 Statistics of high-speed railways (HSRs) in operation in China

No.	Line	Length (km)	Designed speed (km/h)	Construction start date	In operation
1	Qinhuangdao-Shenyang PDL[8]	405	200–250	Aug.1999	Oct. 2003
2	Suining-Chongqing Railway	120	200–250	Feb.2003	Apr. 2006
3	Beijing-Tianjin Intercity Railway	120	350	July 2005	Aug. 2008
4	Wuhan-Guangzhou PDL	1068	350	June 2005	Dec. 2009
5	Zhengzhou-Xi'an section of Xuzhou-Lanzhou HSR	485	350	Sept. 2005	Feb. 2010
6	Shanghai-Nanjing Intercity Railway	300	350	Aug. 2008	July 2010
7	Shanghai-Hangzhou Passenger HSR	202	350	Feb. 2009	Oct. 2010
8	Beijing-Shanghai HSR	1318	350	Apr. 2008	June 2011
9	Guangzhou-Hong Kong HSR	105	350	Apr. 2008	Dec. 2011
10	Harbin-Dalian PDL	904	350	Aug. 2007	Dec. 2012
11	Shijiazhuang-Wuhan PDL	841	350	Oct. 2008	Sep. 2012
12	Beijing-Shijiazhuang PDL	281	350	Oct. 2008	Dec.2012
13	Hefei - Nanjing Railway	166	250	June 2005	Apr. 2008
14	Qingdao-Jinan Passenger Railway	170	250	Jan. 2007	Dec. 2008
15	Shijiazhuang-Taiyuan PDL	190	250	June 2005	Apr. 2009
16	Hefei-Wuhan Railway	359	250	Aug. 2005	Apr. 2009
17	Wenzhou-Fuzhou Railway	298	250	Jan. 2005	Sept. 2009
18	Fujian-Xiamen Railway	275	250	Sept. 2005	Dec. 2009
19	Changchun-Jilin Intercity Railway	96	250	May 2007	Dec. 2010
20	Hainan East Ring Intercity Railway (Haikou-Sanya)	308	250	Sept. 2007	Dec. 2012
21	Xiamen-Shenzhen Railway	502	250	Nov. 2007	Jan. 2011
22	Ningbo-Taizhou-Wenzhou Railway (Yongtaiwen Line)	282	200–250	Dec. 2004	Sept. 2009
23	Chengdu-Dujiangyan Intercity Railway	66	200–250	Nov. 2008	May 2010

(continued)

[8] PDL: Passenger Dedicated Line

Table 1.3 (continued)

No.	Line	Length (km)	Designed speed (km/h)	Construction start date	In operation
24	Nanchang-Jiujiang Intercity Railway	92	200–250	June 2007	June 2010
25	Guangzhou-Zhuhai Intercity Railway	142	200–250	Dec. 2005	Dec. 2010
26	Wuhan-Yichang HSR	291	200–250	Sept. 2008	July 2012
27	Tianjin-Qinhuangdao PDL	261	350	Nov. 2008	Dec. 2013
28	Nanjing-Hangzhou PDL	249	350	Dec. 2008	July 2013
29	Hangzhou-Ningbo HSR	152	350	Mar. 2009	July 2013
30	Panjin-Yingkou PDL	90	350	May 2009	Sept. 2013
31	Xi'an-Baoji PDL	138	350	Nov. 2009	Dec. 28,2013
32	Guangxi coastal intercity Railway: Nanning-Qinzhou HSR, Qinzhou-Fangchenggang HSR and Qinzhou-Beibuwan HSR	259	250	Dec. 2009	Dec. 30,2013
33	Chongqing-Lichuan Railway	164	200–250	Dec. 2008	Dec. 28,2013
34	Hangzhou-Changsha PDL	984	350	Dec. 2009	Sept. 16,2014
	Total	11,683	–	–	–

PDL: Passenger Dedicated Line

1.3 The Impact of High-Speed Railway on Regional Economic and Social Development

1.3.1 Some Cases

The history of human transportation is a history in which technological progress in transportation has continuously promoted human development, social progress and economic development. Only when the level of development of transport meets the needs of a certain level of socio-economic development can economic and social development be realized. Regional economic development go hand in hand with transport and transport has always played an important role in the development of regional economies. Since the birth of the railway, it has been a popular mode of transport due to its better comfort, greater transport capacity and higher operating speed.

Table 1.4 Statistics of China's high-speed railway under construction

No.	Line	Length (km)	Designed speed (km/h)	Construction start date	Status (UC: under construction)
1	Bengbu-Fuzhou HSR	941	350	Jan. 2009	UC
2	Changsha-Kunming PDL	1168	350	July 2010	UC
3	Chengdu-Chongqing PDL	308	250–350	Sept. 2010	UC
4	Guiyang-Guangzhou HSR	857	250	Jan. 2008	UC
5	Nanning-Guangzhou Railway	577	250	Nov. 2008	UC
6	Chengdu-Mianyang-Leshan Intercity Railway	323	250	July 2009	UC
7	Lanzhou-Wulumuqi PDL	1776	250	Nov. 2009	UC
8	Datong-Xi'an PDL	678	250	Dec. 2009	UC
9	Jilin-Hunchun PDL	593	250	Jan. 2010	UC
10	Qingdao-Rongcheng Intercity Railway	299	250	Mar. 2010	UC
11	Shenyang - Dandong PDL	207	250	Mar. 2010	UC
12	Lanzhou-Chongqing Railway	824	200–250	Sept. 2008	UC
13	Kunming-Nanning Railway	716	200–250	Dec. 2009	UC
14	Xiangtang-Futian Railway	636	200–250	Nov. 2007	UC
15	Hunan-Guangxi Railway (from Hengyang to Nanning)	724	200–250	Dec. 2008	UC
16	Nanjing-Anqing Railway	257	200–250	Dec. 2008	UC
17	Wuhan-Xiaogan Intercity Railway	62	200–250	Mar. 2009	UC
18	Qingdao-Lianyungang Railway	197	200–250	Jan.2009	UC
19	Dandong-Dalian Railway	292	200–250	Mar. 2010	UC
20	Chengdu-Xi'an Railway	660	250	Nov. 2011	UC
21	Chengdu-Lanzhou PDL	730	200–250	Feb. 2011	UC
22	Baoji-Lanzhou PDL	400	350	Oct. 2012	UC
23	Chongqing-Wanzhou PDL	245	250	Mar. 2013	UC
24	Hainan West Ring High-speed Railway	344	200	Sep. 2013	UC
25	Beijing-Shenyang PDL	705	350	Feb. 2014	UC
26	Hangzhou-Huangshan HSR	287	200–250	July 2014	UC
	Total	14,806	–	–	–

Table 1.5 Statistics of China's high-speed railways under construction schedule

No.	Line	Length (km)	Designed speed (km/h)	Status
1	Yinchuan-Xi'an Railway	600	350	Planned
2	Wuhan-Xianning Intercity Railway	90	300–350	Planned
3	Beijing-Tangshan Intercity Railway	150	350	Planned
4	Chengdu–Guiyang PDL	633	250	Planned
5	Zhengzhou-Xuzhou PDL	362	350	Planned
6	Nanping-Sanping-Longyan Railway	227	300–350	Planned
7	Beijing-Zhangjiakou Intercity Railway	174	200–250	Planned
8	Qingdao-Ganyu Railway	197	200–250	Planned
9	Yuncheng-Sanmenxia Railway	83	200–250	Planned
10	Fuyang-Yingtan-Shantou Railway	780	200–250	Planned
	Total	3296	–	–

Since the 1950s, people's requirements for transportation have been increasing due to the gradual rise of highways, automobile and the aviation industries. Although the railroad once faced great competition, its role in regional economic development has been non-replaceable. For example, in Germany, ICE trains have been in operation since 1971. There are 14 ICE lines covering nearly 6,000 km, connecting over 80 large and medium-sized cities in Germany. In addition, six lines of high-speed train operation network are closely linked to the ICE network and form a huge passenger network. In addition, suburban (inter-city) railways have been interconnected with underground railways and light rail railways, forming a rapid transport system in large cities, and solving the problem of heavy passenger traffic in areas with high population density. In Australia, railway passenger transportation is dominated by urban railways, and its passenger traffic is 50 times that of passengers in a city without railways.

In addition, the suburban railway is an important part of the urban rail transit system, and also the main form of railways participating in the urban rail transit in many other countries. For example, in Japan, the suburban railway in Tokyo is nearly 2,000 km, with a daily passenger flow of 35 million passengers and an annual passenger traffic of about 5 billion passengers. The French suburban railways, which have an annual transportation volume of 540 million passengers, account for 65% of the total passenger traffic of the French national railways, occupying an important position and role in the transportation system.

Since the 16th National Congress of the Communist Party of China held in 2002 put forward the principle of regional coordinated development, great achievements have been made in the development of the western region, the revitalization of the old industrial base in northeast China and the rise of the central region. This has played an important role in narrowing the regional development gap, expanding domestic demand, improving macro-control and building a harmonious society. This effect is mainly reflected in the deepening of the links within and between regions, of which the development of transportation is most important. Through the continuous innovation and promotion of transportation modes, the connection between regions has been increased; the cost of regional development has been reduced; and the efficiency of regional exchanges has been enhanced. After more than ten years of accumulation and exploration in theory and practice, at the beginning of the twenty-first century China's high-speed railway began to enter a stage of rapid development. The opening of the Qinhuangdao-Shenyang Passenger HSR and the Beijing-Tianjin Inter-city Railway in succession marks a breakthrough in the application of high-speed railway technology in China. The construction of the Beijing-Shanghai High-speed Railway has even enabled China to lead the development and transfer of high-speed railway technology.

Investment in high-speed railways is a direct boost to economic growth. The introduction of labor, capital, materials and intermediary services stimulate regional and local economies. However, while the above actions have important policy implications (for example, as part of a package of economic stimulus packages), they are generally short-term in nature and account for a small part of the overall project cycle and project effectiveness. As can be seen from the example of China's high-speed railway driving economic growth, high-speed railways bring more than just the multiplier effect of investment. Due to the high-speed rail's opening, the transportation resources have been effectively integrated, releasing the railway capacity, easing the tension between the long-term transportation capacity and the volume of transportation, and urging the rapid circulation of production factors such as people flow, logistics, capital flow and information flow. Therefore, cities along the high-speed rail lines have been favored by domestic and foreign investors, who have come to inspect projects and invest in building factories. In addition, the development value of some "resource-exhausted" cities has also been reassessed and revitalized.

After learning about the situation of the high-speed railway in China, the media realized that the high-speed railway greatly shortened the space–time distance between cities, between urban and rural areas and between regions. The train journey from Beijing to Tianjin was shortened from 1.5–2 h to 30 min, from 11 h to 3 h in Wuhan to Guangzhou, from 6 h to less than 2 h in Shijiazhuang-Taiyuan and Zhengzhou-Xi'an respectively, from 7 to 8 h in Hefei to Wuhan to about 2 h, from 11 h in Fuzhou to 90 min in Xiamen, and from 73 min in Shanghai to Nanjing. For example, Hefei is 457 km away from Shanghai, 405 km away from Hangzhou and 156 km away from Nanjing. In the past, it took 6–10 h, 5–7 h and 3–4 h to get to these three places by train from Hefei. Due to the lack of high-capacity, high-speed and high-efficiency railway transportation links, on the one hand, the big cities around Hefei are overloaded, on the other hand, Hefei's own development is relatively backward, and the population outflow is greater than the inflow. After the opening of the Hefei-Wuhan high-speed railway, the running time between Hefei and Nanjing was shortened to one hour, and it took three hours to get to Shanghai, forming a "one-to-three-hour traffic circle" with major cities in the Yangtze River Delta, such as Shanghai, Nanjing and Hangzhou. The radiation and driving effects of developed cities in the Yangtze River Delta on Hefei were rapidly revealed, highlighting the exemplary role of Hefei in the development of the "Wanjiang City Belt".

In addition, the opening of the Hefei-Nanjing HSR and Hefei-Wuhan HSR has shortened the journey from Hefei to Wuhan from seven to eight hours in the past to two hours, greatly enhancing the strength of economic ties with the surrounding areas, and its location advantage of "connecting the south to the north, bearing the east to the west". More population flowed into Hefei and more enterprises settled in Hefei. Hefei and Nanjing formed a one-hour shopping district. Many large commercial enterprises like Nanjing Central Emporium (Group) Stocks Co., Ltd., Nanjing Xinjiekou Department Store Co., Ltd., and Golden Eagle Retail Group Limited all rushed to Hefei for investment, while Anhui's outstanding local enterprises were "going out" to expand new markets.

The "Changsha-Zhuzhou-Xiangtan Metropolitan Area" in Hunan Province is located in the middle of Guangzhou and Wuhan, which used to be called "survival in the cracks". After the opening of Wuhan-Guangzhou high-speed railway, it has become the most important transportation corridor connecting Changsha, Zhuzhou and Xiangtan. It takes 24 min from Changsha to Zhuzhou and 25 min from Changsha to Xiangtan. Fast and convenient transportation greatly strengthens the links among the three cities, and the same city effect is amplified again. It's very convenient for Changsha citizens go shopping in Zhuzhou. People in Zhuzhou can also enjoy delicious food in Changsha comfortably. They can also live in Zhuzhou, which is ecologically suitable, and invest and buy houses in Xiangwan, Changde and other places. Hunan Province proposes to further speed up the construction of railway lines and branch lines between peripheral cities of urban agglomeration, and expand the Changsha-Zhuzhou-Xiangtan city circle to the surrounding five cities of Yueyang, Yiyang, Changde, Hengyang and Loudi so as to make it an urban agglomeration and integration of "Changsha-Zhuzhou-Xiangtan 3 + 5".

Zhengzhou, the capital city of Henan Province, is located in the Central Plains and is the national land transportation hub and the junction of two major trunk lines, Beijing-Guangzhou Railway and Lanzhou-Lianyungang Railway. However, for many years, it has been only the node city through which the railways pass. The proportion of passenger and freight traffic in the trunk railway of the whole province is as high as 71% and 85%, but the development in the region is limited. The opening of the Zhengzhou-Xi'an High-speed Railway in 2010 has provided important support for the development of the Central Plains Urban Agglomeration. Using high-speed rail transportation system as the main support, Henan is going to build a province-wide spatial structure with Zhengzhou metropolitan area as the core and linking peripheral cities along major transport corridors. With Zhengzhou as the centre, it will take half an hour to reach eight cities, including Kaifeng, Luoyang, Pingdingshan, Xinxiang, Jiaozuo, Xuchang, Mohe and Jiyuan, and one hour to reach nine surrounding cities, shortening the space-time distance between northern Henan, southwest Henan and the Huang-Huai Area.[9]

High-speed railways also have significant impact on other modes of transportation in the region. In China, especially between regions with 3-h high-speed rail travel, the passenger flow of intercity highway passenger buses and airlines has dropped considerably, resulting in the reduction of frequency, fare, even shutdown and grounding of airplanes. For example, it takes about two hours for one to traverse from Beijing to Tianjin by bus through the Beijing-Tianjin expressway, while it takes only half an hour to go by the Beijing-Tianjin inter-city high-speed railway, resulting in a 77% increase in railway passenger traffic volume (compared with that before the high-speed railway was opened), a 70% reduction in the number of bus trips by road and a 75% loss of bus passengers. It takes 6–8 h to travel by Zhengzhou-Xi'an highway passenger bus, but it takes only 2 h by the Zhengzhou-Xi'an high-speed rails, making the Xi'an-Zhengzhou flight schedules and the highway bus passenger service basically shut down. With the open of Hefei-Wuhan HSR and Hefei-Nanjing HSR, the number of passenger flights on the Nanjing-Wuhan expressways has dropped from more than 30 to 2–3 daily. Passenger traffic on the Hefei-Nanjing expressways has dropped by 58% and that on the Hefei-Shanghai highway by 80%. Since the Wuhan-Guangzhou High-speed Railway was put into operation, Hainan Airlines grounded Changsha-Guangzhou flights, Wuhan-Guangzhou flights provided by China Southern Airlines were reduced from 8 to 5 flights per day, and Changsha-Guangzhou flights from 9 to 4 flights per day. The opening of the Shijiazhuang-Taiyuan High Speed Railway has made the traffic volume of Taiyuan-Beijing Expressway passenger buses drop by 60–70%, and the number of Taiyuan-Beijing round-trip flights drop from more than 20 to 16 flights per day; the passenger load factor of the flight dropped by 10 percentage points, and the passenger load factor for some flights was less than 50%.

High-speed railway has changed the existing transportation market structure. But local governments along the route don't seem to worry about it. Transportation departments generally believe that in the past, due to the slow development of railways and the shortage of capacity, they were unable to meet the needs of passengers and cargo owners. A lot of traffic that should have been borne by the railway had to be completed by other modes of transportation, which consumed a lot of high-cost transportation resources and caused serious road congestion, environmental pollution and traffic accidents. High-speed railways are accelerating the restructuring of transport, enabling all modes of transport to show their comparative advantages and become a modern integrated transport system that is more conducive to saving resources and protecting the environment. Ultimately, it is the broad masses of the people who benefit. Therefore, the opening of high-speed railways has been taken as a rare opportunity to optimise the allocation of transport resources and adjust the transport

[9] Broadly speaking, Huang -Huai Area includes more than 30 prefecture-level cities in Henan, Anhui, Jiangsu and Shandong provinces. In a narrow sense, Huang -Huai area generally includes central Henan and northern Anhui. Major cities include Huai'an, Xuzhou, Lianyungang, Suqian, Bozhou, Fuyang, Suzhou, Huaibei, Bengbu, Shangqiu, Xuchang, Zhoukou, Pingdingshan, Luohe, Nanyang and Zhumadian.

structure, speeding up the implementation of new transport development strategies, and initial results have been achieved.

On the one hand, the relevant highway departments along high-speed rail lines reduce the frequencies of long-distance passenger lines that run parallel to the high-speed railways. On the other hand, the local government has been working hard to make the best of the multi-faceted, flexible and mobile road transportation and actively expand urban suburbs, rural townships, and tourist attractions so that the coverage of road passenger transport will be expanded in the passenger transport market. Hefei, Wuhan, Nanjing, Zhengzhou, Xi'an, Taiyuan and Fuzhou have greatly increased passenger lines to surrounding areas, counties, townships and towns, extending the highway passenger transport network in depth, further narrowing the distance between urban and rural areas and enhancing the radiation range of central cities. In areas where tourism resources along the Zhengzhou-Xi'an and Wuhan-Guangzhou high-speed railways are abundant, high-speed railways have become an important road development strategy that furthers the "combination of transport and tourism". HSR has become an impetus for booming tourism, which has led to the rising of road passenger transport. The long-distance passenger capacity to be released will be transferred to the development of the chartered tour business, while shuttle buses will be directly linked to the high-speed railway to provide tourists with the transportation services that combine shuttle bus with the public rail transport. In some cities such as Fujian, Wuhan, and Hefei, the tour buses may pull into the high-speed railway station where they pick up travellers and then offer them a one-stop direct access to the scenic area. For example, the Yangchun Lake (in Wuhan) Passenger Transfer Center has launched the commercial express services that integrate the advantages of shuttle buses and chartered buses with the characteristics of business passenger flow. The center has put its focus on the miniaturization of modes of transport, rapid transportation, and regional operation, and expanded the traditional point-to-point transportation into one point-to-points transportation, which provide more thoughtful and convenient services for short-distance transfers of high-end passengers.

Facing new competition in the transportation market, highway, civil aviation and railway transportation enterprises have taken many active and effective countermeasures featuring reform and innovation. One strategy has been to reduce ticket prices and attract passenger flow. Fares for highway passenger transportation along the high-speed railway have generally dropped. After the Shijiazhuang-Taiyuan High-speed Railway was opened, the ticket price for road passengers in Taiyuan-Shijiazhuang and Taiyuan-Beijing had dropped by 22%-38%. After the opening of Wuhan-Guangzhou High-speed Railway, in order to win more passengers, China Southern Airlines also opened "Air Express" in Wuhan, Changsha and Guangzhou, with a minimum fare of only RMB 390 yuan. Another strategy has been to improve service and quality. The relevant departments of the high-speed railway have specially hired professional teachers from civil aviation, schools of foreign languages and transportation schools to carry out intensive training of professional skills ranging from service etiquette to emergency handling for the crew working on the high-speed trains. After the opening of Shanghai Nanjing Intercity High-speed Railway, the railway department made three changes in train schedules in one month so as to satisfy the requirements of passengers. Henan Province has actively promoted the "New Trunk Lines for Green Passenger Transport in Central Plains" project throughout the province. This strategy is to guide enterprises to speed up the renewal of gas vehicles and reduce operating costs. In addition, Henan province has been piloting the "one county, one network, one company" corporate transformation of rural passenger transport to improve the intensive management level of enterprises.

The opening of the Shanghai-Hangzhou High-speed Railway did not impact Hangzhou's civil aviation passengers, but greatly shortened the travel time from Hangzhou to Shanghai and vice versa, making it easier for passengers from the two cities to take the high-speed railway to catch the plane. The Shanghai-Hangzhou high-speed railway, for example, has made Shanghai and Hangzhou, which are 150 km apart, closer together, creating a "same city effect".[10] People in Hangzhou may choose to travel from Hangzhou to Shanghai by high-speed rail and take their flights from Shanghai, and Shanghaiers may be visiting Hangzhou for their flight departures. To account for such travelling choices, the price for air tickets has become a key factor. For most travelers, the ticket price and the margin of the offer are usually important considerations for people's decision-making of where to fly. China Southern Airlines has nine flights from Hangzhou and Shanghai to Guangzhou every day. The air distance and flight time between the two places are similar, but the fares are different. The full-price ticket from Hangzhou to Guangzhou is 1050 yuan, while the full-price ticket from Shanghai to Guangzhou is 1280 yuan. From the nominal point of view, we spend much more on 230 yuan from Shanghai. Even if we count the 82 yuan high-speed railway ticket from Shanghai to Hangzhou, we can save 148 yuan from Hangzhou. If we can buy a low discount ticket from Hangzhou, it will be even more cost-effective. In terms of ticket prices, domestic routes from Hangzhou to Beijing and Guangzhou will be cheaper. In terms of flight schedules, more choices will be available if one chooses to fly to Hong Kong and Macao, or Europe and the United States from Shanghai. For example, KLM Royal Dutch Airlines (or KLM for short) has only 3 flights a week from Hangzhou to Amsterdam. By contrast, the number of KLM's flights from Shanghai to Amsterdam and Paris, if combined with its flights from code-sharing with Air France and China Eastern Airlines, can be as many as 30 per week, 10 times that of Hangzhou. Therefore, in the first half of November after the opening of the Shanghai-Hangzhou high-speed railway, the passenger load factor of KLM from Hangzhou to Amsterdam was 82%, slightly higher than before. With that, some passengers may have chosen to fly to Europe from Shanghai.

High-speed railways have released the railway freight capacity to meet the logistics needs of urban development along the routes. According to preliminary statistics, China's high-speed railway currently in operation can make the existing line to get

---

[10] The same-city effect refers to the important effects and linkage effects that occur in adjacent areas or larger areas. The same city effect, which means "the same city", is a new trend of urban modernization development, and is also an inevitable outcome of the exchanges, cooperation and development between cities under the economic globalization.

an increase of fixed cargo trains to 83 pairs, equivalent to an average of 10,200 cars 620,000 tons, an annual increase in cargo transport capacity of 230 million tons. Among them, the Wuhan-Guangzhou high-speed railway has "freed up" 87.6 million tons of freight capacity. From January to August, 2010, the national railway freight transportation volume reached 2,408.38 million tons, up by 12.8% year-on-year, with an over-year progress of 78.24 million tons the opening of Shanghai-Nanjing intercity railway not only relieves the tension of passenger transport in the corridor, but also frees up a large part of the original Shanghai-Nanjing line capacity to ensure the freight demand. The number of fixed freight trains increased by 32 pairs, equivalent to about 230,000 tons of 3,840 cars per day, and the annual freight transport capacity increased by 83.95 million tons.

In addition, Shanghai Railway Administration[11] has timely launched the small-group, high-density, bus-based container sea-rail intermodal train products. Many local transportation companies immediately expressed their strong desire for joint transport with railways. The railway department has now started developing a rich system of freight product to improve freight service quality and meet different levels of market transportation demand. The opening of Wuhan-Guangzhou high-speed railway effectively eases the transport tension on the existing Beijing-Guangzhou line. At present, the existing Beijing–Guangzhou railway can add 33 pairs of freight trains per day, about 3960 cars. The daily average of 1,300 vehicles originally adjusted to Jiaozuo-Liuzhou Railway and Beijing-Kowloon Railway due to Beijing–Guangzhou railway's tight capacity is gradually returning to Beijing–Guangzhou railway. Electricity, coal, oil, iron and steel, grain and other key materials related to the national economy and people's livelihood have achieved greater growth in transportation capacity. In the past, due to the mixed operation of passenger and cargo on the Beijing–Guangzhou railway, almost all freight traffic in Beijing–Guangzhou railway was suspended during the Spring Festival to ensure passenger transport. In the first China's annual Spring Festival travel rush (or "Chunyun") after the high-speed railway was completed in 2012, 316 direct freight trains were put into operation through the existing Beijing–Guangzhou railway, transporting more than 450,000 tons of key materials such as electricity, coal, oil, steel and so on, with freight volume up 23% year that year. The improvement of railway freight transportation capacity has guaranteed electricity and coal transportation in Hunan and Hubei Provinces, and there has been no switch-off and power restriction since then.

As is shown above, although China's high-speed railway has not been in operation for a long time, they have already led to significant changes in the development of cities along their routes. These impacts have been demonstrated in several ways, including improved transport conditions in cities along the route, enhanced inter-city

[11] Shanghai Railway Administration generally refers to China Railway Shanghai Bureau Group Co., Ltd. It is one of the 18 bureaus of the large-scale railway transportation enterprise managed by China Railway Group Co., Ltd., referred to as "Shanghai Bureau". On November 4, 2017, the State Administration for Industry and Commerce of the People's Republic of China (SAIC) announced that the name change of the Shanghai Railway Bureau under the China Railway has been approved. The Shanghai Railway Bureau was renamed China Railway Shanghai Bureau Group Co., Ltd.

links, growth in the total national economy of each city, promotion of regional industries, optimisation of regional transport patterns, conservation of social resources and enhancement of the cities' economic and social development potential.

In recent years, the focus of high-speed railway on regional economic and social development has shifted. The Beijing-Shanghai high-speed railway set up a station in Dezhou City, Shandong Province, thus attracting surrounding towns to be incorporated into the development of the Dezhou high-speed railway economic circle, and resulting in the rapid expansion of the city scale layout in Dezhou. Meanwhile, Ling County, which is adjacent to Dezhou, seized this favorable opportunity by taking advantage of high-speed rail transportation. It took the initiative to incorporate itself into the expansion layout of Dezhou's city scale to enhance its location value of the "same city economy" between Ling County and Dezhou. Along with Dezhou City, Ling County will be committed to greater economic, social and technological advancement through developing "modern urban agriculture" with its characteristics.

In April 2010, the TOHO RAILWAY (TOHO Create High-speed Railway New Technology Development Co., Ltd.) high-speed rail industrial park, planning to invest 5 billion yuan, started construction in the Guanghan Economic Development Zone in Sichuan Province, and was expected to be built into a well-known high-speed rail product manufacturing base. The high-speed railway has become a new opportunity and new theme for the economic development in Guanghan. A large number of high-speed rail-related product research and development projects have settled in the industrial park, greatly accelerating the industrialization process of high-speed rail technology.

Xingyang is adjacent to the Zhengzhou High-Tech Development Zone. The Zhengxi High Speed Railway runs from Henan Xingyang through the city to Zhengzhou, bringing Xingyang within easy reach of Zhengzhou. The city of Xingyang has seized the new opportunities for urban development brought about by the high-speed railway, and is taking a more active and proactive role in accepting the radiation-driven role of Zhengzhou, integrating into the city's urban layout and making Xingyang a new city and "Western Garden" of Zhengzhou. With the advantage of its location near the Zhengzhou Railway Station, Xingyang attracts many companies. Strategic investors such as the top 500 national enterprises, Central Government-led Enterprises, and multinational corporations are here to vigorously cultivate some core backbone enterprises to promote the formation of pillar-leading industries. According to the person in charge of the Zhengzhou New District Management Committee, in the first half of 2013, there were 59 projects signed in Zhengzhou New District, with a total contract value of 52.1 billion yuan, of which 10.6 billion yuan was actually in place. There were 57 new projects worth over 100 million yuan, including 13 projects worth over 1 billion yuan. In addition, investment in fixed assets in cities and towns reached 20.56 billion yuan, up 57.1%., local fiscal general budget revenue 2.13 billion yuan, an increase of 85.4%. And Zhengzhou New District had a total export value of 280 million US dollars, an increase of 105.3%.

The opening of the Wuhan-Guangzhou high-speed railway has made Changsha the central city in the "1-hour economic circle" of Chang-Zhu-Tan. Taking advantage of the passenger flow, logistics and information flow brought by the high-speed railway,

Hunan Province has undertaken more than 1,600 industrial transfer projects, of which 138 projects have an investment of more than US$10 million. Changsha will be built into four 100 billion industry clusters of engineering machinery, automobile industry, food industry and material industry, and the total industrial output value is planned to reach 800 billion yuan in 2012. Hunan Province has implemented 34 new preferential policies involving taxation, industry and commerce, finance, manpower and many other departments for this purpose. The high-speed rail has made Wuhan the central city of China's "four-hour economic circle". To this end, Wuhan has adjusted its industrial structure and re-planned its industrial development around the new opportunities of high-speed rail, including urban rail transport, modern services, manufacturing and textiles, to achieve new development in the high-speed rail era. A large amount of data shows that regions along the high-speed rails have become the most active and potential areas for China's economic development. We have every reason to be optimistic and foresee that high-speed railways will play a huge role in supporting coordinated regional development, optimising resource allocation and industrial layout, building an efficient integrated transport system, reducing social logistics costs and promoting the integration process of towns and cities and sustainable economic development.

Xuzhou, Jiangsu Province is the only prefecture-level city along the Beijing-Shanghai high-speed rail line to have started trains, sending an average of 11,000 passengers daily. The Beijing-Shanghai high-speed railway has brought more people flow, logistics and capital flow to Xuzhou's economic transformation and development. In the first half of 2012, Xuzhou's gross regional product was RMB 196.07 billion, an increase of 13.2% year-on-year, ranking first in the province, with the output value of high-tech industries increasing by 65.3% year-on-year. In October 2012, our research team learned from our field work in the areas along the Beijing-Shanghai high-speed railway that the railway improved the accessibility of Suzhou and its neighboring areas, making commuting between the regions more convenient. In the same year, a large number of graduates chose to work in Suzhou, but live elsewhere. This provided the necessary conditions for the establishment of the science and technology parks along the Beijing-Shanghai high-speed railway.

As can be seen, high-speed railways have brought with them even more intellectual capital that plays a role in the medium and long term, such as the flow of human capital, science and technology parks that rely on high-speed railways and the new major enterprises that have moved in along the routes. As a developing country, after the reform and opening up, China's regional economic development has been continuously upgraded in strategic planning, gradually forming a more reasonable regional division. The economic development in the eastern and central parts of China has reached a considerable level, and the contribution rate of science and technology to economic growth has also increased. However, compared with the 80% contribution rate of science and technology in developed countries, China's level of 20–30% is far from it, and the gap between less developed areas in western China is even greater.

Li Keqiang, Premier of the State Council and leader of the National Science and Technology Education Leading Group,[12] stressed at the first plenary meeting of the national leading group on science and technology education (on August 29, 2013): "education and science and technology should always be placed in an overall and strategic position and be highly valued." To give full play to the role of science and technology as the first productive force, to promote the deep integration of science and technology with the economy and society, and to be oriented towards innovation and entrepreneurship. Scientific and technological innovation is the key to the sustainable development of regional economies, and knowledge spillover is the power source of regional economic development. Thus, research on regional knowledge spillover has become an important basis for China's underdeveloped regions to achieve technological catch-up, guide governments in formulating appropriate regional policies on science and technology, capital investment and talent flow, improve regional competitiveness in science and technology, accelerate coordinated regional development and maintain sustainable regional economic growth.

Looking at the investment in intellectual capital along high-speed railways, we can conclude that the characteristics of integration have changed from the integration of population and investment to that of technology, knowledge, and human capital. The former is the external economic growth pulled by investment and production elements, while the latter is the intrinsic economic growth model driven by technology and innovation. Therefore, it is of great significance to study how the high-speed railway plays its part in the transformation of the regional economic growth model, and what the mechanisms (the underlying rules and principles) of this role include. This is not only a new theoretical basis for exploring the scientific development of the high-speed railway in China, but a new angle for exploring the coordinated and sustainable development of regional economies and the narrowing of regional gaps.

1.3.2 Literature Review

Regional economic and social development mainly refers to the growth of regional economic aggregates, the upgrading of economic structures and the progress of society. In regards to the theoretical study of the impact of high-speed railway construction on regional economic growth, we can find many arguments from numerous regional economic institutes and theories. For example, from the perspective of regional economics itself, there are theories of economic radiation, development axis and new economic geography. There are also studies and literature on

[12] On August 8, 2018, the General Office of the State Council issued a notice on the establishment of the National Science and Technology Leading Group. In accordance with the deepening of the reform of the party and state institutions and the adjustment of relevant arrangements and work needs according to the deliberation and coordination body, the State Council decided to rename the National Science and Technology Education Leading Group for the National Science and Technology Leading Group.

transportation, transportation channel and other topics in the field of transportation and regional economic development. For instance, according to the "transportation theory" proposed by Rong Chaohe (2001), transport is an important feature of industrialization and an economic process that accompanies industrialization. The above-mentioned theories have been key for discussing the principle of high-speed railways for regional economic growth and social development.

Historically, the rapid economic growth of the United States, Western Europe, and Japan has been in line with the large-scale construction of transportation infrastructure such as railroads. Economists believe that railway transportation infrastructure should be a key factor in promoting economic growth and development in a region. In the ***World Development Report 2008: Agriculture for Development***, the World Bank also pointed out that "The transportation infrastructure represented by railways is one of the best ways to promote the development of underdeveloped areas and an important means to ensure its effective accessibility to a growing and competitive urban market." However, there is also a saying in China that: "transportation infrastructure is prior to regional economic and social development". Further, from a global perspective, the level of economic and social development in developed countries is closely related to their better transportation infrastructure compared to developing countries.

Spiekermann and Wegener (2006) argue that Trans-European transport network (TEN), including high-speed rail, effectively improves the spatial accessibility of EU countries and promotes economic growth, manifesting in an increase in per capita GDP. With the construction of high-speed railways, the cities along the Shinkansen in Okayama, Hiroshima, Oita, Fukuoka and Kumamoto have rapidly adjusted their industrial structures. Traditional industries like the steel and petrochemical industries have been gradually replaced by cutting-edge industries such as the automotive, the electromechanical, integrated circuits, and household appliances, or other processing industries, thus fostering development of the urban industrial structure and regional economic development. Nakamura and Ueda (1989) conducted in-depth studies on population changes in the cities along the Joetsu Shinkansen and the Tohoku Shinkansen from 1975 to 1988. By comparing six cities with Shinkansen and four cities without Shinkansen to the national average, they found that although only three of the six Shinkansen cities had population growth rates higher than the national average, the urban population growth rate of the cities without Shinkansen was lower than the national average. That is to say, in the process of urbanization, the Shinkansen not only attracts the population from the outer regions, but also redistributes the population between the urban areas along the rail. Thus the population center of gravity will gather in cities with more comparative advantages.

Colin Clark (1957) called the high-speed railroad that began in the second half of the twentieth century "the maker and breaker of the city," and Hall and Baniser (1994) thought it opened the "second railway age". Spiekermann and Wegener (1994) believe that high-speed rail is bringing unprecedented shrinkage of time and space, forming the shrinking continent. Blum et al. (1997) argue that high-speed rails are a potential replacement for air transport on the one hand, which will link geographically separate urban individuals to form new urban belts (or urban corridors) and economic zones, and facilitate economic integration within the region. Due to the highly integrated internal economy within the new urban corridor, each economy faces a greater labor market and market share, which brings economic growth opportunities to each city. At the same time, studies on high-speed rail and regional economic and social development have begun to examine the integration of regional and metropolitan transport resources, and suggest that high-speed rail should be more widely accepted as a major means of regional transport (Wang et al. 2014).

Studies on the Shinkansen in Japan by Hirota (1984), Amano & Nakagawa (1990), Brochier (1991), and Sands (1993) found that the effects of high-speed rail are mainly concentrated in tourism-related activities. By studying the first high-speed railway (TGV) in France and Europe, Bonnafous found that, due to the remarkable improvement of communication convenience, it was not necessary for enterprises engaged in specialized service industries in Rhone-Alpes to move from Lyon to Paris, and competitors in Paris felt that it was not necessary to expand their business activities to Lyon (Lin et al. 2010). Taking the Tennessee Railway as an example, Holman (2010) attempted to predict the time-benefit and accessibility benefits of building a high-speed rail to Tennessee by using a traffic demand model. They concluded that the high-speed rail would drive the Tennessee economy.

Similar to the controversy over the role of transportation infrastructure such as railways and highways in regional economic development, economists have different views on the impact of high-speed rail on the spatial economy. Cheshire (1995) believes that high-speed railways link together geographically separated urban individuals to form new urban belts (or urban corridors) and economic zones, promoting the economic integration within the area. Because of the highly integrated internal economics of such urban corridors, each economy faces a greater labor market and market share, which will provide economic growth opportunities for each city. Blum et al. (1997) argue that high-speed rail becomes a potential alternative to air transport and that high-speed rail influences the market structure and market organisation within the corridor, promoting economic integration in the corridor, allowing cities to exploit their comparative advantages through specialised division of labour and trade, increasing productivity and promoting income equality. Peter (2003) believes that the stimulating effect of high-speed railways on regional economic growth can be divided into two types: First, as catalysts, high-speed rails increase new economic dynamism in regions that are in low growth or economic transition, driving regional economic growth. The second is that the high-speed rail has a positive effect on the already prosperous regional economy. However, Puga (2008) holds that high-speed railways will benefit large cities but not small cities, thereby aggravating regional imbalances.

Chen and Hall (2011) argue that the disagreement among economists about the impact of transportation infrastructure on the regional spatial economy is largely due to differences in research methods. In the fields of transportation economics and space economics, mathematical modeling methods are commonly used to predict regional economic development. However, the imperfect competition model of the new economic geography theory points out that lower trade costs may reinforce already existing core-periphery structures and widen the development gap between regions due to the presence of local market expansion effects. Nevertheless, these general modelling approaches, which are cost-oriented and static, do not reflect the quality of dynamic development over time, to the extent that they suggest that transport infrastructure and regional economic development are weakly, or even negatively, correlated (Hall 2009). In the study of the Japanese Shinkansen done by Nakamura and Ueda (1989), Amano and Nakagawa (1990), etc., it was found that cities connected to high-speed rail have higher population and economic growth rates than those not. And the impact of high-speed rail is mainly concentrated on tourism-related activities. However, we found no more details as to the different impacts that high-speed rail has on developed and underdeveloped regions.

The earliest research literature on the regional economic impact analysis of large construction projects in China is found in the preliminary theoretical study by Zheng (1994). They proposed to use the indicator system method, input–output analysis method and other methods to conduct a comprehensive feasibility analysis of the social, economic, and natural factors faced by the super-large engineering projects in a region. Meng (2006) studied the inner mechanism of the role of large-scale motorway construction projects in promoting regional economic development and concluded that motorway construction projects improve regional transport capacity, lay a solid material foundation for regional economic development, make a strong guarantee for the national economy to achieve effective supply, enable transport to meet the needs of regional socio-economic development. This enables transport to meet the needs of regional socio-economic development and promotes the coordination of regional economic structure and transport capacity, thus promoting regional economic development. Ming Libo et al. (2007) briefly reviewed the development and research of evaluation methods for large-scale transportation construction projects at home and abroad. They then carried out an induced empirical case study of regional economy on the assumption that a Wuxi-Nantong river-crossing channel could be built by figuring out the traffic volume and the regional economic aggregate.

The convenience of high-speed rail has greatly increased opportunities for face-to-face interpersonal communication, which is conducive to knowledge creation and business exchange, the development of the knowledge economy and tertiary industries such as commerce and services, and the transformation of industrial structures. Meanwhile, as the main service targets of the high-speed rail, the development of the knowledge-based economy and the tertiary industry provides a broad market space for its operation. Whether spatiotemporal shrinking brought by the high-speed railway can be transformed into the expected spatial economic model depends to a large extent on the synchronization between high-speed rail and economic development. Hu and Shen (1999) systematically analyzed the impact of the Beijing-Shanghai HSR construction project on relevant regional economies regarding the role of it in easing transportation tensions, saving travel time, forming economic integration, promoting the development of the knowledge economy, tertiary industry and tourism, and providing labor and employment opportunities. Zhao and Lin (2010) used the log-linear model and the grey forecasting model to quantitatively analyze the regional economic impact of the Beijing-Tianjin Inter-city Railway from the perspectives of transportation stress relief, travel time saving, economic integration and high-tech industry development. Bai (2010) analyzed the impact mechanism of high-speed railway on regional social and economic development, and believed that the impact of high-speed railway mainly includes direct impact and indirect impact. These impacts are embodied in seven aspects, i.e. HSRs help to improve the railway transportation capacity, regional accessibility, drive regional economic development, optimize industrial structures, increase employment, speed up urbanization, and protect the environment. His research results provide a basis for building an analytical indicator system.

Lin et al. (2010) analyzed the impact mechanism of high-speed railway on the regional economy from three aspects: spatial connection effect, industrial structure effect and employment effect. They calculated the economic impact of Beijing-Tianjin inter-city railway on Beijing and Tianjin by using the grey forecasting method and multiple linear regression model. According to Ding (2012) high-speed railway has both positive and negative spatial spillover effects on regional economic development. The spatial spillover effects of high-speed railway on the macro level include regional economic effects, regional industrial agglomeration, diffusion effects, urban agglomeration effects and social and cultural effects. Based on the data of 287 cities at the prefecture level and above in China from 2006 to 2010, Wang and Nian (2014) used DID(difference in difference) estimation method to examine whether the opening of high-speed railway has significant impact on the development of the regional economy. The research results show that, in the short term, high-speed railways do not lead regional economic growth in the context of the current slowdown in the Chinese economy as a whole.

There are many qualitative analysis documents on the research of high-speed rail on regional economic development in China, with different research angles, but most of them affirm the positive impact of high-speed rail on regional economic growth in the medium and long term. Regarding high-speed railway and scientific and technological innovation, Liu (2002) emphasized the technical problems of high-speed railway and mentioned its role in economic development, but failed to combine technological progress with economic development. Bu and Lin (2007) analyzed the connotation and essence of independent innovation with the help of technology track and technology sub-track theory, differentiated the relationship among its three forms, and then put forward the evolution law of independent innovation. Based on this, they analyzed and summarized the evolution law of independent innovation path of China's high-speed railway. In addition, Cheng (2011) pointed out that, in corresponding to China's national reality, the model of integrating production with learning and research (i.e. the model of cooperative entity with the government's participation) is an ideal model for high-speed railway technology innovation. They also made a triple helix model for the integration of production, learning and research in China's high-speed railway technology innovation.

Compared with other countries, Chinese scholars' research on high-speed railways has developed rapidly, despite its late start. With the successive opening and increasing mileage of high-speed railways in China, more and more research on high-speed railways has been conducted in China, but from different perspectives. For example, some studies focus on people's livelihood, the economic and social significance of high-speed railways for regional economic and social development, while others examine the high-speed rail from a strategic perspective, which is related to military defense, science and technology. Researchers more or less emphasize the positive role of high-speed rail in promoting regional economic development (Li 1998; Hu and Shen 1999; Liu (2002); Zhang and Xu 2005; Lin 2007/2010; Wu 2009; Yang 2010; Zhou 2010; Liu 2011).

In summary, the existing literautures on high-speed railways have the following implications: First, the role of high-speed railways in economic development is significant, although some phenomena often require a certain period of gestation. Second, the establishment of high-speed railway networks and a strict scientific and harmonious organization of transportation and economic development are the preconditions and material basis for the economic take-off of developed countries such as Japan, Germany, and France. Third, the development of high-speed railways is an important driving force for accelerating the development of regional economies and social integration, especially during the downturn of the economic cycle. Fourth, consensus has been reached on the significance and value of building the high-speed railway in China, which will have a positive impact on China's regional economic and social development and the rational distribution of spatial resources. Fifth, it is of great theoretical and practical significance to further explore and refine the impact and role of high-speed rail construction on regional economic and social development.

1.4 Layout of the Book

As a new emerging "green public transport" that is becoming increasingly popular, the high-speed railway has broad prospects for development in China. So, it is of realistic and important significance to study its basic role in influencing regional economic and social development. This book analyzes the significant impact of high-speed railway construction and operation on regional economic and social development by integrating theoretical considerations with empirical studies and cases. Based on the theoretical discussions of the relationship between transportation and regional economies, using system theory as the starting point of the fundamental research, this book makes an attempt to apply transportation economic theories, regional economic theories, and modern organizational management theories to systematically analyze the impact that high-speed railways (as the axis of regional transportation development) have on regional economic and social development. Specifically, this book examines the scope of its impact and its radiation range, its significant influence on the regional economic growth, industrial restructuring, population mobility, urban construction and logistics. Moreover, through building mathematical models to implant real data, the actual benefits brought by high-speed railways in the digital sense will be compared and analyzed. Also, using technical and economic analysis, we will discuss the feasibility and reality of high-speed railway in China's regional economic and social development, its role in China's urbanization development, and the strategic considerations behind its choice of development mode from the perspective of industrial economic development.

The logical structure of each chapter is as follows:

Part I: An Overview of the Birth and Development of High-Speed Railways

Chapter 1 Introduction. This chapter explains the source and research background of this book, including the history and current status of high-speed railways in China and around the world with detailed statistics. Also, this chapter introduces the research results and trends in the field of railways' impact on regional economic and social development. This discussion clarifies the research scope and logical framework, identifying the scientific research questions worthy of study, while formulating a research system to discuss the basic role of the construction and development of high-speed railway on regional economic and social development.

Part II: The Regional Fundamental Role of High-Speed Railways: The Rationale for the Formation and Its Reality

Part Two consists of four chapters. Chapter 2 is Mechanism of the Fundamental Role of High-Speed Railway on Regional Development. Chapter 3 is Technical and Economic Characteristics of High-Speed Railway. Chapter 4 is Time Benefits of High-Speed Railways (HSR) and Calculation of Its Time and Space Competitiveness (TSC). And Chapter 5 is High-Speed Railway and Regional Traffic Structure. The above four chapters focus on the role that high-speed railways play in reshaping space and time due to its unique technical and economic properties. The above four chapters focus on the role of high-speed railways in reshaping space and time due to their unique technical and economic attributes, and are divided into two sections: one on economic theory and mechanism, i.e. Chaps. 2 and 3, and the other on the path to realisation, i.e. Chaps. 4 and 5.

Chapter 2 **Mechanism of the Fundamental Role of High-Speed Railway on Regional Development**. This chapter aims to clarify the theoretical basis to be applied in the following chapters. From the construction and operation of high-speed railway, it looks at the problems of regional economic and social development, and sorts out the intrinsic link between high-speed railway and regional economic and social development. The high-speed railway is separated from the transport economy and regional economy theory, and then the content of high-speed railway is rebuilt in the theoretical system of regional economic and social development from a unique perspective, thus forming the argumentation logic of this book.

Chapter 3 **Technical and Economic Characteristics of High-Speed Railway**. This chapter mainly describes the characteristics of high-speed railway transportation from the beginning of its establishment relative to other modes of transportation, and analyzes the technical and economic characteristics of high-speed railway, namely the flexibility of speed, the complexity of cost, the rigidity of environmental impact, the derivation of demand and the cross-boundary as well as others. Through comparisons with other modes of transportation, this chapter describes the comparative advantages of high-speed railway, namely the volume of high-speed railway, the efficiency of land resource utilization along the line, energy conservation, emission reduction, travel time, safety and comfort, etc. are expounded.

Chapter 4 **Time Benefits of High-Speed Railways (HSR) and Calculation of Its Time and Space Competitiveness (TSC)**. Because of its remodeling of time and space, high-speed railway may rapidly improve regional accessibility, achieve considerable time benefits, and enhance regional space–time competitiveness. This is the fundamental mechanism for high-speed railways to play a key role in promoting regional economic and social development. The construction and operation of the high-speed railway can fundamentally shorten the travel time between the regional node cities and improve the accessibility of the area. This chapter will focus on establishing a set of indicators for measuring the time and space competitiveness of the high-speed rail and evaluate the effect of high-speed railways on the time–space competitiveness of areas along the line. Evidence has it that when the high-speed railway is opened, it not only reduces the travel time between cities, but also increases the accessibility of cities along the route to different degrees, thus improving the space–time competitiveness of the node cities.

Chapter 5 **High-Speed Railway and Regional Traffic Structure**. As a new mode of transportation, the high-speed railway will inevitably bring about great influence when it is introduced into the original regional transportation system. Of course, the soft factors such as market impact, the completeness of the original transportation system, and the embedded system of the high-speed rail will exert their influence. In order to verify these effects, this chapter puts forward the key mechanism of high-speed railways in the regional transportation system by analyzing some cases such as the Shinkansen in Japan, high-speed railways in Taiwan (China) and also those in mainland Hainan Province: that is, competition before coordination. No matter when the high-speed railway enters the regional transportation system, it will help to optimize the original traffic pattern in the long-term, but its evolution path and influence mechanism may be varied in different transportation market structures.

Part III: Performance of the Regional Basic Functions of High-Speed Railways

The basic role of high-speed railways in modern regional economic and social development is embodied in the contents of Chaps. 6 to 9, namely, the High-Speed Railway and Regional Economic Aggregate (Chap. 6), High-Speed Rail Economic Belt and the Improvement of Regional Economic Structure (Chap. 7), High-Speed Railway and Regional Land Development (Chap. 8), High Speed Railways and Regional Service Quality (Chap. 9).

Chapter 6 **High-Speed Railway and Regional Economic Aggregate**. Investment itself is economic growth. High-speed railway construction is a transportation infrastructure investment, which may directly affect economic growth through demand pull and capital accumulation in the short term and long term. This chapter takes the high-speed railway as the axis of economic growth in the region, and quantitatively measures its impact on the total economic growth along the line by analyzing its guiding role in the investment, labor and other factors of production in the regions. It is found that high-speed railways, as an important transportation infrastructure, have a fundamental role in the construction and operation of regional economic aggregates along the line. As a matter of fact, they have been deeply affecting the development activities around the site at the planning stage.

Chapter 7 **High-Speed Rail Economic Belt and the Improvement of Regional Economic Structure**.

Economic restructuring is the key to the sustainable development of the regional economy. High-speed railway reshapes people's production and quality of life because of their high density, high technology and high-quality service. Of course, the degree of its basic effects varies with the quality of the regional economy. This chapter analyzes the optimization of regional economic structure through the case studies of Beijing-Shanghai High-speed Railway, Guangxi High-speed Railway and Beijing-Guangzhou High-speed Railway, we found that the construction and operation of high-speed railway does not only drive the growth of regional economic aggregate along the line, the upgrading of relevant industrial structure, but also produces obvious industrial agglomeration effect, providing rich elements or resources for further adjustment and optimization of regional economic structure.

Chapter 8 **High-Speed Railway and Regional Land Development**. This chapter considers that the corridor effects of rail transit on land quantity and land type are equally applicable in the field of high-speed railways. The high-speed railway not only affects the nature of the land and the inherent law of optimal allocation of land resources, that is, the high-speed railway construction has a far-reaching impact on the nature of land use, the intensity of land development, and the spatial form of regional land use, which can promote the high-density development of land along the line. Such a land development model can create higher land use value and form a multi-win situation. This is in line with China's national policy of land-saving, intensive land use, and improvement of land use.

Chapter 9 **High Speed Railways and Regional Service Quality**. Transportation service quality is an important part of regional service quality. High-speed railway can rapidly improve the quality of transportation service in the region because of its high-quality service. On the basis of an extensive literature review, this chapter mainly discusses the high-speed railway passenger service quality evaluation system and its control technology. We put forward a set of high-speed railway passenger service quality evaluation systems and a model of service quality control selection area. It is hoped that we can study the service quality of high-speed railway from the comprehensive perspective of evaluation and control, and provide useful references for improving the service quality management of the high-speed railway.

Part IV: Strategic Options for More Emphasis on the Significant Role of High-Speed Railway in Regional Development

High-speed railway is a subsystem of transportation. Transportation is a subsystem of the region, and the region the subsystem of the macro. The role of high-speed railways in regional development, regardless of its size or sustainability, is subject to two dimensions. First, on the micro level, the high-speed railway itself has a good organizational management system and a correct development model. Second, on the macro level, the high-speed railway plays an important part and unique role in the national macro strategy. Only by combining the two can the coordinated development of high-speed railway with the main bodies of transportation, industry, industry, city, region and country be formed. Chapter 10 (The Development Models and Policies of High-Speed Railway), and Chap. 11 (High-Speed Railway and National Macro-strategy) correspond to the above two dimensions respectively.

Chapter 10 **The Development Models and Policies of High-Speed Railway**. As the backbone transportation mode in the transportation system, the high-speed railway must provide high-quality transportation services to the society under the dual constraints of existing technology and funds. It is necessary to carefully select its development model to meet the requirements of economic and social development for people and goods, and realize the "balance between supply and demand" between the economy and the social system. This chapter positions the future development model of China's high-speed railway from the dual perspective of industry and enterprise development. That is to say, from the perspective of industrial development, it is necessary to follow the development model of circular economy and pay more attention to achieving sustainable development from a strategic perspective. Meanwhile, it is also necessary to attach great importance to the branding of high-speed railway as transportation enterprises to enhance their market competitiveness, and to regard them as the basic guarantee for China's high-speed railway enterprises to occupy a place in the future international and domestic market competition.

Chapter 11 **High-Speed Railway and National Macro-strategy**. The development of the world high-speed railways has its own unique and profound economic and social background. Countries undergoing economic transformation have regarded high-speed railways as the engines for economic development, since they are mostly located in areas that are economically developed and where there is a relatively large population, and strong demand for travel. In countries with economic maturity, the high-speed railway has been an important means of realizing the transformation of technology and economic growth that focuses on technology investment. In addition, the high-speed railway has become the supporting force for the balanced development of the economy, which has focused on technology export and land development. When environmental issues become a solid constraint on global sustainability, the high-speed railway tends to be constructed as an important starting point for energy conservation and emission reduction and building green eco-cities. This chapter explores the development and the role of high-speed railway concerning national strategic issues like high-end urbanization, technology transfer, and energy conservation and emission reduction.

References

Amano K, Nakagawa D. Study on urbanization impacts by new stations of high-speed railway[C]//Conference of Korean Transportation Association, Dejeon City. 1990.

Andrew Holman. In conversation with David Ellis [J], *British Journal of Learning Disabilities*, 2010, 38(2): 83–85

Bai Yunfeng. Study on the impact of high-speed railway on regional social and economic development [D]. Beijing Jiaotong University, 2010.

Blum, U., Haynes, K. E. and Karlsson, C. (1997): 'Introduction to the special issue The regional and urban effects of high-speed trains', *The Annals of Regional Science, 31*: 1–20.

Brochier, J. E. (1991). Géoarchéologie du monde agropastoral. In Pour Une Archéologie Agraire (pp. 303–322). À La Croisée Des Sciences de l'homme et de La Nature. Armand Colin Éditeur.

Bu Chao, Lin Xiaoyan. Study on Independent Innovation Evolutionary Path of High-speed Railway Based on Technological Trajectory Theory[J]. *Science of Science and Management of S. &T*, 2007, 28(10): 52–57.

Chen C L, Hall P. The impacts of high-speed trains on British economic geography: a study of the UK's InterCity 125/225 and its effects[J]. *Journal of Transport Geography*, 2011, 19(4): 689-704.

Cheng Qinghui. The Study of Model of Production and Research Integration and Cooperation for High-speed Railway Technology Innovation [D]. Changsha: Central South University, 2011.

Cheshire, P (1995) A new phase of urban development in western Europe? The evidence for the 1980s'. *Urban Studies*, 32(7), 1045–1063.

Ding Dongmei. The Research on Regional Economic Development Effect of High-speed Railway [D]. Beijing Jiaotong University, 2012.

Hall P, Banister D. The second railway age[J]. *Built Environment,* 1994, 19(3/4).

Hall, P., Investment-spatially targeted or spatially blind? [J]. *Town and Country Planning*, 2009, 78, 298–300.

Hirota, R.. Present situation and effects of the Shinkansen[R]. International Seminar on High-speed Trains, Paris, 1984(11):15.

Hu Tianjun, Shen Jinsheng. Effect analysis of Beijing-Shanghai High-speed Railway on Regional Economic Development [J]. *Economic Geography*, 1999, 19(5): 101-104.

Li Jingwen. The influence of Beijing-Shanghai high-speed railway construction on the economic development in the areas along the line [J]. *China Railway*, 1998, 10: 44-50.

Lin Xiaoyan, Chen Xiaojun, Zi Yunfeng, Han Xinmei. Quantitative analysis of the impact of Beijing-Tianjin intercity high-speed railway on regional economy [J]. *Railway Economics Research* 2010 (5): 5-11.

Lin Xiaoyan. Important Economic Impact of High-speed EMU [J]. *China Railway*, 2007(4):45-47.

Lin Xiaoyan, Research on Comparative Advantage of Transportation Mode from the Perspective of Energy Conservation and Environmental Protection [J]. *China Transportation Review*, 2010 (06), 25-29.

Liu Jing. Research on Customer Satisfaction Evaluation Based on Analytic Hierarchy Process Method [D]. Shanghai Jiaotong University, 2011.

Liu, Wanming. Research on main engineering economics problems of Chinese high-speed passenger-only guided transport system, Doctor Degree Dissertation, Southwest Jiaotong University, 2002.

Meng Wei. Research on the Evaluation Method of the Impact of Expressway on Regional Economy [D]. Changsha University of Science and Technology, 2006.

Ming Libo, Zhen Feng, Jun Jeong. The construction of Wuxi-Nantong cross-river passage construction and the evaluation of its regional economic impact [J]. *Human Geography*, 2007, 22(04): 105–109+2+6.

Nakamura H, Ueda T. The Impacts of Shinkansin on Regional Development[J]. *Proceedings of WC*,1989,3:95-109.

Peter M. J. Pol, 2003. "The Economic Impact of the High-Speed Train on Urban Regions," ERSA conference papers ersa 03p397, European Regional Science Association.

Puga D. Agglomeration and cross-border infrastructure[J]. *EIB Papers*, 2008, 13(2): 102-124.

Qing, S.H., Li, X.M., & Qing, G.H. Development and Technical Innovation of China's High-speed Railway, *High Speed Railway Technology [J]*, 2014, 5(1): 1–7.

Rong Chaohe. On Scale Economy and Scope Economy of Railway [J]. *Railway Economics Research*, 2001(04):5-8.

Sands, B. (1993). The Development Effects of High-Speed Rail Stations and Implications for California. *Built Environment*, 19(3), 257–284.

Spiekermann K, Wegener M. Accessibility and spatial development in Europe[J]. *Scienze Regionali*, 2006, 5(2): 15-46.

Spiekermann K, Wegener M. The shrinking continent: new time-space maps of Europe[J]. *Environment and Planning* B, 1994, 21: 653-653.

Wang Yao, Nian Meng. Did high-speed trains promote the development of Regional economy? [J]. *Shanghai Journal of Economics*, 2014 (02): 82-91.

Wu Zhaolin. Comprehensive Transportation Planning [M]. Beijing: Tsinghua University Publishing House, 2009.

Yang Weifeng. Analysis of the influence of Beijing-Shanghai high-speed railway on regional spatial structure of China [J]. *Social Sciences of Beijing*, 2010(06):38–43+7.

Zhang Nannan, Xu Yilun. Research on the impacts of Highspeed rail on regional development [J]. *Areal Research and Development*, 2005,24(3):32-36.

Zhao Juan, Lin Xiaoyan. Evaluation of Influence of Beiing-Tianjin Intercity Railway on Regional Economy [J]. *Railway Transport and Economy*, 2010, 32 (01): 11–15+2+6.

Zheng Youjing. Analysis of the current situation and trend of regional economic development in China—Analysis of the development and experience of South China's economic circle (Selected) [J]. *Science & Technology Industry of China*, 1994, (12):15–16.

Zhou Xiaowen. The role of high-speed railway in promoting the coordinated development of regional economy [J]. *Railway Economics Research*, 2010, (06):19-22.

Part II
The Regional Fundamental Role of High-Speed Railways: The Rationale for the Formation and Its Reality

Chapter 2
Mechanism of the Fundamental Role of High-Speed Railway on Regional Development

In recent years, with the improvement of high-speed railway networks and the development of regional economy and society, the role of high-speed railways in optimizing the economic structure of the region along the route, improving the investment environment, and enhancing regional competitiveness has become increasingly important. However, due to the lack of knowledge about the fundamental rules of high-speed railways in the process of regional economic and social development, problems arise in relation to the harmony between high-speed railway planning and construction and regional economic and social development. Wherefore, this chapter looks at the intrinsic relationship and mechanism between high-speed railway and regional economic society, highlighting the factors affecting the role of high-speed railway in regional economic and social development.

2.1 Theoretical Basis

2.1.1 The Interaction (Between Transportation and Regional Development) Theory

The interaction between transportation and regional development generally lies in two aspects. First, traffic push and pull: In a particular transportation system, the development of economy usually calls for new requirements for transportation which are often within the limits of the original modes of transportation in their initial period of gradual accumulation, and thus can be met through the transformation of the existing transportation systems. In this stage, the role that the transportation industry plays in economic development is often invisible, since it primarily functions as a support for economic growth and is in a relatively passive position. If the development of the transport sector can keep pace with economic development, the role of the

X. Lin, *High-Speed Railways and New Structure of Socio-economic Development in China*, Research Series on the Chinese Dream and China's Development Path,
https://doi.org/10.1007/978-981-19-6387-2_2

transport sector seems to be non-existent and often goes unnoticed. But if it fails, it will show its role in economic development in a negative way by hindering economic development. In this case, the economic development may be an impetus for the transportation industry. Then, as time goes on, the limits of the original transport system will be broken, ushering the emergence and development of new modes of transport, which sends a strong material demand to the society and leads to the birth of a great number of new industries. In this period, the transportation industry manifests its visible role in powering up and stimulating the economic growth.

Second, the interplay between transportation and regional development lies in that they come into play with each other and move forward together. On the one hand, transportation links all economic regions together, and is the necessary basic condition for the regional economies to achieve mutual transportation links. Transportation determines the strength of the amount and speed of links between various economic regions, the flow of passengers and goods, and has a positive impact on the economic development and prosperity of various economic regions and social progress, and stimulates the rational allocation of resources. At the same time, economic development has asked for more rational economic structure, the social and regional division of labor and collaboration, and the degree of specialization of production. It also requires that the economic structure be adjusted and reconfigured to ensure the smooth progress of economic development. Changes in the economic structure will bring forth the rapid development of some industries, calling for the reconfiguration of production factors in various industries and changes in transportation demand, thus stimulating the development of transportation.

2.1.2 The System Theory

As a disciplinary reform in the field of science in the twentieth century, the emergence and development of system theory have profoundly influenced the future evolution of the entire scientific community and even human culture through further changes in world outlook, scientific outlook, rubrics of methodology and way of thinking. The so-called system in system theory generally refers to an organic whole composed of a number of interconnected and interacting elements that have specific functions and laws of motion. To put systematic research methods into practice, it is necessary to pay attention to the essential characteristics of systematic ingredients such as integrity, structure, connectivity, dynamics, orderliness and dynamic adaptation to environmental conditions. Among them, integrity means that the overall function is not equal to the sum of the functions of each part; a system may consist of several elements, but the sum of the functions of each part does not add to the overall function. Usually, it can be seen in the following two contexts: when individual segments come up to a harmonious and orderly structure, the overall function is greater than the total functions of its segments; if all segments form a structure that interferes and conflicts with each other, the overall function may be smaller than the sum of the parts. Structure here can be said as the forms of organizing various elements within the system, and function means the certain effects it has under different circumstances. The structure of a system determines the function of the system, and different structures may function in different ways. Therefore, in the case that the elements have been determined and the environmental impact is unchanged, the skillful arrangement of temporal and spatial structure of the system is key to make the system come into play. Connectivity means that if one certain part is always found in a certain system, it should be an element of the system. Everything is self-contained and has its internal structure. For a particular system, the other system is the external environment in which the system exists. Therefore, the system, the elements and the environment are organic, unified, and also interrelated and mutually confined. Dynamics means that any system is not something absolutely closed and static but is found in a specific environment, and is seen in its exchanges with external world in terms of energy, matter and information. Moreover, as each system is constantly in a process of changing, its processes and stages of changes should be analyzed from a dynamic point of view.

When it comes to high-speed railways and their adjacent areas, the system can be understood as an organic whole with specific functions formed by a number of mutual constrained elements, such as human resources, materials and sources, packaging equipment, loading and unloading machinery, delivery vehicles, facilities for warehousing and communication systems in a certain period of time and space. Therefore, it is of paramount to study the role that the high-speed railways play in regional economic and social development from the perspective of the system theory in modern transportation economy.

2.1.3 Regional Economic Theory

As a science established in the 1920s–1940s, regional economics deals with "the differences between different economic regions in the practice of uneven distribution of resources and incomplete free flow within regions" from the economic point of view. It looks into "individual or group industries that effectively engage in production activities and make profits in some regions". It is also concerned with "how to establish the principle of the division of labor and cooperation within a region, organize the industrial development and layout of each region in the system to form a regional economic system that combines aggregation and dispersion, maximizes regional comparative advantages, and enhances regional division of labor and cooperation."

Regional economics enables us to focus on the "temporal-spatial compression" effect of high-speed railways and the geographical distribution of resource like people and materials from the perspective of the geographical distribution of economic activities and the organization of space transportation. On this basis, it helps to analyze the fundamental role of high-speed railways in the regional economic and social development, and explore the characteristics and rules of regional high-speed rail operations in different practical circumstances, as well as the interaction and dependence between high-speed railways and regions, thus transforming the layout of the economic and social development and transport industry in the region. Moreover, it invites the principles of the division of labor and cooperation, helps to realize the optimal allocation of regional resources, and establish high-speed railway industries that conform to regional advantages, and a multi-level and coordinated development of a regional economic and social development system.

2.1.4 Transportation Economic Theory

As an intermediate economic discipline that integrates the industrial analytic framework and the basic characteristics of the transportation industry, transportation economics is based on microeconomics and gradually adds such factors as information, transaction costs, barriers to entry, and government regulation in the perfect competition model, making its theoretical model closer to the reality and more explanatory. Transport economics provides strong theoretical support for studying the basic role of high-speed railways in regional economic and social development. It may be applied to a wide scope of topics, issues and concerns with regard to the development of the high-speed railway industry (the evolving process, life cycles, the general laws and major influencing factors of high-speed rail industry development, and so on), the subjectivity of high-speed railway industry (the impact of high-speed rail enterprises, competitions in transportation market, high-speed rail enterprises' goals and development), the industrial structure (such as its development model, its balanced or unbalanced growth, and its transformation potential), the interconnectivity between high-speed railway industry and other related industries (e.g. the analysis in their relations, structures, and vibrations); and the management, organization, policies, security, profit, and sustainable development of the high-speed railway industry.

2.1.5 Modern Organization Management Theory

Generally speaking the construction and operation of high-speed railways is all about transport organisation, where transport organisation management is fundamental to all transport activities. The management of the high-speed rail system in a region is the art of a balanced evolution of the system, which directly affects the operational efficiency of the high-speed railway in the region. The management of the regional high-speed rail system looks at the operational efficiency and policy-making process from the perspective of organization management of government departments and railway enterprises. Some management theories on the macro level have thus become important theoretical bases for regional high-speed rail system management. For example, when explicating the function of the regional high-speed rails under the framework of urban management theories, we are actually looking at how the municipal government is coordinating the layout of urban functions with the planning, construction and operation of high-speed railway infrastructure to meet new demands for the survival and development of the city.

In short, taking the interaction theory (the interrelationship between transportation and regional economy) as the basis, and the system theory as the starting point, this book aims to elaborate on the fundamental role of high-speed railway in regional economic and social development.

2.2 Factors Affecting the Fundamental Functions of the HSR

In general, the basic factors affecting the role of high-speed railways in regional economic and social development include the technological base of high-speed railways and the economic and social foundations of the areas along the line. Among them, the technical basis of high-speed railway refers to the technological system that provides displacement services for people, information and material in the region within a certain transportation network by using certain infrastructure and vehicles within the area along the line. The technical system is mainly concerned with road network, carriers and people, information and materials that use these road networks and vehicles. Therefore, the technical base of high-speed railways is the technical or the carrier basis for the movement of people, information and materials in a region. The economic foundation refers to the value created by those who make use of the regional resources to engage in activities such as the production, circulation, distribution and consumption of materials. The regional economic foundation is to some degree the material basis or "hard" basis that affects the essential role of high-speed railways. The social foundation of the regional development can be understood as the cultural literacy, ideology and spiritual temperament of those who are involved in or organize these management activities. This may include the management level of government sectors and railway enterprises, and people's cognitive knowledge about the market, laws, rules and other aspects. Therefore, the social foundation can be seen as the management basis or "soft" basis that may release the role of high-speed railways.

The reasons why the above elements are put on the same level are as follows: first, the main theme in this book is to discuss the fundamental role of high-speed railway in regional economic and social development. To make it clear, it is necessary to discuss its essential elements and the relationship between them. The technical foundation, regional economic foundation, and social development base of the high-speed railway are precisely the decisive factors that determine the above-mentioned effects. Second, the development of any of the above-mentioned decisive elements may push or impede another, which will further restrain the high-speed railway from playing its fundamental role in the development of the regional economy. Therefore, only by coordinating the relationship between them will it be possible to promote

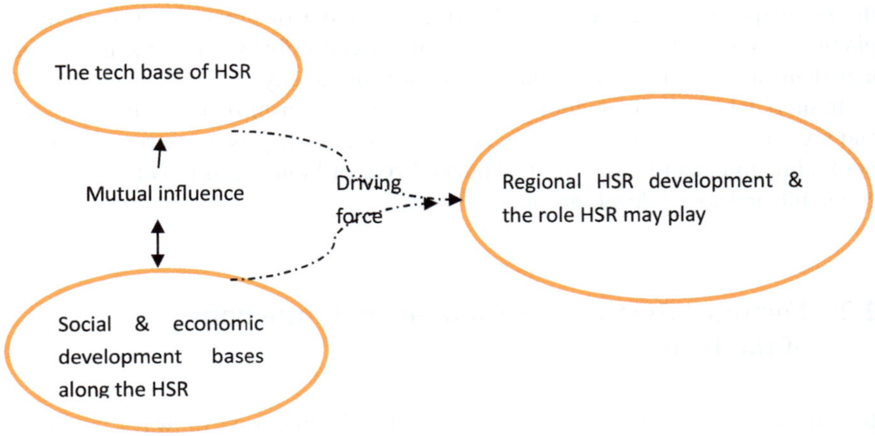

Fig. 2.1 The essential elements and their driving effects of regional HSR development

regional economic and social development to a more advanced stage. Third, from the perspective of the development of the high-speed railway industry, the construction and operation of high-speed railways, the development of its equipment manufacturing and related industries have also played an invaluable role in promoting regional economic and social development.

As can be seen, the technical foundation of the high-speed railway and the economic and social basis of the region play important roles and work as the driving force for developing high-speed roads (Fig. 2.1).

In addition to the high-speed railway technology and the quality of the economic and social development of the region along the highway as shown in Fig. 2.1, that is, the two basic factors of technology and economic and social infrastructure, organizational management department plays a key part: first, the high-speed railway as a new type of transportation method can be integrated into the established transport system, which involves the coordination and fit between the various modes of transport. Second, it is feasible to imbed an organizational management department between the high-speed railway and the regions alongside. This will be further elaborated in the following sections focusing on high-speed railway and transportation structure and the development model of the high-speed railway.

2.2.1 Differences in the Impact of Regional Socio-economic Development on the Fundamental Role of the HSR

From the perspective of urban development practices at home and abroad, there exist great differences in the impact of high-speed railways on cities within the radiating region. Cities that have successfully benefited from the development of high-speed railways tend to share some similar characteristics. For example, they may enjoy better soft environment good for urban development, better public service facilities and infrastructure, and reasonable positioning, and appropriate development strategies.

High-speed rail cities with better soft environment are easy to get development opportunities. Some scholars believe that high-speed railway reduces the resistance of factors of production flowing into passenger station cities, and creates conditions for strengthening the resource attraction of passenger station cities and forming accumulation effect. Since the opening of Tokaido Shinkansen, the net migration population of Tokyo metropolitan area is much larger than that of Osaka and Nagoya metropolitan areas. Correspondingly, local enterprises in Osaka and Nagoya have automatically moved closer to Tokyo and established corporate headquarters in Tokyo. Thus, the agglomeration effect directed to Tokyo was triggered.

Whether the internal public service facilities and infrastructure in the city are developed or not will affect the speed of development of the city with the help of high-speed railway. Producers tend to choose places with convenient transportation conditions, where they are more likely to approach the raw materials and commodity markets of products. In this regard, central cities often have more advantages than small and medium-sized cities. If the public service facilities and infrastructure of small and medium-sized cities are not developed, the convenience brought by the high-speed railway outside the cities will make producers feel the inconvenience inside the small and medium-sized cities more easily, which will lead to the relocation of small and medium-sized enterprises to central cities, the outflow of population and industry in small and medium-sized cities, and the clustering in big cities. This is confirmed by the experience of Lille, a city along the TGV high-speed railway line in France. Lille has taken advantage of the improved access to the railway to improve or develop and build services in parallel, and the industrial area of Lille has been rapidly transformed into a new industrial area. The other small and medium-sized cities around Lille did not improve or develop their services in the same way as Lille. As a result, tertiary industry continued to move out of the surrounding small and medium-sized cities and into Lille, and the surrounding small and medium-sized cities suffered a severe blow.

High-speed rail cities that are well positioned and have appropriate development strategies are more likely to achieve rapid development. When the positioning of different cities in the region effectively complements each other, the central city may drive small and medium-sized cities and even towns to accelerate their development together, forming a city cluster with positive interaction. In this respect, the German high-speed railway is a good example. ICE, a high-speed railway system in Germany, came into being after the socio-economic development in Germany reached a very high level. Its high-speed railway was not designed to relieve the pressure of passenger transport, nor to allow cities along the route to gain more development opportunities, but to further reduce the time distance between cities, making travel and business activities within and across Germany more comfortable and convenient. Kassel, Hessen, Germany, a typical city with a population of less than 200,000, has benefited from the transference of industrial expansion and extension. Because Hanover-Fort Ulcz high-speed railway passes through it, and it takes less than one hour from Kassel to Frankfurt and Hanover by ICE. The high-speed railway has given Kassel the opportunity to take advantage of the shortened time available to the city to (i) revitalise the old city centre, (ii) build a new high-speed railway station to expand the city's space, and (iii) make for the transfer of industries from neighbouring cities. One lesson in this respect is the French town of Le Crusoe. Le Crusoe, France, is an area which is trapped in economic reconstruction due to the closure of coal mines. Although TGV high-speed railway has greatly shortened the distance between the town and Paris, Le Crusoe has not done a good job in infrastructure construction and environmental improvement, nor has it rationally positioned itself and formulated its development strategy by determining its role in the regional industrial chain according to local conditions. As a result, its industry did not benefit from the construction of high-speed railway. After TGV has been running for 6 years, there are only two small companies near the station in this town.

2.2.2 Differences in Short-Term and Long-Term Effects of HSRs

Time is also an important factor affecting the difference in the basic impact of high-speed railways on regional economic and social development. Due to the limitation of social capital to support the production and development capacity, the process of high-speed railways affecting regional economic and social development is longer than the relatively limited traffic system in service areas such as urban rail transit. Different durations will have different effect. In the short term, the impact of the high-speed railway is mainly reflected in its influence on the urban market of production factors. It is an immediate effect that can be seen after the high-speed railways are put into use. While in the long term, the influx of labor, capital and technology will continue to promote the differentiation and integration or the market, the reduction of the relative prices of production factor; and the increased competitiveness of urban products will ultimately build to the adjustment of the economic and social structure. Such effect will not be seen within a short period of time but after a relatively long period of development.

HSR's short-term impact on the market. In the short term, high-speed railway will have an impact on the factor markets of a city. First of all, high-speed railways bring comfortable, convenient and fast transport services to travellers, triggering a change in their choice of transport and competition between different transportation means, thus impacting on the transport market. Secondly, high-speed railways improve urban traffic conditions, shorten the time distance between the originally segmented markets, and reduce the social transport costs of direct production factors such as labor pools, capital, and technology in cities, thereby exerting an impact on the direct market of production factors in cities. Specifically, they are: (1) HSR's impact on the transportation market: experience from the practice of foreign high-speed railways has made it clear that HSR's competition with civil aviation usually ends with the victory of high-speed railways. For instance, the Japanese Shinkansen has doubled the speed of the regular trains and reduced their operating time from 6.5 h to 3.2 h. The superior price/performance ratio has attracted a large portion of air travellers so that the Tokyo-Nagoya flight has been suspended. The Paris-Lyon TGV high-speed train is running every hour. It takes only two hours for the entire trip so that it has taken away most of the domestic civil aviation market; the Eurostar train linking London, Paris and Brussels and the Thalys high-speed train between Paris and Brussels has also taken a a considerable portion of the market share of civil aviation; ICE high-speed train in Germany has resulted in Lufthansa Airlines' cutting prices and gradually attracted passengers on medium and short-haul routes. As a new popular transportation, high-speed railways provide efficient services to attract a large number of passengers by improving traffic conditions between cities, which has led to a rapid increase in passenger traffic while boosting economic growth. In turn, the rapid development of the urban economy has brought about more employment opportunities, attracting a large number of immigrant residents for the city (circle), and the increasing population has further magnified the traffic demand. In addition, high-speed railway has strong technical and economic advantages. In its competition with road and air transport, a large number of tourists are attracted and transferred, resulting in the redistribution of the transport market, and a great change in the composition of the transport modes, which has an important impact on the transport market.

(2) Its impact on the market of production factors: high-speed railways improve the traffic conditions between cities, speed up the flow of labor, capital, and technology, shorten the space–time distance between cities, and at the same time, expand the scope of cities to attract production factors, thereby significantly reducing the transportation and transaction costs for a city's access to production factors, which may bring the growth of economic trade between cities and market efficiency, thus creating conditions for market integration and development. For example, the French TGV reduced the travel time from Nantes to Paris from three hours to two hours. The Korean high-speed railway KTX reduced the travel time from Seoul to Busan from 5 h to 2.7 h, both of which facilitated regional economy and trade, formed and expanded products, services and labor markets.

HSR's impact on the structural adjustment of urban development in the long term. In the long run, the high-speed railway attracts abundant production factors such as labor, capital and technology for the cities around the pathways, which makes the cities more attractive to advantageous industries and enterprises. However, as the strong industries and enterprises tend to gather in the central cities with better production conditions, and the industries in the central cities are upgraded, the competition for production conditions and markets gradually raises the factor prices of the central cities and drives the adjustment of industrial layout. Industries and enterprises with higher requirements for information, capital, high-level talents and production technology remain in the central city, while manufacturing industries with more labor-intensive and land use are constantly shifting to neighboring cities with lower labor and land costs, which leads to the redistribution of industries and population in cities (circles) and the adjustment of urban industries and development structure. For example, the efficient and fast Shinkansen in Japan has caused a large number of talents to flock to developed areas, and the development of the tertiary industry, which is mainly high-tech industry, has been accelerated in cities in the radiation area of high-speed railway, and the urban economic structure has been adjusted accordingly. By reducing travel time, Shinkansen has changed the social and economic development potential and spatial distribution of social and economic activities in cities along the route, thus having a long-term impact on the social economy.

It can be seen that the high-speed railway plays an important role in improving the traffic conditions and the accessibility of cities in the radiation area, and reducing the social transportation costs and transaction costs in the area. Its existence will make it possible for cities (clusters) to obtain resources in a larger scope and bring together a large number of production factors such as labor, talents, capital and technology for cities at a lower cost. On the one hand, the prices of factors of production have been reduced; on the other hand, the influx of low-cost and high-quality factors of production will make the original factors of production reallocated. Under the action of market rules, production factor markets are continuously integrated and new markets are formed; the prices of production factors also continue to fall, creating conditions for cities to expand production and continuously adjust the capital factor ratio, leading to regional industrial restructuring and upgrading.

2.3 The Relationship Between HSR and Regional Economic and Social Development

A review of the available literature shows that the earliest documentation on transport and regional economic and social development is the classical location theory, represented by Alfred Weber's industrial location theory. In essence, location theory was the first theoretical recognition of the fundamental role and influence of transport in regional economic and social development, which focused on the role of transport factors in location choice and the resulting impact of transport costs on product costs. Since 1950s, with the rise of large-scale regional economic and social development, western researchers of regional science and transportation economics began to attach importance to the role of advanced transportation technology in regional economic and social development, and linked the construction of new transportation system with regional economic and social development. Therefore, the economic and social benefits of the construction and operation of the new transportation system are gradually valued. In these studies, the important parameters of high-speed railway efficiency are mainly about the cost and time saving, while the research on the structure, characteristics and operation laws of high-speed railway is rare. By far, Japan, France and Germany may be the countries that have studied the underlying principles of high-speed railways and their regional economic and social benefits.

As China's high-speed railways are currently in a period of rapid development, local governments at all levels actively support the development of high-speed railways and have introduced relevant plans and policies fit for local interests. However, if the coordination and integration of the economic and social development level and the existing transportation resources is not well considered, it will inevitably lead to regional conflicts and repeated construction under the new economic conditions. Therefore, in order to solve the basic problems of the function of high-speed railway in regional economic and social development, it is necessary to clarify the relationship between high-speed railway and regional economic and social development.

First, the influence of the degree of regional economic and social development on the development of high-speed railway: (1) the developed regional economic society is the guarantee for the construction and operation of high-speed railway. Without it, it is difficult to complete the construction and operation of large-scale modern high-speed railway. (2) The level of regional socio-economic development also determines the flow rate of people, information and materials in the region, which may greatly affect the operation efficiency of high-speed railway. It can be said that without the developed regional economy and society, it is unlikely that high-speed railway will exert its huge benefits. The above relationship can be simply summarized as follows: the development of high-speed railway must be based on certain regional economic and social development, as the degree of regional economic and social development determines the role of high-speed railway.

Secondly, the impact of high-speed railway on regional economy and society. Similar to the impact of regional economic and social development level on the development of high-speed railway, the impact of high-speed railway on regional economic and social development can be seen from the following two aspects. On the one hand, the increase in demand for labour, materials, machinery and information during the construction of the high speed railway network will cause an increase in demand from other industries in the regional economic and social sphere, which in turn will have a pull on the regional economic and social development. On the other hand, the economic and social benefits brought about by the construction and operation of high-speed railways, such as savings in travel costs, increased frequency of information and knowledge exchange and the release of freight capacity, will contribute to regional economic and social development.

To sum up, the relationship between high-speed railway and regional economic and social development is as follows: regional economic and social development will definitely show an increase in demand for high-speed railway transportation, which will inevitably drive the construction of high-speed railway infrastructure to meet the transportation demand derived from regional economic and social development. Meanwhile, the construction and operation of high-speed railway will promote regional economic and social development. However, blind high-speed railway construction may become a burden for local economic development. The relationship between high-speed railway and regional economic and social development mentioned in this book means that the demand of construction and operation of high-speed railway should be in line with the stage of regional economic and social development, and the construction and operation of high-speed railway should be conducive to regional economic and social development, while there is no oversupply of high-speed railway.

2.3.1 The Impact of Regional Economic and Social Development Demand on the Development of HSR

Generally speaking, the regional economic and social impact on the development of high-speed rail can be seen in the following areas:

The demand of population mobility has an important impact on the development of high-speed railway. Regional economic and social development can bring more demand for population mobility, which makes the development of high-speed railway necessary. The remarkable characteristics of regional economic and social development are the expansion of market demand, the increase of spatial distance and the more frequent interpersonal communication among economic units in a region. Therefore, with the further development of regional economy and society, higher requirements will be put forward for the infrastructure supporting population mobility. To a certain extent, this will require the continuous improvement of high-speed railway infrastructure to meet the growing demand. Conversely, a weakening of the regional economic and social base will lead to a reduction in demand for mobility, which in turn will constrain the development of high-speed rail infrastructure.

Capital supply has certain influence on the development of high-speed railway. With the rapid development of regional economy and society, various forms of funds available for investment will gradually increase, among which the funds that can be allocated to high-speed railway infrastructure will also increase, thus speeding up its construction speed and improving its network. On the contrary, a weakened regional economic and social base will result in less funds being available for developing high-speed rail infrastructure.

Technology has a technical impact on the development of high-speed railway. Regional economic and social development will accelerate the division of labor within the region, which in turn leads to the improvement of the original technology or the innovation of new technology. The construction and operation of high-speed railway infrastructure is undoubtedly a high-tech industry, so the breakthrough of technical difficulties will have a positive impact on the construction and operation of high-speed railway infrastructure. On the contrary, backward technology will restrict the development of high-speed railway infrastructure.

Institutional supply has a positive impact on the development of high-speed railway. With the development of regional economy and society and the further improvement of market mechanism, a series of new management alternatives are produced to meet the actual needs in the region. For example, in order to solve the problem of insufficient funds in the development of high-speed railway infrastructure, BOT, franchising and other methods have been gradually introduced, and great achievements have been made in practice. On the contrary, imperfect institutional arrangements may make the development of high-speed railway infrastructure lack of innovation vitality and enthusiasm.

2.3.2 *The Basic Role of HSR in Promoting Regional Economic and Social Development*

The construction and operation of high-speed railway will promote the sustainable development of regional economy and society. It is mainly manifested as follows: the construction of high-speed railway will drive the adjustment of industrial structure in the region, further unify the urban–rural structure, deepen the degree of urbanization, increase employment opportunities, improve the efficiency of regional economic and social production, the frequency of exchange of knowledge, technology and culture, and the quality of life of people in the region to a higher stage. The development of high-speed railway has the following two effects on regional economic and social development: one is to push and the other is to restrict.

2.3.2.1 The Push Effect of HSR on Regional Socio-economic Development

There are two ways in which the construction and operation of high-speed railways can contribute to the regional social economic development: First, the increase of output of high-speed railway construction and operation will directly lead to the increase of total output of regional economic and social foundation; Secondly, the increase in investment in high-speed railway infrastructure flows or the increase in the size of the stock will cause an increase in investment, foreign trade and output, indirectly stimulating the growth of the total output of the regional economy and society.

(1) The construction and operation of high-speed railway will directly promote the regional economic and social development. Simon Kuznets (1901–1985), an American economist, used the historical statistics of developed western countries to analyze the internal structure of industry and service industry carefully, and thought that "the share of transportation and communication sector, which is one of the richest productive components in modern economic growth, is obviously rising continuously, especially in GDP and total labor force". His research results mean that the added value created by the transportation and communication infrastructure sector is increasing in the national economy. That is to say, the output value created by transportation and communication industry is increasing faster than the average economic growth rate, and its role in contributing to economic growth is strengthening.
(2) The indirect contribution of the construction and operation of high-speed railways to regional economic and social development is in the following areas:

First, the construction and operation of high-speed railway will help reduce the production and transaction costs in the region, optimize the structure of the transportation market, and drive the rapid development of regional economy and society. On the one hand, the construction and operation of high-speed railway will improve the infrastructure in the region, thereby reducing the production cost of enterprises and improving the labor productivity. As the construction investment of high-speed railway is a prerequisite for technological innovation and production development, its impact on technological innovation and production activities is mainly reflected in reducing the employment cost of talents and improving the technological innovation ability of enterprises, thus a higher production efficiency of enterprises. On the other hand, the construction and operation of high-speed railway improves the transaction cost and efficiency in the region. Transaction costs in this context are the fees paid to obtain accurate market information in the region, as well as the costs of negotiating and entering into recurring contracts. It may include: (i) ex-ante transaction costs, i.e. the cost of entering into a contract setting out the rights, responsibilities etc. of the parties to the transaction; and (ii) the cost of entering into a contract and then changing the terms to resolve problems with the contract itself, from changing the terms to withdrawing from the contract. One of the most important reasons for transaction cost is that there is asymmetry of information, or the amount of information owned by both parties to the transaction is not equal, since both parties need to obtain as much information as possible in order to be in a more favorable position in the transaction. Good high-speed railway infrastructure is conducive to more frequent exchanges between the two parties, thus obtaining more transaction information and reducing transaction costs. Therefore, with the further improvement of high-speed railway infrastructure and the increase of its services, it will help to reduce the cost for traders to obtain transaction information.

Second, the construction and operation of high-speed railway will help to improve the accessibility and interconnectivity of a region, and enhance its competitiveness in time and space, thus stimulating the rapid development of regional economy and society. The most important impact of the construction and operation of high-speed railway on the areas along the line is to improve the accessibility of various areas and cities along the line, and to change their economic geographical position and traffic geographical status, and improve their location advantages, thus promoting the rapid development of local economy and society. Further, the improvement of regional and urban accessibility means that the transportation conditions for the external contact of the region or city become better, and the transportation distance between it and other regions is shortened, the time is shortened and the convenience is improved. All these improvements in transportation accessibility may bring an increase of the total output of the regional economy and society, and be beneficial to regional spatial expansion, which is extremely important for the development of areas with backward transportation conditions (more specific is shown in Fig. 2.2). In addition, from the perspective of "input–output", the connectivity in the regional economic, social and industrial system can be expressed as the size of "flow", that is, "population flow", "technology flow", "capital flow", released "material flow", and so on. These different forms of "flow" must be carried out with a certain transportation infrastructure as the carrier for regional spatial displacement. Without doubt, the continuous construction and operation of high-speed railway will facilitate the displacement of such "flows", thus strengthening the contact frequency between industries and creating greater connectedness.

Third, the construction and operation of the high-speed railway may help to form the high-speed railway economic belt along the line, which may speed up the regional trade development, thus bringing about rapid regional economic and social development. From the perspective of regional aggregate demand (including investment, consumption and import and export), the increase of import and export will undoubtedly drive the development of regional economy and society. For example, Japan's rapid economic development from World War II to 1970s largely benefited from the export pull of Shinkansen. In addition, from the perspective of supply, the construction and operation of high-speed railway is conducive to the development of foreign trade and the formation of economies of scale along it, which helps to improve the efficiency of resource allocation such as manpower, capital and knowledge. Plus, according to New Growth Theory prosed by Paul Romer and Robert Lucas in 1980s, transportation mode with advanced technology may promote economic growth by

<table>
<tr><td colspan="8">The construction and operation of high-speed railway</td></tr>
<tr><td colspan="8">Improved accessibility and interconnectivity and competitiveness in time and space</td></tr>
<tr><td colspan="2">Differences in development opportunities caused by differences in enterprises, information and interpersonal contact</td><td colspan="2">Differences in awareness, education and culture caused by differences in people's motives</td><td colspan="2">Industrial agglomeration and changes in the surrounding natural environment caused by the improved accessibility</td><td colspan="2">Institutional environment and policy differences caused by spatial differences in resource endowments and population mobility</td></tr>
<tr><td>Relatively fast-growing areas</td><td>relatively slow-growing areas</td><td>areas with relatively progressive social structure</td><td>areas with relatively conservative social structure</td><td>Areas with relatively high environmental load</td><td>Areas with relatively low environmental load</td><td>Areas with relatively more policy support</td><td>areas with relatively less policy support</td></tr>
</table>

Fig. 2.2 The effect of construction and operation of high-speed railway on the improvement of regional economic and social accessibility and its connectivity

improving factor productivity. Therefore, it can be said that the construction and operation of high-speed railway will greatly help to improve the service level of infrastructure, accelerate the flow of production factors, thus boosting the rapid economic and social development in the region.

Fourth, the construction and operation of high-speed railway may help to improve the competitiveness of the regional market and related enterprises and optimize the regional economic structure. From the historical development experience, high-speed railway, as a regional basic industry, plays significant role in stimulating the progress and competitiveness of modern industries in a region. The reason is: during the continued construction and operation of high-speed railway, if an enterprise or region is far away from the market because it is not accessible to good transportation infrastructure services, its competitiveness will be weakened because of the high transportation cost, resulting in its disadvantage in the development process. On the contrary, those enterprises or regions accessible to good transportation infrastructure services may have more opportunities in economic development, thus the burgeoning of new production enterprises.

Fifth, the construction and operation of high-speed railways may lead to an increase in the efficiency of land use in the region, i.e. changing the nature of land use, increasing the intensity of land use and the value of land use. To be specific, in the first place, the construction and operation of high-speed railway will have a great impact on the function of land use. Accordingly, the land use types along the high-speed railway will also change according to the market rules. Not only will the functions of finance, trade and service in the city center be highlighted, but also the nature of land use will be changed. Also, in the area around the station, with the opening of high-speed railway, the nature of land use in the surrounding area will change greatly. Due to the increase in land price brought by high-speed railway, some high-profit industries such as commerce, service industry and office buildings are often arranged in reasonable walking areas, which will prompt other industries to gradually change to these industries. In view of the sensitivity of various industries to the change of location, the reasonable walking area of high-speed railway station has the greatest influence on the commercial use of the location, followed by office, residence and industry. Theoretically speaking, within 500 m from the traffic station, the benefits from business and office are the greatest. In the second place, the construction and operation of high-speed railway will enhance the intensity of land use. High-speed railway is characterized by speed, punctuality and large capacity. The small car drives the development of the area, but if its density is too high, it will inevitably bring a lot of traffic, which will cause traffic congestion due to the limited road capacity, thus reducing the attractiveness of this area and is not conducive to the development of the area. In contrast to the small car, the high-speed railway brings more people rather than traffic flow and does not create congestion or regional pollution, so it may greatly improve the accessibility of its traffic rational area, thus contributing to the high density of regional land use. In the third place, the construction and operation of high-speed railway will increase land value. Generally speaking, the reason why the construction of high-speed railway has an impact on the land price along the line can be summarized into three aspects: (1) The regional high-speed railway can greatly improve the traffic conditions of the land and the accessibility and convenience of the area along the line, shorten the space distance between the land along the line and the regional center, and save the traveler's traffic time and cost; (2) The development of high-speed railway provides an opportunity to adjust the planning and modify the property of land use; Land use planning may be guided by transportation, and its purpose of use may be justified by planning, which directs to an intensive use of land and an increase of profit from land use. (3) High-speed railway can improve the relative location conditions, strengthen the mutual connection of the land along the line, and the complementary effect of the land, which makes the land along the line become an organic whole supporting each other, links the land values of adjacent regions, and adds to the land price.

Sixth, the construction and operation of high-speed railways may promote energy conservation and environmental protection in the region and promote social development. The history of railway development in the world proves that the adoption of high-speed railways as the backbone transport mode in the regional economy and society is an inevitable choice for environmental protection and energy conservation. From the perspective of energy efficiency, in meeting the same transport needs, the per capita energy consumption of high-speed railways is only 3.64 degrees per 100 kilometres, which is 1/12 of passenger aircraft, 1/8 of small cars and 1/3 of medium-sized buses, making it undoubtedly the most energy-efficient mode of transport in comparison. Further, from the perspective of environmental protection, high-speed railway has the least environmental pollution, which is mainly due to the advantages of its energy sources. Since the high-speed railway is powered by electric power, no exhaust gas is discharged during the running of the train, and the pollution of dust, soot and other exhaust gas is basically eliminated. It is a clean and green means of transportation. Even compared with ordinary railways, high-speed railways have great comparative advantages in environmental protection and sustainable development.

Seventh, the construction and operation of high-speed railway will help to improve the local people's travel habits and transport service quality. Since the advent of high-speed railway, the high-speed and safe travel brought by high-speed railway is gradually changing people's lifestyle and consumption concept. During holidays, high-speed railways often carry more than five times the usual passenger flow, creating a time for tourism and leisure in the regions along the route and bringing convenience to long-distance off-site travel. Furthermore, because of its fast-running speed, the high-speed railway attracts the traffic demand of remote and backward areas to the high-speed railway station in its economic center and quickly transports it to other more developed high-speed railway station areas. This not only effectively expands the application scope of advanced social resources and services in developed areas, and plays an obvious role in helping the poor, but also provides stable and continuous support for improving the travel quality and transportation services of local people, and further enhances the regional social value.

2.3.2.2 Reverse Effect of High-Speed Railway on Regional Economic and Social Development

The reverse effect of high-speed railway on regional economic and social development is that when its construction and operation fail to match with the foundation of regional economic and social development, it will become an obstacle to local development, thus being detrimental to regional economic and social development. According to previous studies, there are two kinds of negative effects: backward bottleneck effect and the excessive prepaid effect. Backward lagging effect refers to the fact that if the construction investment of high-speed railway is seriously insufficient, the operation and management technology is backward or an effective network is not formed, its service standards will be far behind the needs of regional economic and social development. In other words, when the construction and operation of high-speed railway is slower than the demand of regional economic and social development, it will cause the flow of regional production factors to be tight, the transportation cost to be raised and the service standard to be lowered, which will seriously restrict the regional economic and social development and lead to the decline of economic and social benefits. This is commonly referred to as the bottleneck effect. Like the bottleneck, the excessive prepaid expense of the construction and operation of high-speed railway will also bring serious constraints to the regional economic and social development. On the one hand, if the construction investment of high-speed railway is seriously excessive, or there is a decision-making error in the investment direction, the construction and operation level of high-speed railway far exceeds the demand of regional economic and social development and the carrying capacity of environmental resources, eliciting the low operational efficiency of regional economic and social development. On the other hand, excessive investment in high-speed railway construction leads to insufficient investment in other sectors in the regional economy and society, making the overall development unbalanced and inefficient.

To sum up, the research on the interaction and coordination between the construction and operation of high-speed railway and regional economic and social development needs to be re-examined, because the interaction and coordinated development between them may actually constitute a complete positive feedback cycle. When they adapt to each other in scale, structure and function, they form a positive feedback cycle (solid line) that benefits each other, so that they can develop in harmony and evolve to a higher level together. However, if the two fails to adjust to each other, the development of either party may become an important obstacle to its counterpart, and then form a negative feedback cycle (dashed line) that restricts each other. This positive feedback loop is shown in Fig. 2.3.

In a word, there is an interaction between high-speed railway and regional economic and social foundation. The regional economic and social foundation acts on the high-speed railway, while the high-speed railway reacts on the regional economic and social foundation. This mutual influence constitutes a positive feedback loop. It is also on the basis of the overall circulation mechanism composed of high-speed railway and regional economy and society that multiple positive feedback circulation subsystems will be derived. The development of each link or subsystem will affect

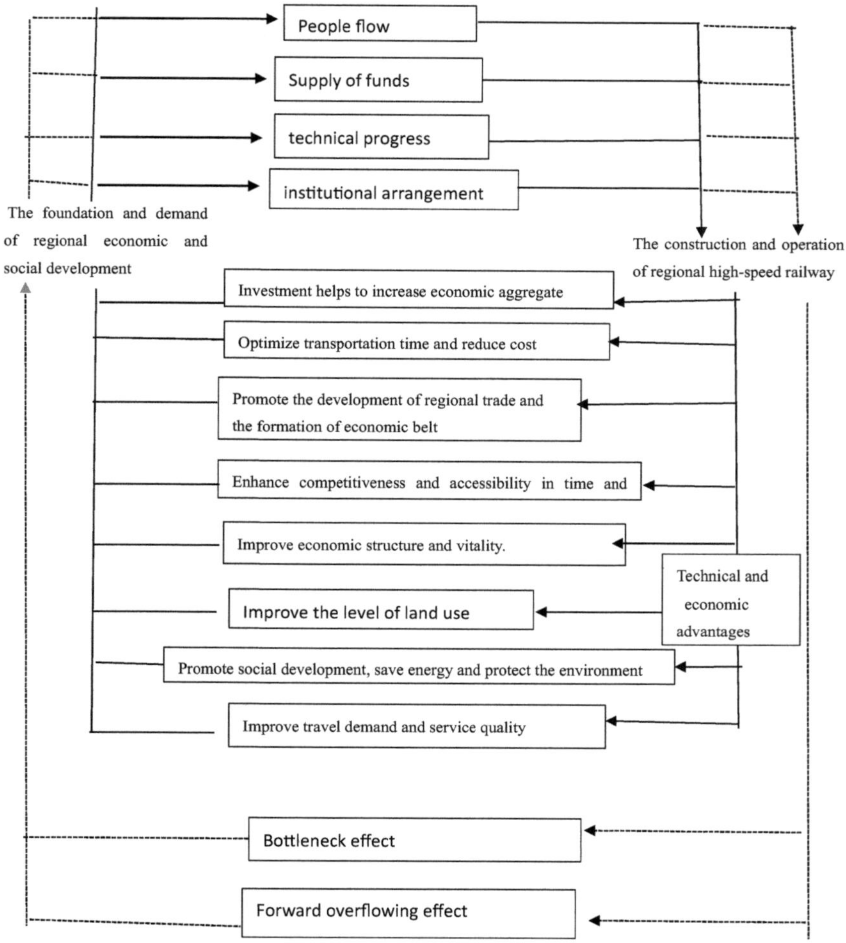

Fig. 2.3 The effect of HSR on regional economic and social development

the next one or other subsystems, and then the functions of the whole will be continuously enhanced. As this positive feedback cycle continues, the organic combination of high-speed railway and regional economy and society will form a more extensive regional economic and social network.

Chapter 3
Technical and Economic Characteristics of High-Speed Railway

In the process of social development and progress, the market share of high-speed railway is generally affected by its technical and economic characteristics when the total social demand is fixed. Since the high-speed rail is basically used for passenger transport, and it is still railway transport in essence, this chapter mainly summarizes the characteristics of railways, expressways, waterways and air transport respectively (Wu, 2009) (Table 3.1).

3.1 Technical and Economic Characteristics of High-Speed Railway

Historically, there are three reasons for the emergence of high-speed rail system. First of all, in order to overcome the limited transmission capacity of conventional routes, some new investments and more effective solutions are needed. This is the fundamental reason for the Tokaido Shinkansen in Japan and TGV Southeast Line in France. The same is true of Korea, China's Taiwan Province. Second, the high-speed rail is designed to increase the speed of the particularly slow part of conventional lines, which cannot be improved due to the huge cost and low rail transit technology. This is the case with Germany's high-speed rail. Third, high-speed rail is a way to improve accessibility in remote areas. The most typical examples are the Sanyo Shinkansen running through Osaka and Fukuoka and the first AVE route in Spain, from Madrid to Seville.

X. Lin, *High-Speed Railways and New Structure of Socio-economic Development in China*, Research Series on the Chinese Dream and China's Development Path,
https://doi.org/10.1007/978-981-19-6387-2_3

Table 3.1 Technical and economic characteristics of main transportation modes

Rail transport	Strong adaptability: continuity and reliability, two-way transportation Large transport capacity: The transport capacity depends on the weight of trains and the number of trains passing through the line day and night High safety: high-tech applications such as electronic computers and automatic control to prevent accidents and passenger casualties and reduce damage High speed: 60–80 km/h for conventional railways, 140–160 km/h for some railways and 210–310 km/h for high-speed railways Low energy consumption: Compared with highways, the friction resistance between wheels and rails of high-speed rail is less than that between automobile wheels and the ground, and its energy consumption is only one tenth of that of highways Low environmental pollution Low cost: higher than waterway and pipeline transport, and several times to ten times lower than highway
Highway transport	Good technical performance indicators: improved power performance and reduced fuel consumption Mobility and flexibility: door-to-door transportation is possible; Has strong adaptability to the size of passenger traffic and freight volume Fast delivery speed: the delivery speed of short-distance passenger transport is generally faster than that of railway; The delivery speed of long-distance bus passenger transport relying on expressways is often higher than that of railways when the distance is similar Less initial investment and short payback period: According to relevant data in the United States, for every $1 in revenue, road freight enterprises only need to invest $0.72, while railways need $2.7. Highway transportation capital can be turned around three times a year, while railway capital can be turned around once every 3–4 years The unit cost of highway transport is high, and it is harmful to the environment, especially for long-distance transportation
Waterway transport	The initial capital construction investment is low, and the construction investment, maintenance and management costs of the waterway are relatively low Large transport capacity: With natural waterways, the transport capacity is almost unlimited; Though restricted by inland waterways, waterways with good transportation conditions can still reach a large scale in capacity Low unit transportation cost: the waterway has a large carrying capacity, so the unit cost is lower than that of railway and highway. In particular, sea transportation is the least expensive transport means, which is suitable for the transportation of raw materials and bulk materials with low transportation cost tolerance Long distance: the transportation distance can be as long as several thousand nautical miles to tens of thousands of nautical miles; may reach any accessible port The slow delivery speed: First, the power and fuel cost of the main engine are saved; Second, water transport is often an intermediate transport link, so ports at both ends have to rely on other means of transport The external operating conditions are complex, with many uncertain conditions and great operational risks: the depth and velocity of inland waterways vary greatly in seasons, and some river sections also have dangerous beaches and reefs; Ocean shipping routes means long voyages, often passing through different climatic zones, with ever-changing climate. In addition, due to the multi-link requirements of water transport, it needs the cooperation of ports, ships, communications and navigation, agencies, inspection agencies, customs and other departments throughout the travel. The management work involved is complex and international

(continued)

Table 3.1 (continued)

Air transport	Very high-tech High speed: greatly shortened travel time for passengers, reduced inventory and operating costs Better mobility and flexibility: Air transportation is not limited by topography and landforms, and air routes can be opened with the support of airport and air route facilities. Greater mobility if helicopters are used Safety, reliability and comfort: measured by unit passenger turnover or death per unit flight time. With the use of wide-body aircraft, comfort can be brought into full play. Advanced equipment and thoughtful ground service create a comfortable travel environment for passengers Short construction period and less investment: Short capital construction cycle and low investment: shorter construction cycle, lower investment and faster payback than railways and highways High transportation costs: small volume and load capacity, large energy consumption per unit of transport turnover, high mechanical maintenance costs, highest transportation costs

3.1.1 The Elasticity of Speed

It is customary to think of high-speed rail as a technical concept and define it with maximum speed. Generally speaking, high-speed railway refers to the railway running at a speed of more than 200 km/h. It is a comprehensive integrated system of modern high-tech that combines infrastructure, fixed and mobile equipment suitable for high-speed operation, scientific security configuration and transportation organization. For newly built lines, the speed must reach 250 km/h. The speed of high-speed rail lines upgraded on traditional lines should reach 200 km/h. If the upgraded route is affected by topography or urban planning, the speed should be adjusted appropriately to adapt to new situations. Theoretically, the term 'speed' may define all railway facilities that provide high-speed rail services. But in fact, speed is not the best indicator, because in fact speed should be controlled. For example, in densely populated areas, or at places where there are viaducts and tunnels, the speed of high-speed rail cannot meet the prescribed requirements. In densely populated areas, noise and accident rate must be reduced, while in places with viaducts or tunnels, the speed must be controlled at 160–180 km/h for safety reasons.

The engineering principle of high-speed rail is the same as that of ordinary railway: the track provides a very smooth and hard track surface, which makes the train generate the least friction and energy consumption when traveling on it. However, the traditional railway is technically different from the high-speed railway. In terms of operating system, the traditional railway relies on external (electronic) signal lights and automatic signal systems, while the high-speed railway can be operated in the cab. For example, CBTC (Automatic Train Control System Based on Wireless Communication) is used to control the operation of high-speed rail in China. The outstanding advantage of CBTC is that it has made the two-way communication between vehicle and ground a reality. In addition, due to the large amount of transmitted information and fast transmission speed, the mobile automatic block system is formed, which greatly reduces the interval cable laying, and the one-time investment and daily maintenance work, so it does not only greatly improve the interval passing capacity, flexibly organize bidirectional operation and unidirectional continuous departure, but also makes the train operation control of different speeds, different traffic volumes and different types of traction possible.

3.1.2 The Complexity of High-Speed Rail Costs

The construction of a high-speed rail requires a special technique to eliminate all technical obstacles that limit the speed of the railroad to 250 km/h–300 km/h. This technique includes road intersections, emergency stops and sharp turns at high speeds. Sometimes, new signal mechanisms and more powerful power systems are also needed. Moreover, in some joint development models, it is also necessary to build some connections and specialized rails so as not to use rails together with trucks and trains running at normal speed. These common design features do not apply to all high-speed rail projects. As a matter of fact, due to the different technologies used in different high-speed rail projects, the cost of high-speed rail construction is also different.

According to UIC (International Union of Railways) (2005), there are three main costs for building a high-speed rail: (1) planning and land price, including feasibility study and technical design, and others (legal and administrative fees, licenses, business licenses, etc.), which account for about 5%–10% of the total investment cost. (2) Infrastructure construction costs, including all costs related to terrain preparation and platform construction, generally account for 10% and 25% of the total investment. In some special cases (such as viaducts and tunnels), the cost will account for 40%–50%. (3) The cost of superstructure, including track, signal system, pantograph, electrification mechanism, communication and safety facilities, accounts for about 5%–10% of the total cost. Although these three costs are included in any project, their proportions are different in different projects. According to UIC (2006), five types of high-speed rail construction can be identified: (1) Large corridors separated from other high-speed rail lines, such as AVE in Madrid-Seville section. (2) The reticular and complete large corridor, such as the Paris-Lille section, which makes the Paris-Lyon section fully integrated into the French high-speed rail network. (3) Minor expansion or supplement to existing corridors, such as Madrid-Toledo and Lyon-Valance. (4) Large-scale projects, such as European Tunnel, the Grand Belt, and Mexico Sea-crossing Bridge. (5) Small projects built on traditional railway networks, such as Germany or Italy.

Once a high-speed railway is in operation, it also incurs operating costs for infrastructure and rolling stock. The operating costs of the infrastructure are primarily its use and maintenance costs. The line maintenance cost of high-speed rail includes five types: track maintenance, power costs, signalling costs, communications and other costs. Some of these costs are fixed, and the operation path conforming to technical and safety standards is adopted. Other costs, such as track maintenance, depend on the frequency of track use. Similarly, the cost of electronic traction devices and pantographs depends on the number of trains passing by. According to the statistics of five European countries (Belgium, France, Italy, the Netherlands and Spain), on the whole, the maintenance cost of infrastructure and track accounts for 40–67% of the total cost, the signal cost accounts for 10–35% of the high-speed rail cost, and the maintenance cost of a single track of high-speed rail is between 28,000 and 33,000 euros/km.

Operating costs of rolling stock can be divided into four main types: shunt and train operating costs (mainly labor costs), and maintenance costs of all vehicles, which depend on the technology adopted by the train, and equipment, energy and operating costs which ultimately depend on the operator's expectation of traffic volume. In Europe, almost all countries have their own technical systems to solve their own traffic problems. For example, France used TGV Reseau and Thalys to provide international services, but in 1996 it introduced TGV duplex with double capacity. ETR-500 and ETR-480 are used in Italy, AVE is used in Spain, and five types are used in Germany: ICE-1, ICE-2, ICE-3, ICE-3 polycourant and ICE-T. Each high-speed train has different technical characteristics, namely length, structure, integration, weight, energy, traction, and inclination. If the vehicle type is not considered, the transition cost depends on the distance between the main station and the sub-station and the time when the train stops at the main station. On average, the cost per seat is 53,000 euros. For a 500-km high-speed rail line, assuming full load, the operating and maintenance cost of the train per kilometer per seat is 41.3(500*0.0776 + 500*0.005) Euros (French double-deck high-speed railway) to 93 Euros (German ICE-20)[1].

However, among various factors affecting the construction and operation cost of high-speed railway, the construction mode of high-speed railway also plays an important role to a certain extent. It is generally believed that there are four development modes of high-speed rail in the world (Yan et al., 2012): the first one is the exclusive development mode. High-speed railways are completely separated from traditional railways, such as the Shinkansen in Japan. The existing traditional railway in Japan is a narrow track with a gauge of 1.067 m, which has reached its capacity. That is to say, the Shinkansen needs to build a new track suitable for high-speed rail, because the standard gauge of high-speed rail is 1.435 m. The main advantage of this model is that the market operation of traditional railway and high-speed rail is separated. When Japan National Railways (JNR) went bankrupt in 1987, the complete high-speed rail services and facilities could be privatized. There are exceptions, of course. The Shinkansen in Japan has not reached the maximum speed, because some sections still retain the original narrow gauge, which is shared by high-speed rail to reduce the land cost. In addition, in places like Tokyo and Osaka, high-speed trains have to slow down and wait and make way. The second mode is the hybrid high-speed, whereby the high-speed train runs on the newly-built line as well as the upgraded traditional railway. For example, TGV in France. The advantage of this model is that it reduces the cost of railway construction. The third mode is the mixed traditional, whereby ordinary trains run on high-speed rail lines. For example, AVE of Spain applies a technology that can adapt to these two railways to trains. The biggest feature of this

---

[1] HSR Database, data in 2002 values.

mode is that it not only saves the purchase cost and maintenance cost of vehicles, but also enables ordinary trains to continue to serve on a high-speed rail line. The fourth is the full mixed mode, running trains of two or more speed classes, with high-speed trains mixed with ordinary coaches and even express freight trains. Examples include the German ICE and the Italian (Rome-Florence) high-speed trains, which use upgraded conventional railways during the day and ordinary wagons at night. This makes up for the maintenance cost of infrastructure.

The environmental cost of high-speed railway cannot be ignored and are largely dependent on the choice and use of technology. The infrastructure construction or service operation of high-speed rail will generate environmental costs, including cost of land occupancy, barrier effect, visual intrusion, noise, air pollution and its impact on global warming. As for pollution, the amount of polluted gas emitted by high-speed trains depends on the energy consumption in a given trip, while air pollution may come from the electric devices in high-speed trains. As we all know, high-speed rail is more environmentally friendly than private cars or airplanes. If converted into gasoline, the initial energy consumption of high-speed rail is 2.5 L per 100 km (6 L for cars and 7 L for airplanes). Similarly, the carbon dioxide emissions per 100 km are 17 tons for airplanes, 14 tons for private cars and 4 tons for high-speed trains.

Energy saving has always been one of the most important operation, organization and management contents of Shinkansen in Japan, and its main means is to update energy-saving vehicles. Take JR Tokaido as an example. The area of the three metropolitan areas connected by this line accounts for more than 24% of Japan's total territory, with a population of about 59% and a GDP of about 65%, which is definitely the main artery of Japan's traffic. After its establishment in 1964, starting with the 0-series vehicles at 210 km/h, the 100-series vehicles at 220 km/h, the 300-series vehicles at 270 km/h, the 700-series vehicles at 285 km/h and the N700-series vehicles at 300 km/h were successively updated. Among them, energy-saving vehicles refer to 300 series, 700 series and N700 series vehicles, also known as the second -generation vehicles, which mainly adopt technical means such as light weight of the car body, car body tilting system and electric power regeneration. Through the renewal of energy-saving vehicles, Tokaido JR has achieved remarkable energy-saving and emission-reduction effects (Sun et al., 2011). Train noise depends to a great extent on the technology used. Generally speaking, the noise generated by high-speed railway is wheel-rail noise. Although the sound of high-speed trains is 80–90 decibels, it is considered as noise pollution in urban areas. According to the calculation, if the speed is 280 km/h and the noise is controlled at 55db (this value is tolerable), then a 150 m sound-proof corridor is needed. It is costly to build such a soundproof corridor. For example, the Paris-Lyon Line has brought noise problems to the neighboring towns, causing many complaints from the public, so a large number of sound insulation screens have to be built to reduce the impact of noise on the surrounding residents.

3.1.3 The Inherent Nature of Environmental Problems in High-Speed Rail

Different from the noise emitted by ordinary trains, high-speed trains use slag-free tracks, which eliminates the noise caused by the friction between wheels and tracks, but its noise is much higher than that of ordinary trains in the past, because of its fast-running speed. As high-speed railway lines generally pass through many large and medium-sized cities with dense population, noise pollution should be the most important effect that high-speed railway has on urban environment. Table 3.2 shows the noise generated at different train speeds. The faster the speed, the greater the noise generated.

The main sources of railway noise pollution in the areas along the high-speed railway are the running noise of trains, the overhaul and preparation operations of various sections and multiple units, and high-noise fixed equipment. Among them, aerodynamic noise accounts for a large proportion in the high-speed running of high-speed railway trains. In view of the sources and characteristics of high-speed railway noise, noise prevention should be considered from the acoustic performance of running high-speed trains and the planning and layout of towns and cities. For instance, it is suggested to choose high-speed trains with excellent acoustic performance, avoid building sensitive functional areas such as residential areas, schools and hospitals in areas where high-speed trains pass, and build isolation and protection barriers. The main measures of noise control include demolition and transfer, setting up sound barriers, green forest belts and sound insulation walls. According to the analysis of noise and vibration data measured at survey points along the line during the construction period of high-speed railway, the over-standard rate of noise data in hospitals, schools and nursing homes is 100%, and the average rate in residential areas is 70%. Therefore, the noise pollution of high-speed railway is an environmental problem that cannot be ignored.

Table 3.2 Noise levels generated at different speed conditions

Train speed (km/h)	Noise value (dBA)	Train speed (km/h)	Noise value (dBA)
160	79.5	240	86
170	80	250	86.5
180	81	260	87.5
190	81.5	270	88
200	82.5	280	88.5
210	83.5	290	89
220	84.5	300	89.5
230	85.5		

Location of measuring point: 25 m from the center of train running line and 3.5 m above the track surface

Vibration produced by trains on high-speed railway will disturb people's daily life, such as disturbing residents' sleep, rest and study. Because vibration is closely related to people's life, long-term exposure to vibration will also lead to physical diseases. High-speed railways that don't operate at night have relatively little influence on residents' sleep. In addition to its effect on people, vibration will also affect the normal use of precision equipment and instruments, and even cause damage to buildings. Although environmental vibration will not cause the collapse of buildings like earthquake and explosion, it may still do great harm to some ancient and old buildings, and even cause damage. When the frequency of vibration is 20–2000 Hz, noise will be generated and the environment will be worsened.

According to the evaluation standard of vibration level, the vibration level on both sides of the railway trunk line, that is, residential areas on both sides of the outer rail 30 m away with daily traffic volume of not less than 20 trains, should be less than 80 dB. When the train speed exceeds 160 km/h, the vibration level (NVH) measured at the observation point of 15 m has exceeded 80 dB, and with the increase of train speed, the vibration level also rises. The running speed of high-speed railway is far greater than 160 km/h, and its vibration level is between 70 and 90 dB in the area 30 m away from the outer rail of the railway. According to the data of environmental monitoring, the over-standard rate is 30.6% indoors and 57.9% outdoors.

To sum up, the environmental problems of high-speed railway mainly include vibration, noise and electromagnetic interference, among which vibration and noise are getting more and more public attention. High-speed train noise brings great disturbance and harm to the living, learning and working of residential areas along both sides of the line, causing noise in many functional areas to exceed standard values. For this reason, the government needs to pay more attention to noise and vibration issues. It is expected that the further maturity of high-speed maglev transportation technology will bring hope for noise treatment in high-speed environment. See Fig. 3.1 for an overview of the comparison of noise made by different types of high-speed rails.

Noise pollution around the airport is more serious. In the airport, the area with aircraft noise level higher than 75 dB is about 50 square kilometers, excluding the area occupied by airport buildings and areas with noise level higher than 90 dB. In terms of noise pollution, high-speed railway produces less noise than other modes of transport. For example, in France, the measured noise level is 65 dB(A) at a distance of 25 m from TGV high-speed railway line. The average noise level of a highway passing through 30,000 vehicles every day is 75 dB(A), which is 10 dB(A) higher than that of high-speed railway. While the noise near the airport is higher, and the noise level in the area 300 m away from the airstrip reaches 100dB(A). In contrast, due to the improvement in operation characteristics and dynamic design, the noise pollution of the new generation of high-speed trains is decreasing. For example, the noise of TGV 3rd generation trains is reduced by 8 dB(A) compared with that of the 1st generation trains.

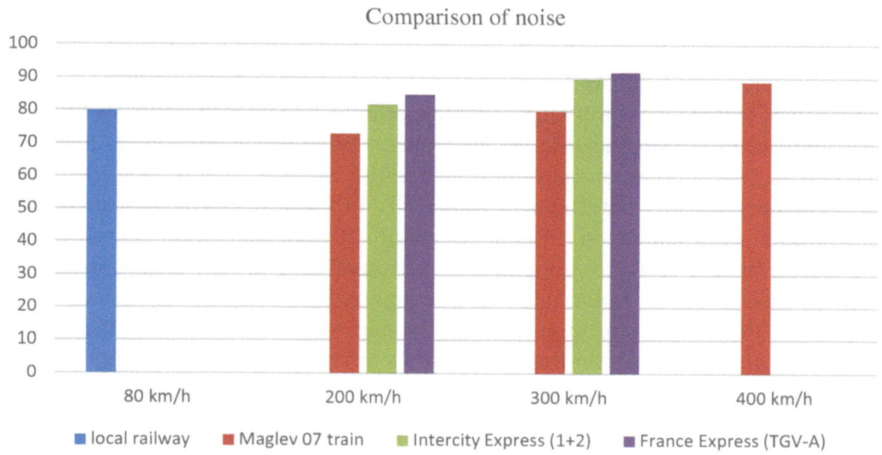

Fig. 3.1 Noise comparison diagram of different types of high-speed rails (The sound level/decibel (A) at a distance of 25 m when the train passes by)

3.1.4 Derivation and Cross-Border of High-Speed Rail Market Demand

Since the commercialisation of railway projects in the early 1970s, high speed railways, which meet demand and generate revenue, have been seen as a success story. In many countries it is seen as one of the key factors in the revival of passenger rail transport, which had lost significant market share due to fierce competition from road transport and air transport. For example, in France and Spain, high-speed rail is the only business of railway companies that can make up for operating costs. Estimates of the demand for high-speed rail are highly controversial. By 2005, the Shinkansen in Japan had accumulated 150 billion passengers. In South Korea, the high-speed rail line opened in 2004 has severely hit the domestic air transport sector in just two years, transporting 40 million passengers per year. As for Europe, it reached a record 17 billion people/kilometer in 2005. Between 1994 and 2004, the transportation sector grew at an average annual rate of 15.6%, with double-digit growth in early years and some decline in recent years. In addition to the demand drivers such as price, quality and income, its growth depends heavily on the construction of high-speed rail. From 1994 to 2004, the largest share of European high-speed rail growth was the Paris-Lyon line in France, which was 70% at the initial stage (and then decreased to 55%). France's high-speed rail has developed faster in the hub zone of Paris, while other lines, especially the older ones, have not increased significantly.

Further, the high-speed railway does not rely solely on its technical characteristics to obtain good market demand automatically, so the derivation of its transportation demand still exists. Moreover, in view of its higher cost than the existing railway, the high fare level is a strong preference of the high-speed rail operating department under the conditions permitted by government policies. If the competition is sufficient, the competitive advantage of high-speed railway needs to be realized by seeking other comprehensive means to improve quality service besides fare. It can even be obtained by leading the comprehensive transportation system to optimize its structure and improve the comprehensive service quality. What needs attention is that from the perspective of comprehensive development of railway land, in fact, the timing of high-speed railway development (especially in China) is closely related to the acceleration of urbanization. In reality, even if the high-speed rail project has not started, its impact on the development of surrounding land has already started the moment it is included in the nationally recognized planning or other documents. Therefore, the demand for high-speed rail comes not only from the transportation factor, but also from the non-transportation factors which rely on the improvement of transport facilities to gain huge external benefits. This is the cross-bordering of high-speed rail demand.

3.2 Technical and Economic Comparative Advantages of High-Speed Railway

Compared with other modes of transport and traditional railways, the technical and economic comparative advantages of high-speed railways are generally reflected in speed, transport capacity, time, safety and comfort, land resource utilization, and energy conservation and environmental protection.

3.2.1 Advantage in Speed

Fast speed, short departure interval and high traffic density are the main technical advantages of high-speed railway, and other advantages are basically derived from them. Research and practice have shown that the reason why the traditional railway is gradually at a disadvantage in the competition of the existing transportation market lies in that the improvement of its technical speed has reached the bottleneck, especially the economic and safety bottleneck. Before the large-scale construction of high-speed railway in China, the existing lines have been speeded up, and remarkable achievements have been made. See Table 3.3.

In order to overcome the speed limitation of existing railways, countries all over the world have committed to the development of high-speed rail technology and have repeatedly set new test speeds. See Table 3.4.

There is still much room for improvement in the speed advantage of high-speed railway. With the maturing of high-speed maglev transportation technology, especially the official opening of the 30-km high-speed maglev airport line in Shanghai on

Table 3.3 List of 200 km/h speed-up lines

No	Lines	Speed-up section	mileage/km	Line mileage/km	Reconstruction project
1	Beijing-Harbin Railway	Tongzhou–Fengrun	242.4	1106.4	Electrotechnical improvement, special reconstruction
		Shanhaiguan–Huanggu Tun	756.0		Remediation of Diseases at Subgrade
		Caijiagou–Wujia	108.0		Reconstruction for Special Purposes
2	Beijing-Shanghai Railway	Zhoulizhuang–Qingxian County	112.8	694.6	Electrotechnical improvement
		Jiedi–Changzhuang	205. 6		Electrotechnical improvement
		Gaojiaying–FuLiji	96.8		Electrotechnical improvement
		Suzhou–Tangnanji	82.0		Electrotechnical improvement
		Zhenjiang South–Benniu	102.8		Electrotechnical improvement
		Kunshan–Shanghai	94.6		Electrotechnical improvement
3	Beijing-Guangzhou Railway	Doudian–Caohe	185. 8	952.8	Special transformation
		Yuanshi–Xingtai	136. 6		Special transformation
		Hebi–Weihui	79.2		Special transformation
		Guanting–Luohe	167.4		Reconstruction for Specific Purposes
		Luohe–Changtaiguan	270.0		Reconstruction for Specific Purposes
		LiJiazhai–Chenjiahe	113.8		Reconstruction for Specific Purposes
4	Guangzhou-Shenzhen Railway	Xintang–Honghai	43.2	43.2	Electrotechnical improvement

(continued)

Table 3.3 (continued)

No	Lines	Speed-up section	mileage/km	Line mileage/km	Reconstruction project
5	Lanzhou-Lianyungang Railway	West Xuzhou–East Zhengzhou	660.8	828.4	Reconstruction for Specific Purposes
		Xianyang–Changxing	167.6		Electrotechnical improvement
6	Shanghai-Kunming Railway	Bailutang–Tangya	244.2	1507.0	Electrotechnical improvement
		Bailongqiao–Guixi	540.4		Electrotechnical improvement
		Yingtan–Binjiang	564.8		Electrotechnical improvement
		Liangcun–Baiyuan	47.8		Electrotechnical improvement
		Yaojiazhou–Wulidun	109.8		Electrotechnical improvement
7	Qingdao-Jiaozhou Railway	Loushan–Licheng	644.0	644.0	Electrotechnical improvement
8	Wuchang-Jiujiang Railway	Heliu–Yangxin	227.4	227.4	construction of the second line
Total			6003.8	6003.8	–

Source Adapted from Qing Sanhui et al. (2014)

Table 3.4 The highest speed of test for high-speed railways in the world from 1988 to 2011

Countries	Railways	Locomotive type	Locomotive model	Test speed/(km/h)	Test date
Italy	–	–	TAV	319	1988
South Korea	–	–	KTX	352.4	2004
Germany	–	–	ICE	406.9	1988–05
Japan	Sanyo Shinkansen	Electric locomotive	JR500	443	1996–07
China	Beijing-Shanghai High-speed Railway	EMU (electric multiple unit)	CRH380BL	487.3	2011–01
France	TGV	EMU	TGV-V150	574.8	2007–04

December 31, 2002, the high-speed maglev train with the highest speed of 430 km/h has been initially accepted by the market.

3.2.2 *Advantage in Traffic Volume*

The volume of transport is the amount of passengers and goods carried by the transport sector in a given period, expressed in terms of volume and turnover. The volume is the actual number of passengers and tonnes of goods transported by the transport sector, and is accordingly divided into passenger and freight volumes, measured in 'persons' and 'tonnes' respectively. Turnover is a transportation production output index that indicates the transportation volume and transportation distance, which is divided into passenger turnover, cargo turnover and conversion turnover. Passenger turnover is the product of the number of passengers transported by the transportation department and their transportation distance in a certain period of time, expressed in passenger kilometers. Turnover of goods means the product of the tonnage of goods transported by the transportation department and its transportation distance in a certain period, expressed in freight ton-kilometers. It can be divided into total turnover, turnover of goods by various modes of transportation and turnover of goods by category. Converted turnover is the amount of passenger turnover and cargo turnover converted into the same unit of measurement, measured in converted tonne kilometres. In China, 1 person-kilometre is usually converted to 1 tonne-kilometre for railways and water transport enterprises directly under the Ministry of Communications, 10 person-kilometres to 1 tonne-kilometre for motor transport and 13.33 person-kilometres to 1 tonne-kilometre for international routes and 13.89 person-kilometres to 1 tonne-kilometre for domestic routes in civil air transport. Here, the absolute numbers of capacity and turnover are mainly applied to compare the capacity of various modes of transport.

Railways, due to their technical characteristics, can take on a large number of passenger and cargo transport tasks, and have long been the absolute mainstay of long-distance cargo transport in large volumes. The railways are also a major market for medium and long distance passenger transport during peak holiday periods. High-speed railway, as a new mode of transportation, has been rising for nearly half a century, and its comparative advantage in traffic volume is one of the main reasons for building high-speed railways in various countries. From the Shinkansen in Japan to the pioneering countries of high-speed rail such as France, to the countries and regions where modern high-speed rail is developing rapidly such as Taiwan Province and mainland China, the first high-speed rail line is always built in the most densely populated and economically developed regions. This is a clear indication that high speed railways are primarily driven by high density and year round huge demand for travel and logistics.

An important technical and economic comparative advantage of high-speed railways is their large transport capacity, with an annual capacity of up to 160 million passengers on a two-line high-speed railway. Especially, when high-speed railway runs high-density, public transportation and flexible grouping EMU trains between cities, its passenger capacity is unmatched by highway and civil aviation. At present, almost all the high-speed railways in the world can meet the requirements of driving interval of 4 min or less. During the peak hours of Shinkansen in Japan, the minimum interval between trains can reach about 3 min, and the departure density of trains can basically meet the demand of passengers. After the completion of Beijing-Shanghai high-speed railway in China, one-way annual passenger transport can reach 80 million passengers, and its passenger turnover far exceeds that of the same line of highway and civil air transport.

3.2.3 Advantage in Time

Although there is a correlation between time and speed advantage, they are not the same. Speed advantage is basically a purely technical attribute, while time advantage includes the advantages of transportation organization and operation management. With the continuous improvement of labor productivity, the value of passengers' time increases, and there is a requirement for further saving of travel time, so the number of passengers who choose higher-speed means of transportation is increasing. In the early stages of economic development, this change manifests itself in an increase in the market share of high-speed rail as travellers shift from lower speed conventional rail and road to high-speed rail. At higher levels of economic development, the public can afford to seek higher travel speeds and a proportion of long-distance travellers move to air, so the market share of high-speed rail tends to decrease. As a result, travel time, one of the most important concerns of travellers, is largely determined by the travelling speed of the transport on which the journey is made. High-speed railway is the longest and fastest transportation tool on land. In recent years, the operating speed of high-speed railways that have started construction one after another is over 300 km/h, far exceeding the speed of highway traffic. Compared with the fastest air transportation, the high-speed railway also shows great time competitiveness by virtue of its advantages such as convenience for passengers to travel, short departure interval, and on-demand travel. Take Beijing to Shanghai as an example. Taking into account factors such as check-in and waiting time at the airport, as well as the time spent travelling to and from the airport (which is generally built at a distance from the city), a total of 5 h is required for a flight, which is not a time advantage compared to high-speed rail.

High-speed railways can form an economic belt or economic corridor from isolated, scattered economic areas within 800–1000 km of each other, making the economic and social activities in this area more closely linked and developing at a faster pace. The resulting social and economic benefits are very significant and incalculable. The following is an example of the high-speed railway in Guangxi, China, to verify the time advantage of high-speed railways over a certain spatial distance.

In general, the value of travel time saved is an important parameter in determining consumer travel behaviour and transport choice. Consumers consider the value of travel time savings that can be derived from choosing a particular mode of transport and weigh this against the price they would have to pay if they chose that. The total value of travel time saved is the sum of the value of travel time saved by all consumers who choose this mode of transportation. Generally speaking, the traffic construction project is feasible only when the total value of saved travel time exceeds the corresponding total cost.

The basic idea of this analysis is to first analyse the utility function that influences consumers' choice of transport mode for long-distance travel and the main factors that determine the value of travel time savings, and to explain the impact of the value of travel time savings of different transport modes on their market shares, using three different transport modes: road, rail and civil aviation as examples. Consumers often choose among various modes of transportation for long-distance travel, and their choices are influenced by many factors. Assume that consumers mainly consider three factors: the value of saved travel time (VTTS), the price paid (P_i) and the preference of consumers for a certain mode of transportation (Pre_i). Only when the benefit of travel time saved by choosing a certain mode of transportation to consumers exceeds the price paid by consumers will they choose this mode of transportation. The utility function for the consumer's choice of the "i" mode of transport can therefore be expressed as (3.1).

$$U_i = U\{VTTS_i(T_i,\ T_0,\ aI),\ P_i,\ \Pr e_i\} \tag{3.1}$$

Among them, T_i is the total travel time of the "I" mode of transportation; T_0 is the total travel time for another mode of transportation; $VTTS_i(T_i, T_0, aI)$ represents the value of saved time by choosing the "i" mode of transportation relative to another; The value of time is related to the multiple *a* of consumer's income level I. The higher the income level, the higher the economic value of time. If there is a linear relationship between the value of saved travel time and income level, it can be assumed that $VTTS_i(T_i, T_0, aI) = a(T_0\text{-}T_i) \times I$. Consumers with high incomes may afford to pay for faster and more expensive modes of transportation. Consumers may compare and contrast the benefits $VTTS_i(T_i, T_0, aI)$ brought by the travel time saved by different modes of transportation with the cost (P_i) paid for it, and then tend to choose one that maximizes their benefits.

Travel time refers to the total travel time, that is, the sum of the time spent from the starting point to the destination. For example, the total travel time by air consists of at least three parts, namely: (1) the time for passengers to walk from their residence, take the subway or bus to the airport, pass the security check and board the plane (T_{a1}); (2) The flight time (T_{a2}) of the aircraft, including the time of accelerated lift-off, air flight, deceleration and landing; And (3) the time from getting off the plane, leaving the airport and arriving at the destination (T_{a3}). Use Ta to express the total travel time of civil aviation travel, then $T_a = T_{a1} + T_{a2} + T_{a3}$. The total travel time by train is also composed of three parts, namely, the time when passengers arrive at the station from their residence and get on the train (T_{t1}), the train running time (T_{t2}), and the time from getting off, leaving the station and arriving at the destination (T_{t3}). If Tt is used to express the total travel time by railway, then $T_a = T_{a1} + T_{a2} + T_{a3}$. Then the total travel time difference between the plane and the train is (T_t-T_a) = (T_{t1}-T_{a1}) + (T_{t2}-T_{a2}) + (T_{t3}-T_{a3}). Here, (T_{t1}-T_{a1}) indicates the time difference between passengers arriving at the station from their residence and arriving at the airport, security check and boarding. In general, T_{t1} is less than t_{a1}, because the train station is generally closer to the city center than the airport; (T_{t2}-T_{a2}) indicates the difference between train travel time and plane travel time in the same city. Because the travel speed of train is less than that of plane, $T_{t2} > T_{a2}$; (T_{t3}-T_{a3}) indicates the time difference between the passengers from the railway station to the destination and from the airport to the destination, generally $T_{t3} < T_{a3}$. If we don't consider the time by plane or train, the travel time that passengers can save by choosing railway can be expressed as Δt: $\Delta t = (T_{a1}\text{-}T_{t1}) + (T_{a3}\text{-}T_{t3})$. If S represents the distance between two cities, the time for passengers to take railway or civil aviation between two cities is $T_{t2} = S/V_t$, $T_{a2} = S/V_a$ respectively. V_a and V_t respectively represent the average travel speed of airplane and train. Although the travel speed of airplane is much higher than that of train, the total travel time of traveling by civil aviation or train may be equal, that is, $\Delta t + S/V_a = S/V_t$. From this formula, we can get the distance S* equal to the total travel time of choosing civil aviation travel and choosing railway travel, as shown in formula (3.2).

$$S^* = (\Delta t \times V_t \times V_a)/(V_a - V_t) \tag{3.2}$$

From the point of view of the value of saved travel time and the choice of transportation mode, combined with the actual situation of transportation industry in Guangxi, we can further explain with typical data of high-speed rail and civil aviation, and estimate the impact of Guangxi high-speed rail on passenger transportation in its economic belt. Generally speaking, it takes more time to get to the airport and check in at the security check than to get to the railway station. Assuming (T_{a1}-T_{t1})≈1.5 h, it usually takes one and a half hours to leave before going to the railway station. It is assumed that it takes more than half an hour to get to the city center from the airport than from the railway station, that is, (T_{a3}-T_{t3})≈0.5 h. The sum of them is 2 h, $\Delta t = 2$ h. The average domestic travel speed of airplane is about 600 km/h, and the average travel speed of EMU on high-speed railway with a speed of 300 km/h can reach 225 km/h, so according to Formula 3.2, we can calculate the distance S* with equal total travel time by civil aviation or high-speed railway.

$$S^* = (\Delta t \times V_t \times V_a)/(V_a - V_t) = (2 \times 225 \times 600)/(600 - 225) = 720\ (\text{km})$$

As stated above, when the distance between the two cities is less than 720 km, people who choose to travel by high-speed railway can reach their destination faster than by plane. Of course if the HSR station is far away from the city centre, or if there is no metro or convenient public transport directly to the HSR station, S^* is significantly smaller and the value of the HSR in terms of travel time savings is significantly reduced.

Similarly, according to formula 3.2, we can calculate the distance with equal travel time by high-speed rail or bus. Although HSR runs at a higher speed than a car, a car allows for door-to-door transport. If we assume that the time to the station saved by travelling by car compared to travelling by HSR is 1 h, i.e. $\Delta t = 1$, and the average travel speed of a car, Vv, is 100 km/h, the distance S** which is equal to the total travel time of traveling by car and high-speed rail can be calculated:

$$S^{**} = (\Delta t \times V_v \times V_t)/(V_t - V_v) = (1 \times 100 \times 225)/(225 - 100) = 180\ (km)$$

The calculation shows that although the high-speed rail runs faster than a bus, the car may reach its destination faster than the HSR over a travel distance of 180 km.

According to formula 3.2, we can also calculate the distance with equal travel time by high-speed rail or ordinary railway. Although the running speed of high-speed trains is higher than that of ordinary railways, high-speed trains generally only set up stations in large and medium-sized cities, with relatively few stations. Suppose that by taking the ordinary train one can save 0.5 h, that is $\Delta t = 0.5$, and the average travel speed Vo of the ordinary railway train is 160 km/h. Then according to Eq. 3.2, we can calculate the distance S*** which is equal to the total travel time of taking the ordinary train and the high-speed train:

$$S^{***} = (\Delta t \times V_o \times V_t)/(V_t - V_o) = (0.5 \times 160 \times 225)/(225 - 160) \approx 277\ (km)$$

The calculation shows that although the high-speed railway runs faster than the ordinary railway, within the travel distance of 277 km, we may reach the destination faster by using the ordinary railway than the high-speed railway.

Under the above conditions, if we only consider the shortest total travel time, in the medium distance range of 277 km to 720 km, we can get to the destination faster by high-speed rail than by civil aviation, bus or ordinary train. In Guangxi, for example, most of the high speed trains currently under construction in Guangxi are within 720 km of each other, with only the Guizhou-Guangxi High Speed Railway (857 km) and the Hunan-Guangxi High Speed Railway (1013 km) being greater than 720 km in length. Of course, with the changes of the above parameters, the competitive advantage of high-speed rail will also change, enabling it to compete with civil aviation in a farther distance.

3.2.4 Advantages of High-Speed Railroad in Land Use

From an ecological point of view, transportation should occupy as little land as possible. Therefore, if a transport mode has the same capacity as other modes of transport but occupies less land, the mode of transport will have comparative advantages in the use of land resources. Among the various transports to make the most efficient use of the limited land, high-speed railways undoubtedly have obvious comparative advantages.

Compared with the highway, the infrastructure required by the high-speed railway is only 25% of the area required by the highway to transport an equal number of passengers. The width of subgrade of TGV high-speed railway in France is about 14 m, while the expressway with the same capacity is 28 m (4 lanes) to 35 m (6 lanes), and the land area used by a TGV high-speed railway is only 50% of that of a two-way 4-lane expressway.

Compared with the high-speed rail, the air transport infrastructure such as airport, landing runway, passenger terminal station, hangar, service building, radio radar device and special line to the airport occupy a large amount of land. There is also a large amount of land that loses its value due to the impact of strong noise; for example, the total area of the residential building range is restricted as a result of the noise impact. In addition, as the land required for the construction of airports has to meet certain requirements, the airports in some cities are located far away from the city, so that the relevant city-to-airport highways and other auxiliary facilities need to be built. Nonetheless, high-speed railways can be built in sparsely populated areas where land values are low and, in some cases, existing railway strips can be used. High-speed railways can run directly through cities, making it convenient and using less land. In addition, the areas along the suburban railway can be used for the development of individual villas and manors.

The evaluation of land use in monetary terms shows that it is 3.8 times more expensive to build two airports near St. Petersburg and Moscow, Russia than to build a high-speed railway line between the two cities. It should be pointed out that building a new airport near London can't solve the land problem, which is one of the reasons for building tunnels in the English Channel in Great Britain. Another example is Charles de Gaulle Airport near Paris, which occupies an area of 3000 hm^2, equivalent to 1/3 of the urban area of Paris. By contrast, the total area of Paris-Lyon high-speed railway line is only 2400 hm^2 (Guo and Zeng, 2000).

In addition, the advantages of high-speed maglev transportation technology in terms of land use are technically recognized. Compared with the ICE data of German high-speed maglev train and its intercity express train, the ICE covers an area of 14 m^2/m, and the high-speed maglev transportation covers an area of 12 m^2/m, while the high-speed maglev transportation for overhead construction only covers an area of 2 m^2/m.

3.2.5 *Advantages of High-Speed Railway in Energy Saving*

Transport is a major consumer of energy, and energy consumption standards are an important technical indicator of the merits of transports. In developed countries (Germany, for example), energy consumption for transport accounts for more than 30% of the country's final energy consumption, and is on the rise. Personal cars and airlines account for 70–74% of the total energy consumption in the transport sector, with railways being the least energy-intensive form of land transport. In addition, due to the finite nature of primary energy resources, social development requires for better use of renewable energy sources and new energy sources such as nuclear and solar energy. At present, cars and airplanes can only use disposable mineral oil (gasoline, diesel oil and aviation kerosene) as energy. The railway energy is mainly electricity (high-speed railways are all driven by electricity), so it is possible to use different forms of energy, such as nuclear energy, hydropower and other renewable energy. Therefore, guiding passenger flow to railways and high-speed railways that occupy less land, consume less energy and make full use of secondary energy and renewable energy is an important trend of transportation policy making in various countries. See Table 3.5 for the comparison of transportation means in energy consumption indicators of various countries (Lin, 2010).

According to Lin (2010), there are two energy-saving indicators: first, the energy consumption coefficient of passenger and freight transportation by various modes of transportation, and second, the total energy consumption of passenger and freight transportation in the whole society by a single mode of transportation throughout the year. Energy consumption coefficient is then divided into passenger energy consumption coefficient and freight energy consumption coefficient. The former refers to the energy value consumed by each mode of transportation to complete the unit passenger turnover, while the latter refers to the energy value consumed by each mode of transportation to complete the unit freight turnover. Assuming that the total energy consumption index of the whole society's passenger and cargo transportation in the whole year is completed by a single transportation mode, the calculation formula is as follows:

Total energy consumption of passenger transport in all modes of transportation = annual passenger turnover × passenger transport energy consumption coefficient in all modes of transportation.

Total energy consumption of freight transportation by various modes = annual freight turnover × freight energy consumption coefficient by various modes of transportation.

The results of comparison and calculation are shown in Tables 3.6 and 3.7.

Tables 3.6 and 3.7 show that the passenger energy consumption coefficient and freight energy consumption coefficient of railway are far lower than those of other modes of transportation. Assuming that the whole social transportation volume is

Table 3.5 Comparison of energy consumption of transportations in developed countries in the 1990s (Unit: %)

Country	Percentage of transportation industry in total energy consumption	Transportation method				
		Railways	Highways	Inland river	Aviation	Maritime transport
United States	34.8	2.0	76.7	1.0	14.3	9.0
Canada	27.4	5.1	80.3	4.6	10	–
United Kingdom	30.6	2.4	79.5	2.8	15.3	–
France	28.2	2.0	81.7	0.1	9	6.7
Germany	25.2	3.5	84.1	1.2	8.0	3.0
Italy	30.1	1.9	84.2	1.1	5.9	6.9
Japan	13.8	2.8	51.5	15.2	29.2	1.3

Table 3.6 Energy consumption coefficient of various transportation modes in China in 2007 and energy consumption of annual transportation turnover of each mode

Mode of transport	Total fuel consumption (tons)	Energy consumption factor for freight transport (kg/10,000 ton-km)	Passenger energy consumption coefficient (kg/10,000 person kilometers)
Highway	97,079,140	776.3	77.63
Civil aviation	11,298,913	3093.1	237.92
Water transport	7,723,515	120	40

Note Total oil consumption = passenger transportation energy consumption coefficient × passenger transportation turnover in the current year + freight transportation energy consumption coefficient × freight transportation turnover in the current year

Table 3.7 Annual total energy consumption by a single transport mode in China in 2007 (Unit: ten thousand tons)

	Railway	Highway	Waterway	Civil aviation
Passenger traffic	379	1197	864	3677
Freight	1781	56,219	8690	224,500
Total	2160	57,416	9554	228,177

Note As the pipeline transportation only includes freight, it is omitted

completed by one mode of transportation, the energy consumption advantage of railway is more prominent. The energy consumption of high-speed railway is lower than that of ordinary railway, so it has more obvious advantages in energy saving. China's railway energy consumption accounts for only 18% of the total consumption of the national transportation industry, while the turnover after conversion has reached more than 50%. High-speed railway adopts electric locomotive traction, which has great energy-saving advantages compared with transportation tools using liquid fuels such as petroleum. Compared with road and air transportation, it is the most economical mode of transport in medium and long distance transportation. The fuel consumption per kilometer of high-speed railway is 1/3 of that of automobiles and 1/5 of that of medium-range passenger aircraft. Therefore, the unit energy consumption of high-speed railway, mainly secondary energy, is the lowest, which has a comparative advantage compared with gasoline used by automobiles and airplanes.

Because of the contactless technology, the high-speed maglev system is more economical. At the same speed, Maglev consumes 20–30% less energy than the already "energy-efficient" high-speed wheel tracks. A small car uses three times as much energy as a maglev and an aircraft five times as much for the same distance. According to German experimental data, when the speed of an intercity express train is increased from 125 km/h to 185 km/h, its energy consumption increases from 29 to 51 watt-hours/tonne-kilometre. On the other hand, when the high-speed maglev train runs at 200 km/h, 300 km/h and 400 km/h, the corresponding energy consumption indexes are 22 W hours/ton km, 34 W hours/ton km and 52 W hours/ton km.

3.2.6 Advantages of High-Speed Railway in Pollutant Emission Control

An OECD survey in China found that cars are the largest source of urban air pollution. For example, during the warm season in Beijing, 92% of carbon monoxide (CO), 94% of hydrocarbons (HC) and 68% of nitrogen oxides (NO) come from automobiles. Similarly, in Guangzhou and Shanghai, on average over 80% of CO ~ 140% of NOx (nitrogen oxide) emissions come from automobiles, followed by aviation, and relatively little from trains (Lin, 2010).

The indicators are divided into two categories, namely, the annual per capita emissions of air pollutants and the emissions of air pollutants per 1,000 US dollars of GDP produced. In turn, the three main types of air pollutants are carbon oxides, nitrogen oxides and sulphur oxides. Establishing indicators based on previous statistics on the coefficients of air pollutants emitted by various modes of transport, the corresponding data are calculated according to the proportion of the total emissions from each pollutant, which gives us Table 3.8.

Table 3.8 shows the annual per capita emissions of air pollutants and the emissions of air pollutants per 1,000 US dollars of GDP produced by various modes of transportation. The annual per capita emissions of carbon oxides, nitrogen oxides and sulfur oxides in China's railways are 0.031, 0.53 and 4.02 kg/cap respectively, while the emissions of air pollutants per 1,000 US dollars of GDP are 0.005, 0.10 and 0.67 kg respectively. In the comparison of China's transportation pollutant emissions, the carbon oxide emissions of aviation and highway are 58 times and 55 times that of railway respectively, the nitrogen oxides 4.96 times and 11.5 times, and the sulfur oxides 1 time and 2.3 times that of railway respectively. The value of GDP converted from carbon oxides emitted by aviation and highway is 62 times and 60 times that of railway respectively, nitrogen oxides 4.8 times and 11.8 times, and sulfur oxides are 1 time and 2.31 times that of railway respectively. In the comparison of the three modes of transportation in the United States, in terms of the annual per capita emissions of aviation and highway are 58 times and 57 times that of railways, and the emissions of nitrogen oxides and sulfur oxides are 5 times and 11.5 times, 1 time and 2.3 times that of railways respectively. And for the emission index of air pollutants produced by aviation and highway for every $1,000 of GDP, the emission of carbon oxides, nitrogen oxides and sulfur oxides are 54 times and 52 times, 5 times and 12 times, 1 time and 2.27 times that of railway respectively. In the comparison of the three modes of transportation in OECD countries, as far as the annual per capita emission index of aviation and highway is concerned, the emissions of carbon oxide, nitrogen oxide and sulfur oxide are 58 times and 57 times, 4.9 times and 11 times, 1.003 times and 2.3 times that of railway respectively. In the air pollutant emission

Table 3.8 Comparison of pollutant emissions of three transport modes with two indicators from 1999 to 2004

		Railway			Highway			Aviation		
		COx	NOx	SOx	COx	NOx	SOx	COx	NOx	SOx
China	Annual per capita emissions (kg/cap)	0.031	0.53	4.02	1.76	6.12	9.25	1.81	2.63	4.02
	Average annual emissions per 1,000 US dollars of GDP (kg/l000 USD GDP)	0.005	0.10	0.67	0.30	1.18	1.55	0.31	0.48	0.67
U.S	Annual per capita emissions (kg/cap)	0.170	3.69	11.49	9.62	42.11	26.43	9.90	18.08	11.49
	Average annual emissions per 1,000 US dollars of GDP (kg/1000 USD GDP)	0.005	0.10	0.33	0.26	1.18	0.75	0.27	0.51	0.33
OECD	Annual per capita emissions (kg/cap)	0.096	1.98	6.45	5.43	22.6	14.86	5.58	9.72	6.47
	Average annual emissions per GDP of $1,000 (kg/1000 USD GDP)	0.004	0.08	0.26	0.22	0.92	0.60	0.22	0.40	0.26

Source Adapted from Lin Xiaoyan, Research on comparative advantages of transportation mode from the perspective of energy conservation and environmental protection, *Comprehensive Transportation*, 2010(6). *Note* The data in this table is calculated from the pollutant discharge coefficient table and the pollutant discharge amount table (see Appendix). Kg/cap is annual per capita emissions, and the unit of discharge is kg. Kg/1000 USD Annual average annual GDP per GDP of US$1000 GDP. OECD's average level of emissions within the organization (1999–2004) index of every $1,000 GDP produced by aviation and highway, the carbon oxide emission and sulfur oxide emission are 5 times and 11.5 times, 1 time and 2.3 times that of railway respectively.

High-speed railways are able to operate at more than double the speed of ordinary railways and still meet the environmental requirements of ordinary railways, or even outperform them in some environmental indicators, thanks to the use of high technology and the adoption of corresponding environmental protection measures. According to the data of Shinkansen in Japan, the per capita carbon dioxide emissions of automobiles and airplanes are 5.5 times and 6.3 times that of high-speed railways. Most of the high-speed railways use electric traction, which basically eliminates the emission of dust, oil fume and other waste gases, and realizes zero emission of locomotive without pollution to the air, thus reducing the air pollution of cities and towns along the line. For a large jet airliner, the average amount of waste emitted into the air is 46.8 t of carbon dioxide, 18 t of water vapour, 635 kg of carbon monoxide, 630 kg of nitrogen oxide and 2.2 kg of hard particles, while the aircraft requires about 15 t of aviation petrol and 625 t of air for one hour of flight. 386 g, which is 642 times greater than a similar indicator for high-speed electrified rail transport. German data show that high-speed maglev transport has more advantages in emission reduction than intercity express, as shown in Fig. 3.2.

From the perspective of the cost of pollution control, according to the statistical data of 1991 from the International Railway Union on the cost of 17 European countries for traffic impact on the environment, it is shown that the extra social transportation cost for environmental protection of three forms of transportation,

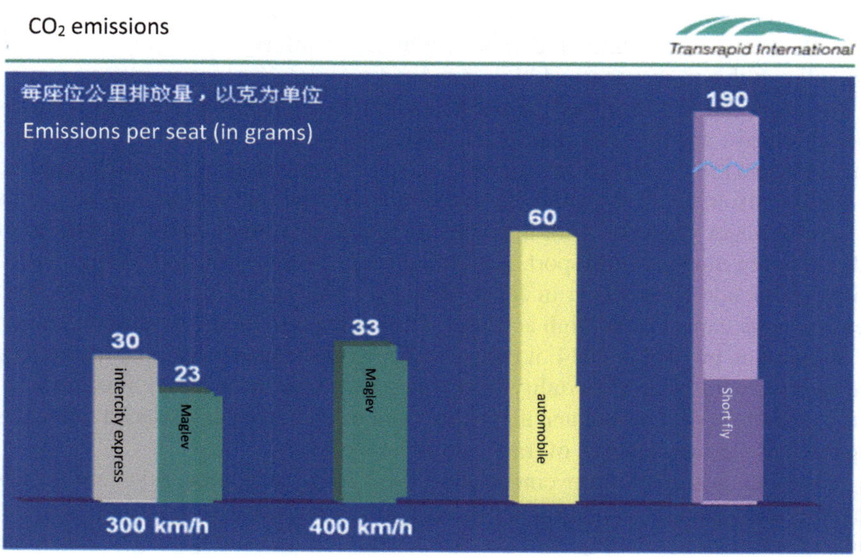

Fig. 3.2 Advantages in emission reduction of high-speed maglev transport, intercity express train and other modes of transportation

Table 3.9 Comparison of the cost of various modes of transport to combat environmental pollution

Mode of transport indicators	Automotive	Aviation	High- speed rail
Cost (100 million ECUs)	1942	124	28
Percent	92%	6%	1.7%

such as aviation, automobile and train, is 209.4 billion European currency units (ECU), equivalent to 31.54% of the GDP of these countries that year. Liao (2006) summarized the cost (100 million ECU) and proportion of various transportation modes to control environmental pollution, see Table 3.9 for details.

3.2.7 *Advantages of High-Speed Rail in Safety and Comfort*

In terms of safety and comfort, the ever-increasing quality of life makes the passengers have higher requirements for the comfort of the travel process. This is reflected in the fact that some passengers will no longer be willing to tolerate long-term bumpy rides. Therefore, they will choose a vehicle with a better ride and a shorter journey. In the early stage of economic development, this change shows that passengers are transferred from conventional railway trains and highway buses with poor comfort conditions to high-speed railway trains, thus increasing the market share of high-speed railways. At higher levels of economic development, road travel will become smaller and more comfortable, and some travellers are able to afford to travel by air with shorter travel times and relatively higher comfort levels. This results in a reduction in the market share of high speed rail.

However, safety is the most important factor in people's choice of transport. High-speed railways are far safer than other modes of transport due to their automated operation in a fully enclosed environment and a range of safety measures. According to statistics from countries with high-speed railways, the accident and casualty rates in several major countries with high-speed railways are much lower than those of other modern modes of transport. According to the statistics of Japan's high-speed railway, Tokaido Shinkansen in Japan adopted a safe operation system with high density, large capacity and high accuracy. During the 50 years from the completion of Shinkansen in 1964 to 2014, it transported 5.6 billion passengers, but maintained the safety record of zero mortality in passenger transportation. Such a record is an unimaginable myth for road and air transportation. Therefore, high-speed railway is regarded as the safest means of transport in the world.

Above all, with the improvement of people's living standards, ride comfort has become an important condition for people to choose transportation. High-speed railway trains run smoothly, with the least vibration and swing range. The space shared by each passenger is much larger than that of cars and civil aviation. Coupled with the advanced facilities and well-equipped features, it is indeed very comfortable to ride on high-speed railway.

References

Guo Wenjun, Zeng Xuegui. The significance of high-speed railway to the sustainable development of transportation [J]. *China Railway*, 2000, 03:25–27+31.

Liao Hong. Analysis of the technical and economic advantages of high-speed railway [J]. *Theory Studying and Exploration*, 2006(1):42.

Lin Xiaoyan, Research on Comparative Advantage of Transportation Mode from the Perspective of Energy Conservation and Environmental Protection [J]. *China Transportation Review*, 2010 (06), 25–29.

Lin Xiaoyan, Chen Xiaojun, Zi Yunfeng, Han Xinmei. Quantitative analysis of the impact of Beijing-Tianjin intercity high-speed railway on regional economy [J]. *Railway Economics Research* 2010 (5): 5–11.

Sun Qipeng, Feng Xuesong, Bian Kai. Operation and Organization Management of High-speed Railway in Japan [J]. *Journal of Transportation Systems Engineering and Information Technology*, 2011 (05): 11–16.

UIC, Estimation des Resources et des Activite´s Economiques Lie´es a la Grande Vitesse. Prepared by CENIT (Center for Innovationin Transport, Universitat Politecnica de Catalunya), Paris October 2005.

UIC, Railway Time-Series Data 1970–2004. UIC Publications, Paris February 2006.

Wu Zhaolin. Comprehensive Transportation Planning [M]. Beijing: Tsinghua University Publishing House, 2009.

Yan Ying, Cui Yanping. Characteristics and concepts of foreign high-speed railway trains [J]. *China Railways*, 2012, (09), 88.

Chapter 4
Time Benefits of High-Speed Railways (HSR) and Calculation of Its Time and Space Competitiveness (TSC)

4.1 Time-Saving Benefits of High-Speed Railways

4.1.1 The Term Time-Saving Benefit *and Its Indicators*

The analysis of time-saving efficiency in social evaluation lies in the fact that when a high-speed rail project is completed and in operation, it brings travelers great social benefits by cutting down their travel time, whereby its societal adaptability is checked and evaluated.

With the rapid development of the Chinese economy over several decades, transportation has entered a more scalable, speedier, and longer-lasting stage of development by leaps and bounds. With the increase of average income and improved living standards, Chinese citizens are expecting more efficiency, wherefore there has been a robust demand for the development of high-speed railway systems. Speed being the hallmark of high-speed rail technology, many countries are driving hard to improve the speed of trains. Beijing-Shanghai high-speed rail had reached 416.6 km per hour in its pilot run. Since then, the testing speed of high-speed railway has exceeded 500 km/hour, and is sure to be higher in the near future. In contrast, the maximum construction speed for automobiles is limited to 240 km/h, and normally, 200 km/h, since the average speed on Chinese highways is confined to 120 km/h. Further, although aeroplanes are much faster than high-speed trains, their speed is falling to between 300–700 km/h, if we take passengers' time spent on transit and boarding into consideration. In this sense, among these transportation means (automobiles, aeroplanes and high-speed rail), high-speed rail still enjoys more advantages in speed. Compared to aeroplanes, high-speed rail is more punctual, safer, and reasonable in price. Compared to conventional railways, and highways, high-speed rail shortens the time needed to get from one place to another. Being one of the basic infrastructures which serves the populace, high-speed rail has made great computed social value, a large part of which is realized by the savings of travelling time. Therefore, it is

X. Lin, *High-Speed Railways and New Structure of Socio-economic Development in China*, Research Series on the Chinese Dream and China's Development Path,
https://doi.org/10.1007/978-981-19-6387-2_4

extremely necessary to figure out the saved time and its social benefits in a scientific way, whereby a comprehensive evaluation is done to a high-speed rail system.

One indicator for measuring saved-time efficiency is value of time (VOT). Travel time can be described as the time a traveler usually spends on a journey from one place to another. Philosophically, time and space, which constitute the basic existence of matter, is limitless for the whole material world. But in terms of economics, time is limited for everything, and so time is a type of resource, which is endowed with opportunity cost and value. It is in this sense that a traveler's VOT is different from time value in a general sense, where a person has made a contribution to society during his working time spent on social and economic production activities, which is seen as a factor of production and thus has value realised in the value of goods. Although a traveler does not help to add more wealth to society, the time saved during his traveling will have the same value as other resources do. So traveling VOT can be obtained by calculating the time value of the increased time for other activities resulting from the decrease of the traveling time. Subsequently, VOT refers to the added value of benefits resulting from the passage of time, and is the monetary expression of benefit loss resulting from the non-productive passage of time. Thus, for different travelers, their VOT is different.

Generally speaking, VOT is influenced by such factors as a traveler's income, his purpose of traveling, the specific time he chooses to travel, and the means of transportation. The saved VOT in mega-cities is quite different from that in second and third tier cities. By the same token, the saved VOT between cities is far different from that in the countryside. The saved VOT in economically developed regions is also different from that in relatively underdeveloped areas. Normally, a traveler's VOT will increase with his personal income. The VOT for people from developed areas is higher than those in relatively underdeveloped areas, the VOT for people who travel for work purposes is higher than those for non-work purposes, the saved VOT in an emergency is higher than that in a non-emergency, and the VOT on workdays is higher than that on non-workdays. In summary, it is essential that the accurate analysis of VOT should be done on more abstract, simplified, and generalised bases, considering the diversified factors influencing the value of time and its complexity in nature.

Time value has long been a concern for both scholars from China and abroad. The two main methods of calculating VOT are **opportunity-cost method (OCM)** and **willingness-to-pay method (WTPM), including** Logit Model and Multinomial Logit Model. For a project, its VOT can be analyzed by doing direct and indirect estimation. Methods that can be applied to direct estimation include product method, profit method, time and cost method, earnings and outlay method, and production-outlay method, whereas indirect estimation methods are mainly made up of revealed preference analysis and stated preference analysis.

The principle underlying the opportunity-cost method is that the saved VOT used for work purposes is the profit made from one's work time minus the profit from one's loss of leisure time. If a person's activity can be divided into production activities and recreational activities, then his time can be cut into production time and leisure time. If a traveler's saved time of value can be understood as VOT used for work and VOT for non-work purposes, we should calculate VOT by multiplying the corresponding probability of time saved.

It is easy and operational to figure out the VOT by using the **opportunity-cost method (OCM)**. But this method fails to include the invisible or unobserved spiritual loss of a traveler suffering from the extended travelling time. And this violates the principle of willingness to pay, which is seeking to reveal a traveler's choice preference by looking at the market behavior of sampled travelers' choices for different transportation through regression analysis and parameter estimation.

4.1.1.1 Models of Calculating VOT

Logit Model is the earliest Discrete Choice Model (DCM), and the most widely used model. It is derived from the IIA (the Independence on Irrelevant Alternatives) and first put forward by Luce (1959). Later on, Marschak (1960) showed that the Logit model is inconsistent with the maximum utility theory. Based on his study of the relationship between the shape of models and the distribution of indefinite items, Marley (1965) showed that the extreme distribution helps to deduce models like Logit. Conversely, McFadden (1974) found out that a utility model in the form of a Logit must conform to the extreme distribution. With these discussions, the Logit model has evolved into a sophisticated discrete choice model system involving many other discrete models such as Probit model, NL model (Nest Logit model), Mixed Logit, and so on, and has been widely applied in other fields such as psychology, sociology, economics and transportation. One shared commonality between these models is that it is based on an observable variable χ and an unobservable variable ε (taking the influence of observable variables into consideration), so we can get a function value of **y** which denotes a certain decision-making behavior.

Let the above assumed ε be an independent variable with an extreme in a normal distribution. The logistic probability of a certain transportation i from a substitution set of transportation alternatives can be represented by (4.1):

$$P_{ni} = \frac{e^{V_{ni}}}{\sum e^{V_{ni}}} \tag{4.1}$$

Here, P_{ni} denotes the probability i of an individual n's choice of transportation from a set of transportation alternatives, while e is the natural logarithm, V_{ni} the observable effect of an individual n's preference for a certain alternative i.

The Logit model is applicable only when a traveler has made the traveling plan. Once the traveler is setting off upon a random trip, this model doesn't work very well, because it is limited to calculating the effect of a single trip, but fails to calculate some consecutive travelling plans.

Compared to the Logit model, Multinomial Logit model is more sophisticated and more flexible. Because it gets away from the restraint coming from the independence and variance hypothesis, the Multinomial logistic model can be applied to random trips. Equation (4.2) provides a description of the function:

$$P_{ni} = \int L_{ni}(\beta) f(\beta) d\beta \tag{4.2}$$

where $L_{ni}(\beta)$ denotes the probability formula of the Logit model, and $f(\beta)$ the density distribution, whereby the most likelihood alternative is relied on.

4.1.2 Different Methods of Calculating Time Value

A. **The Product method**

When applying **product method** to calculating time value, first, the labor of travelers should be considered as one of the production factors, because the reduction of travel time will release this part of the resource, therefore increasing the output, promoting the Gross National Product (GNP) and National Income (NI). Nonetheless, the prerequisites of applying the product method are: saved travel time helps to release production factors, and is used for work; released production factors can be utilized in the production process. In addition, the shortage of labor is also the required condition of applying the product method. That is, there should be the existence of full employment or insufficient labor force.

When applying the product method, the value of travel time is measured by its hourly output. So the time value of an hour's trip can be represented as in (4.3):

$$V_p = GNP/(2000P) \tag{4.3}$$

where V_p represents the hourly value of travel time; GNP is the yearly gross national product; P denotes the average employment in a year; 2000 is the mean of one's yearly effective work time (365 days a year, during which there are about 250 statutory working days, and approximately 2,000 h if one works eight hours each day). But one weakness of the production method is that although it may measure a traveler's time of value objectively from the perspective of an enterprise, it fails to take into account the traveler's preference and his personal income.

B. **The Earning method**

The Earning method, also known as the income method or salary method, calculates the value of saved travel time according to a certain percentage of different travelers' income. The presumptions of this method are: travelers can freely allocate their disposable time to work or leisure; travelers regard travelling time as work time, because travelling occupies a part of their disposable time, preventing them from seeking for jobs and gaining income; the more travelers spend their saved time working, the more income they will get.

To a large extent, the value of leisure time can be said to be a function of social development. A higher development level means more emphasis on leisure time, therefore enticing people to pay more for that. As a result, it is argued that the earning method overestimates the value of saved travel time, and that the value of leisure time and time during extended traveling cannot be calculated on a total wage level, but with a certain proportion of it.

C. **The Time–cost method**

Assuming that the travel time and travel cost are mutual substitutes (meaning travelers can apply either slower but more affordable, or faster but more expensive travel modes), time can be indirectly viewed as a substitute for consumption. Therefore, the proportion of the cost gap to the time gap between these two travel modes, also known as the cost contribution value of travel, is the value of saved travel time, which is demonstrated in (4.4):

$$V = \frac{C_2 - C_1}{t_1 - t_2} \tag{4.4}$$

In this formula, V refers to the hourly value of travel time, C_2 and t_2 refer to the travel cost and travel time of transportation mode No. 2 (faster but more expensive) respectively, while C_1 and t_1 are the cost and time of transportation mode No. 1 (slower but more affordable). This formula is not completely applicable, given that it is more important for travelers with low income to reduce travel cost rather than travel time. Under this circumstance, a slower but more economic mode of transportation might be a preferred choice of traveling.

D. **The Income-expense approach**

The income-cost approach focuses only on employee income and leisure time, while the expense approach focuses only on passenger costs and travel time, without taking into account all the factors that affect travel time. Williamson and Masten (1999) proposed the income-cost approach, whose expressions are (4.5):

$$V = S + \frac{C_2 - C_1}{t_2 - t_1} \tag{4.5}$$

where, S stands for one's hourly payment.

This approach underestimates the potentially increased income or decreased income of travelers who have a preference for leisure time, while it overestimates the income of travelers who prefer income.

E. **The Product-cost method**

This method combines the value of work time that is generated by the time saved from travel and used in production, with the discrepancies between different transport modes utilized. As is shown in (4.6):

$$V = \frac{GNP \times (t_2 - t_1)}{2000 \times N_p} + (D_1 - D_2) + E - (C_2 - C_1) \tag{4.6}$$

where GNP refers to gross national product; D_1,D_2 refers to the travel expenses caused by transportation mode No.1 and No.2 (exclusive of transport fares) respectively; $\boldsymbol{E}$ refers to the incurred loss of slower transportation mode; 2000 is the yearly work hour; N_p is the aggregate national population.

F. **The Indirect estimation method**

The indirect estimation method calculates VOT. It includes the analysis of revealed preference and stated preference.

Revealed preference analysis reckons time value by recording or surveying the frequency of travel. It may well explain how travelers make their decisions (of choosing between two different transportation modes mentioned above, for example). However, it is not cost-efficient and often comes with huge uncertainty, because even if there is only one option, empirical surveys could only be conducted with the selected travel plans, not the abandoned plans.

Stated preference analysis reckons the value of behaviors through questionnaires or direct inquiries and conversations. The advantages of this method are: it is more transparent and economic; it can be conducted within a wide range; questionnaires often ensure a considerable amount of reliable information; it is relatively affordable in a simple empirical survey. The results show that the value of time savings imputed by stated preference analysis is generally lower than that estimated by revealed preference analysis, but the trend is consistent.

G. **The Cost-profit method**

The analysis of cost and profit is one method of evaluating a project by comparing the total cost and its profit. As one way of doing economic decision-making, the cost-profit method has been applied in some government's projects on public affairs which ask for quantitative assessment in order to ensure the maximum profit out of the minimum cost. With a reference to Delucchi (1995), we have come up with the formula of calculating social value of saved traveling time as represented in (4.7):

$$V = PT_d \times (C_{hd} + C_{pd}) \tag{4.7}$$

where V shows the lowering of social cost due to the shortened travel time, i.e. the increase of social benefit, PT_d is denoting the total amount of saved time represented by an individual-hour, C_{hd} the opportunity cost of a time unit, and C_{pd} the implicit cost of a time unit.

As is noted by Button, assuming that the VOT for travelers from different jobs is varied, a traveler's social value of time savings can be represented as in (4.8)

$$V = \sum_{i=1}^{N}\sum_{j=1}^{N}\sum_{m=1}^{M} 0.5 \times (C^0_{ijam} + C^1_{ijam}) \times (T^0_{ijam} - T^1_{ijam}) \times VOT_{am} \tag{4.8}$$

where C^0_{ijam}, C^1_{ijam} represents the frequencies that a traveler whose job is α commutes between the two cities i and j by choosing transportation mode m before he can have other traffic alternatives, while T^0_{ijam} and T^1_{ijam} represent the length of time needed for travel, and VOT_{am} the VOT in a unit of time that the traveler α can create.

4.2 The Saved Time Value of High-Speed Rail (HSR)—A Case Study of Beijing-Shanghai High-Speed Rail

One major function of high-speed rails is to save the time of travelers. Taking the Beijing-Shanghai high-speed railway as an example, it takes a normal overnight sleeper about 14 h from Beijing to Shanghai, but by high-speed rail, it takes him about 5 h. In other words, high-speed rail will undoubtedly be the first choice for those who are endowed with a higher value of time, because, as compared to regular sleepers, it saves 9 h for them. The total length of the Beijing-Shanghai high-speed railway is 1,318 km, along which there are 7 provinces or cities and 24 stations. Table 4.1 below lists 4 cities (Jinan, Xuzhou, Nanjing and Shanghai) and compares the cost and time of different transport to Beijing, without taking the time for transit and other negligible factors into account.

Here, it is necessary to point out that the self-driving distance from Jinan, Xuzhou, Nanjing and Shanghai to Beijing is 431, 713.5, 1065.9 and 1261.4 km respectively. Assuming that the average fuel consumption of a car is 8 L/100 km, the total cost of self-driving can be obtained by adding the cost of fuel to the toll fee on the highway. Here, the ticket charge for Beijing-Shanghai high-speed rail refers to the fee charged for second-class seats, the ticket for normal trains, the fee for hard seats (marked

Table 4.1 Time and cost of different transport modes

Unit: Yuan

Transportation modes from Beijing	To Jinan		To Xuzhou		To Nanjing		To Shanghai	
	Cost	Time	Cost	Time	Cost	Time	Cost	Time
By aeroplane	710.14	1 h	652.57	1 h 20 min	547.43	2 h	508.71	2 h 10 min
By the Beijing—Shanghai HSR	185.00	1 h 32 min	310.00	2 h 44 min	445.00	3 h 39 min	555.00	4 h 58 min
By train	73.00	5 h 50 min	106.00	9 h 53 min	150.00	14 h 2 min	158.00	20 h 12 min
By bus	124.00	6 h 24 min	250.00	10 h 30 min	300.00	15 h 23 min	340.00	18 h 8 min
By Self-driving	406.36	6 h 24 min	694.56	10 h 30 min	1031.90	15 h 23 min	1226.38	18 h 8 min

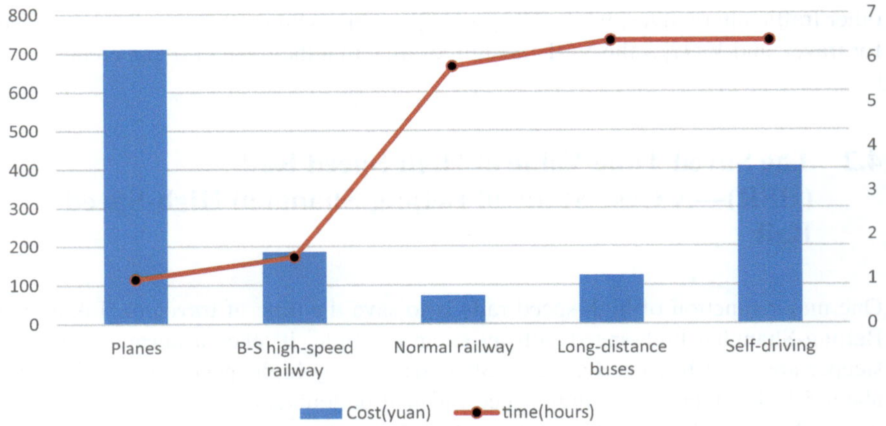

Fig. 4.1 Cost of different transportation modes and their corresponding traveling hours

with the initial letter "K"), while the air ticket is the average air ticket price from October 22nd to October 28th, 2014.

The fare rates of different transportation modes on the Beijing-Shanghai line are: RMB 0.24 yuan for each person every kilometer by trains, RMB 0.40 yuan for each person every kilometer by high-speed rail; for highway buses, RMB 0.25 yuan for each person every kilometer by bus (the express); RMB 0.93 yuan for each person every kilometer by car (self-driving). Our survey shows that most complaints about the high-speed rail are concentrated on the ticket price, where 65% of the population agrees that the ticket is a little bit overcharged. For example, as is shown in Fig. 4.1, in terms of the cost of different transport modes and time from Beijing to Jinan, the traveling time required for Beijing-Shanghai high-speed rail is slightly longer than that for planes, but far shorter than that for regular trains, long-distance buses and self-driving.

The high-speed railway has greatly reduced the distance between places and accelerated the development of China's regional economic integration. The Beijing-Shanghai high-speed railway has shortened the train journey between Beijing and Shanghai to about 5 h, with 90 round-trip train sets each day, running at a mixed speed of 300 km/h and 250 km/h, with 76 round-trip G trains and 14 round-trip D trains, departing every 16 minutes on average, connecting 24 stations in 7 provinces and cities along the way. This not only effectively improves the economic and social situation of the provinces and cities along the way, but also saves passengers' travel time to a considerable extent. The lengths of the three sections of the Beijing-Shanghai high-speed railway, namely Beijing to Dezhou, Dezhou to Xuzhou and Xuzhou to Shanghai Hongqiao, and the time saved compared to the fastest ordinary K trains are shown in Table 4.2.

To be specific, the average time savings per passenger per 100 kilometres from Beijing to Dezhou, Dezhou to Xuzhou and Xuzhou to Shanghai Hongqiao are 0.74, 1.08 and 0.83 h respectively (i.e. 44 min, 1 h 5 min, and 50 min). In other words,

Table 4.2 The length of HSR among cities and its time savings

Segment	Beijing to Dezhou	Dezhou to Xuzhou	Xuzhou to Hongqiao, Shanghai
Distance (kilometers)	419	301	618
Time savings (hours)	3.08	3.25	5.12
Passenger turnover volume (thousand-kilometers)	11,092,849	11,820,366	19,799,045

Note The number of travelers transported was taken from statistics up to the end of September, 2009

among the 24 stations of the Beijing-Shanghai high-speed railway, the average time saved per 100 kilometres varies from station to station due to the different distances and running speeds. By the end of September, 2009, Beijing-Shanghai HSR had an aggregate turnover volume of 42,712,260 thousand-passenger-kilometers, among which Beijing to Dezhou took up 11,092,849 thousand-passenger-kilometers, Dezhou to Xuzhou contributed 11,820,366 thousand-passenger-kilometers and Xuzhou to Shanghai was responsible for 19,799,045 thousand-passenger-kilometers. If we multiply the above figures by the minimal average time saved for each person every 100 km of these three segments respectively, we can infer that the Beijing-Shanghai HSR has saved at least 82 million hours, 128 million hours and 164 million hours (total: 374 million hours) from June, 2011 to September, 2012, meaning that every resident in this country saved 15 min.

In terms of travel costs, the cost of the Beijing-Shanghai High Speed Rail is only a quarter of that of civil aviation, less than half of that of self-drive, and slightly higher than that required by coach and regular rail. The results of a questionnaire survey of travellers on the Beijing-Shanghai High Speed Railway found that 87% of all passengers were public officials and business operators travelling for business purposes, and the majority of the remaining passengers were travelling to visit relatives. Therefore, for almost all Beijing-Shanghai HSR passengers, it is far more important to reduce travel time than travel cost, and under this circumstance, time–cost method can be applied to roughly estimate whether the Beijing-Shanghai HSR brings considerable social benefits. Assume the Beijing-Shanghai HSR as transportation No. 1 and planes as No. 2, according to the time–cost method, the hourly cost of travel is demonstrated in Table 4.3.

The HSR will be more socially beneficial than planes, if the time value of passengers from Beijing to Jinan, Xuzhou and Nanjing is respectively lower than 99, 245

Table 4.3 The VOT of an Hour

Unit: Yuan

Transportation modes (from Beijing)	To Jinan	To Xuzhou	To Nanjing	To Shanghai
Planes	991	245	62	−17

Table 4.4 The hourly time value of travelling by different transport modes

Unit: Yuan

Transportation modes (from Beijing)	To Jinan	To Xuzhou	To Nanjing	To Shanghai
Trains	26	29	28	26
Long-distance buses	13	8	12	16
Self-driving	−45	−49	−50	−51

and 62 yuan, while planes have advantage over the HSR from Beijing to Shanghai, because of their higher speed and lower cost. Assuming the Beijing-Shanghai high-speed railway as travel mode 2, then let the coach, self-drive and ordinary train as travel mode 1 respectively, we can use the time cost method to initially calculate the cost of travel per hour, as shown in Table 4.4.

Table 4.4 indicates that if the time value of a traveler is higher than 29 yuan/hour, as compared to any other transport modes, the Beijing-Shanghai HSR can reduce travelling costs. If the time value of travelers exceeds 16 yuan/hour, the HSR has an edge over long-distance buses. In summary, if the value of passenger time is less than RMB 8, the HSR has an advantage over driving only; if the value of time is RMB 8–16, the HSR has an advantage over driving and coaches in some zones; if the value of passenger time is RMB 16–26 per hour, the HSR can reduce the cost of travel compared to all other modes of travel except ordinary railways. Therefore, the Beijing-Shanghai HSR dramatically reduces a passenger's travel time, and decreases his travel cost to some degree. Specifically, the relationship between time value of travelers and whether the HSR reduces one's travel cost, and the segments where HSR has an absolute advantage, are listed in Tables 4.5 and 4.6.

As is shown in Tables 4.5 and 4.6, for passengers whose hourly VOT is between 26 and 991yuan the Beijing-Shanghai HSR has an absolute advantage in all different segments. When the hourly VOT is less than 26yuan or more than 991yuan, the HSR does not have an advantage in reducing travel cost and raising social benefits, but compared with other transportation modes, it has an obvious edge in raising social benefits. In short, the HSR greatly diminishes travel cost and brings considerable social benefits.

To calculate the social benefits brought by saved travel time since the HSR was put into operation over a year before, first, the VOT difference of various passenger groups should be considered. When the running time and total turnover are known, passengers can be categorized based on the results of questionnaires. Therefore, a cost–benefit method by comparing the aggregate cost and benefits of a project is relatively more applicable at this point. Assume that when there is no HSR, a new transportation tool, passengers who will travel by a high-speed rail will have to go by train. Then, the formula of cost-profit method can be simplified as in Eq. 4.9:

Table 4.5 Transportation modes that have a higher cost than HSR under different hourly time value

Unit: Yuan

VOT	Beijing to Jinan	Beijing to Xuzhou	Beijing to Nanjing	Beijing to Shanghai
0–8	Self-driving, planes	Self-driving, planes	Self-driving, planes	Self-driving
8–12	Self-driving, planes	Self-driving, planes, long-distance buses	Self-driving, planes	Self-driving
12–13	Self-driving, planes	Self-driving, planes, long-distance buses	Self-driving, planes, long-distance buses	Self-driving
13–16	Self-driving, planes, long-distance buses	Self-driving, planes, long-distance buses	Self-driving, planes, long-distance buses	Self-driving
16–26	Self-driving, planes, long-distance buses	Self-driving, planes, long-distance buses	Self-driving, planes, long-distance buses	Self-driving, long-distance buses
26–28	Self-driving, planes, long-distance buses, trains	Self-driving, planes, long-distance buses	Self-driving, planes, long-distance buses	Self-driving, long-distance buses, trains
28–29	Self-driving, planes, long-distance buses, trains	Self-driving, planes, long-distance buses	Self-driving, planes, long-distance buses, trains	Self-driving, long-distance buses, trains
29–62	Self-driving, planes, long-distance buses, trains	Self-driving, planes, long-distance buses, trains	Self-driving, planes, long-distance buses, trains	Self-driving, long-distance buses, trains
62–245	Self-driving, planes, long-distance buses, trains	Self-driving, planes, long-distance buses, trains	Self-driving, long-distance buses, trains	Self-driving, long-distance buses, trains
245–991	Self-driving, planes, long-distance buses, trains	Self-driving, long-distance buses, trains	Self-driving, long-distance buses, trains	Self-driving, long-distance buses, trains
Above 911	Self-driving, long-distance buses, trains	Self-driving, long-distance buses, trains	Self-driving, long-distance buses, trains	Self-driving, long-distance buses, trains

Table 4.6 Segments that HSR has absolute advantages of value for different time units

Unit: Yuan

VOT	Beijing to Jinan	Beijing to Xuzhou	Beijing to Nanjing	Beijing to Shanghai
0–26	–	–	–	–
26–28	HSR	–	–	–
28–29	HSR	–	HSR	–
29–62	HSR	HSR	HSR	–
62–245	HSR	HSR	–	–
245–991	HSR	–	–	–
Above 991	–	–	–	–

$$V = \sum_{i=1}^{N} \sum_{i=1}^{N} C_{ija}^{1} \times (T_{ija}^{0} - T_{ija}^{1}) \times VOT_{am} \tag{4.9}$$

The total length of Beijing-Shanghai is 1,318 km, and the full transit time is 5 h. While for normal trains, we need 20 h, which indicates that by taking a HSR train, we can save 1.16 h (1 h and 10 min) per kilometer. For different passengers, the unit of their VOT is different. Surveys show that 87% of passengers are public officials on business and proprietors, and the travel purpose of the rest of passengers is basically for family visits. Therefore, the formula above can be further simplified as in (4.10):

$$V = 0.87 \times 1.16 \times VOT_{a1} \times \sum C_{ija} + 0.13 \times 1.16 \times VOT_{a2} \times \sum C_{ija} \tag{4.10}$$

Under this circumstance, to calculate the saved VOT of the Beijing-Shanghai HSR, we only need to determine the VOT of different types of passengers.

Normally, due to the current pervasive regional social-economic differences, and disparities among the achievable GDP of different professions in a unit of time, to reckon the saved VOT on a national scale, it is more justified to apply the production method to calculate the VOT of the Beijing-Shanghai HSR passengers, but these passengers' occupations, to a very large extent, fall into secondary industries and tertiary industry. Hence, as is shown in Table 4.7, based on the regional GDP and employment along the Beijing-Shanghai HSR, it is estimated that the per capita GDP of secondary industries and tertiary industry in these regions is 92,274 yuan. Given that passengers on business trips can achieve a much higher GDP than a normal local employee does, the value that these passengers create can be presumed to be 5 times as large as the GDP per person in 7 cities and provinces alongside the Beijing-Shanghai HSR. At approximately 250 effective workdays per year and 8 work hours per day, the VOT of passengers on business is about 243 yuan/hour. However, it should be taken into account that China ranks top in per capita work time (about 275 work days), and passengers on business usually have more hours of work time than the average. Therefore, their yearly work time can be estimated to be 300 h, and based on that, their VOT is estimated to be around 203 yuan/hour. To be more precise, assuming that these passengers work 365 days a year and 8 h a day, their VOT, accordingly, is about 166 yuan/h.

The Income method, i.e. salary method, reckons the saved VOT of travel according to a certain percentage of passengers' income. The advantage of such method is that we can assess both the unit VOT on work and the VOT on leisure. Only a relatively small proportion (13%) of target passengers of the Beijing-Shanghai HSR travel for business and, for the majority, their purposes of travel are various. On a national scale, it is more justified to apply the income method to calculate the saved VOT of travel. Most passengers have to travel by the HSR during their work time because of its night maintenance and daytime operation, and its relatively high fares means that the time value of these passengers should be reckoned at the annual average income for high income groups in the 7 cities along the Beijing-Shanghai HSR. As Table 4.8 demonstrates, peoples' average income in these cities and provinces is about 25,000yuan per month. Domestically, the average income of high income group is twice as high as the national per capita income. With the method applied in the calculation of the VOT of passengers on business, at this proportion, when there are 250, 300 and 365 work days a year and the work time per day is 8 h, the VOT of these passengers, as is calculated, is 37 yuan, 31 yuan and 26 yuan/hour respectively.

If we plug the above VOTs into the formula, given that 42,712,260 thousand-passenger-kilometers were completed from June, 2011 to September, 2012, the social value of the Beijing-Shanghai HSR can be estimated:

From Table 4.9, it is known that the social value created by Beijing-Shanghai HSR during its first 15 months in service was between 70 to 100 billion yuan (around 90 billion yuan).

4.3 Time and Space Competitiveness of a High-Speed Railway

4.3.1 The Term "Time and Space Competitiveness" *and Its Measurement Indicators*

Time and space competitiveness **(TSC)** of a region refers to the extent on which this region can draw the eyes of, and have mutual interaction with, other regions. The more interaction and attraction a region has with other regions, the more TSC it has. Or vice versa. Rather, a city's TSC often depends on its own "quality"–social economic resources such as GDP, population density and resources endowment–and the cost required for its accessibility over resistance and obstacles. So, if its "quality" is better, the cost for its accessibility is much lower, wherefore its TSC will be stronger, and it may enjoy more advantages in social-economic development.

The term ***resistance*** is understood as the possibility and dynamics of the coverage and dominance on location and space in an economic process. Resistance happens

Table 4.7 The regional GDP of 7 cities and provinces along the Beijing-Shanghai HSR

Region	GDP (billion)	Primary industry	Secondary industry	Tertiary industry	Employment (ten thousand)	Primary industry	Secondary industry	Tertiary industry
Beijing	1411.358	12.436	338.838	1060.084	1317.7	65.1	275.8	976.8
Tianjin	922.446	14.538	484.023	423.865	520.8	75.9	213.5	231.4
Hebei	2039.426	256.281	1070.768	712.377	3790.2	1469.6	1261.1	1059.5
Shanghai	1716.598	11.415	721.832	983.351	924.7	36.4	347.4	540.9
Jiangsu	4142.548	254.010	2175.393	1713.145	4731.7	883.3	2141.9	1706.5
Anhui	1235.932	172.902	643.662	419.368	3846.7	1538.5	1132.4	1175.8
Shandong	3916.991	358.828	2123.849	1434.314	5654.7	2004.4	1839.9	1810.4
Total	15,385.299	1080.430	7558.365	6746.504	20,786.5	6073.2	7212.0	7501.3

Note Statistics are between June 2011 and September 2012

Table 4.8 The per capita GDP of 7 cities and provinces along the Beijing-Shanghai HSR

Region	Per capita income (yuan)	Population (ten thousand	Total GDP (ten thousand yuan)
Beijing	33,360.42	1520.75	50,732,858.72
Tianjin	26,942.00	931.55	25,097,820.10
Hebei	17,334.42	1891.81	32,793,429.10
Shanghai	35,378.51	2032.89	72,652,459.59
Jiangsu	25,115.40	3323.38	83,468,018.05
Anhui	17,626.71	1589.84	28,023,648.63
Shandong	21,736.94	3654.38	79,435,038.80

Note Data above were from June, 2011 to the end of September, 2012

Table 4.9 Social value generated by time savings from the Beijing-Shanghai HSR

Work days	VOT of passengers on business (yuan/hour)	VOT of passengers for non-business purposes (yuan/hour)	VOT of distance (yuan/100 km)	Saved VOT (billion)
250	243	37	250.82	107.178
300	203	31	209.54	89.315
365	166	26	171.45	73.409

when the flow of economic factors are held up or blocked by the competition and confinement of natural and human factors. **Temporal and spatial resistance** here refers to resistance against the geographical distances resulting from the flow of economic factors. Such resistance is everywhere, so the cost of temporal and spatial resistance mainly comprises of money needed and time needed for dealing with geographical distance. Further, if the zoning of a place is different, making it belong to different zones, the same place may differ greatly in terms of its TSC. Hence, determining factors for TSC can be represented as in Fig. 4.2:

Fig. 4.2 Determinants for a city's time and space competitiveness

As is shown in Fig. 4.2, the cost of overcoming time and space resistance determines the TSC of a region, while its cost depends on the distance between the two places. Subsequently, the same region may have a very different TSC.

In terms of regional economics, TSC is closely related. The Gravity Model, also called the model of gravitation, was introduced by Reilly (1931) to calculate the gravity at a breaking point between two places where customers will be drawn to one or another of two competing commercial centers. It is one of the easiest but most important mathematical models which focus on the interaction between pairs of spatial points. Reilly (1931) expanded Isaac Newton's law of gravitation to describe spatial interactions between cities based on his observation of the trade area, market boundary and service area. But results of computation by the gravity model are more or less different from what is turning out to be. So, some other scholars have been working on to improve this model, among which the following modified version proposed by Haynes and Fotheringham (1984) has been widely used (as in 4.11):

$$I_{\mathrm{ij}} = K\frac{P_{\mathrm{i}}^{\lambda} P_{\mathrm{j}}^{\alpha}}{r_{ij}^{\beta}} \tag{4.11}$$

where K represents a dielectric constant, which helps to make the whole equation match "proportional features" of the modeling object; α and λ are elasticity coefficients applied to moderating the variable *population size* and its impact on gravitation. P_{i}, P_{j} usually refer to the population size of the two cities: city i and j, but these two variables can also be replaced by other indicators representing the actual scale of a city. $\mathrm{r_{ij}}$ denotes the distance between the city i and j(normally, kilometers are used to express it). As a matter of fact, our assumption about the word "distance" has been challenged with the rapid development of modern transportation, which means the cost of distance may not be an indication of its geographical distance. In other words, time and monetary cost of transportation can also be used as indicators measuring the distance between two cities.

β is an indicator used to measure friction effects. It is put on the independent variable "distance" to see if the distance has been changing in proportion to its interactiveness, because changes in mode and conditions of transport affect the value of β. For instance, in freight transport by air, the cost for each unit of distance will decrease with the increase of the distance between the pairs of points. In this case, the effect of distance on mutual interactiveness is not in inverse proportion to the mileage of the airline, which means we need to cut down the negative impact of distance on its interactiveness. Moreover, the more advanced the transportation between the paired regions is, the smaller the coefficient β will be, and correspondingly, the lower the impact of distance on its interaction will be. So, when $\beta \to 0$ occurs, the variable "distance" does not have any effect. Conversely, if the transportation is less advanced or underdeveloped, the coefficient β will be bigger; its influence on mutual interaction will be greater. But when $\beta \to \infty$ occurs, it means there is completely no relation between the paired two cities; therefore, its interactiveness is zero. However, the opening of high-speed railway does have an obvious influence on the coefficient β, which makes the shortening of distance have less effect on the interaction compared to the less developed transportation, where the decrease of distance greatly impacts on the interaction between the paired places.

The Gravity Model is a mathematical model built on spatial interaction theory, which is used to analyze spatial interaction between two paired cities, where flows of population, material, information and technology are considered to be means of spatial interaction. Aside from these five elements, three more conditions are needed and cannot exist without each other for such spatial interaction: the degree of its complementarity, the status of being intermediary and its accessibility to transport. Specifically, the complementarity, being the impetus for the flow of population, materials, business, funds, and information, serves as the foundation for mutual interaction. Of course, in the course of spatial interaction, there may be the intervention of the third party or place which may incur a substitution for the original starting and ending points of the above-mentioned flows, which may result in a halt on its existing spatial interaction. It is the accessibility of transport, an important means of linking cities together to become a system, that makes mutual interaction possible. Without such a link, a city is isolated. With the rise of high-speed railway systems, the density of material, population, and information flows (indicators for the intensity of spatial interaction) have been changing at a faster speed. Therefore, by figuring out the extent of change in spatial interaction between neighboring cities both before and after the operation of a new mode of transport, we will see the impact that the new transportation has on the growing development of city networks.

In brief, the Gravity Model is introduced to calculate the time and space competitiveness. It can be used as a means to measure the degree of TSC between two paired points of distance before and after the operation of the HSR. But it is worthwhile to notice that the gravity model is applicable only when we are looking at attractiveness between the two places. It is not effective enough to examine the impact of such an integrated HSR system like the Beijing-Shanghai high-speed railway on the TSC of a group of cities along the route. And because we should not do such calculations on only paired cities separately, we need a more complicated and complex measuring model, which we will call "***Accessibility of a region***", to estimate a region's TSC before and after the running of a high-speed railway.

4.3.2 Regional Accessibility Calculations

4.3.2.1 The Term *Accessibility*

The term *accessibility* can be defined simply as the ease of reaching a desirable destination from a typical location. This concept was first proposed by Hansen (1959), and used to express the chances that each node can interact with each other in a transportation network. Goodall (1987) defined accessibility as the degree of how easily a spatial location could be accessed relative to other locations rather than physical distance, while Deichmann (1997) considered that accessibility was the ability to engage or interact with a specific economy, factors of social opportunities and their locations. Nowadays, accessibility is being used as a comprehensive index to evaluate whether the regional transportation network can carry out the transportation task efficiently, whether the basic requirements of plenitude, efficiency, balance and harmony can be achieved.

Passengers or freight in places that are highly accessible can reach the destination point from a departure point quickly. The value of quality transportation mainly depends on the time factor. Meanwhile, travel time is one of the important conditions for traffic network users to consider when they chose their route, which is also one of the indicators of a traffic network layout from the perspective of sustainable development of the transportation system. Accessibility is calculated as follows (4.12):

$$A_i = \frac{\sum_{j=1}^{n}\left(T_{ij} \times M_j\right)}{\sum_{j=1}^{n} M_j} \tag{4.12}$$

where

$$T_{ij} = \sum_{k=1}^{n}\left(T_{ij}^{k} \times W_k\right)$$

In the previous equation, A_i refers to the weighted average travel time of node i in the traffic network, that is, the weighted average travel time that each node in the traffic network needs to reach the node i. A_i can value the time structure of connectivity among nodes. The smaller the value of A_i is, the relatively higher-level of its accessibility is T_{ij} is the weighted average travel time from node i to node j, which reflects the travel cost between nodes. M_j indicates the attraction of node j that shows the quality of node j, such as variables of population, regional gross domestic product (GDP), or other setting values. T_{ij}^{k} is the time needed from j to i by all possible transportation means, k (k ranges from 1–4, and represents highway, railroad, waterborne, and aviation, respectively). W_k states the weight of k occupied in the comprehensive traffic network.

This model is a simple calculation method for the passage time between nodes. We can easily get the numerical variation of the regional accessibility by computing the difference of time round about the construction of traffic facilities. Accessibility can be improved by shortening the passage of time.

4.3.2.2 Accessibility and Regional Development

Distance plays a critical role in the development of a region in the long run. The construction of traffic infrastructure not only helps to build a city's accessibility but also speeds up its development. The relationship between a region's development and its accessibility can be seen from Fig. 4.3.

	Differences in development opportunities between regions, sectors and companies due to differences in raw materials, markets, information and human interaction	Slow growth area Rapid growth area
Variation in accessibility	Educational, cultural and ideological differences due to differences in population mobility between regions	Relatively conventional social structure Flexible and progressive social structure
	Differences in production agglomeration due to changes in accessibility and differences in environmental loads	Landscape with a relatively stable and lower environment pressure Landscape with a high environment pressure
	Spatial differences in government policies due to spatial differences in income, capital, population mobility and agglomeration procedures	A comparatively stagnant social structure A relatively more sensitive and flexible political structure and policy

Fig. 4.3 The relationship between regional development and accessibility

The improvement of a city's accessibility may bring the following benefits:

First, it helps to ameliorate conditions of transportation and communications, thus upgrading its status and influence over other regions. To be specific, accessibility can accurately reflect a city's potential energy in its interaction between two regions, with other regions, or even with all of the regions concerned. In practice, such potential lies in the strength or weakness between regions in terms of its social and economic connectivity and its influence (of, for example, leading, facilitating or competing, and inhibiting) over other regions. As long as the road blocks into the region are removed, making it accessible, and opening it up to the outside world for more communication, its transport situation and conditions will be changed and improved, making it more reachable, and thus more strategically important for its regional development. If a region is relatively not very large in space, the disparity of the spatial distance between each geographical point and the other points in the same region may be less obvious. But accessibility makes things different. Because accessibility is crucial for social economic entities (e.g. enterprises) and their development, they often build their factories or companies near the hub of communications and transportation that are more readily accessible. In short, the accessibility of a region determines, to some extent, its social and economic structure in space.

Second, it helps to strive for more comprehensive accessibility so as to accelerate a region's potential capacity for more speedy development. The building of traffic arteries and lines of communications brings on many other constructions of infrastructure and facilities along the way such as the highway, airport, water transportation channels, branch tracks of railroads, energy and power grids, post and telecommunication industries, and so on. And these things will help to build up a region's accessibility as a whole, and make it more inviting for investors. In addition, when a region has become more available by both means of transportation and communications, it will be easier for the region to get involved in more economic activities on a larger scale and in a wider space. Then, cities on the different points of the regional map are open to more opportunities for social economic or commercial intercourse while narrowing down their physical distance and economic gaps with other developed areas.

Third, it is conducive to the development of accessibility-sensitive industries and industrial restructuring. As a matter of fact, accessibility has a different impact on different industries. In modern society, time is the key factor for successful cooperation between industrial development and economic and technological development in different countries and regions. As manufacturing and markets are changing constantly, it is urgent for enterprises or institutions concerned to keep up with the changing markets, to organize manufacturing and sell properly, and to make sure that the source materials needed, semi-finished products, and finished products are ready for use. Good accessibility is required for making products such as semi-conductors, integrated circuits, clothing, automobile parts and assembly, and toys. The same is also true for the tertiary industry businesses such as enterprises of financing (e.g. banks, stock exchanges, and insurance companies), business and commerce (firms, enterprises, supermarkets, and futures), holidaying and travelling facilities and services, and high-grade hotels. Also, in agriculture, accessibility is essential for fresh and living products, and the flower business. So the above mentioned can be regard as accessibility-sensitive industries, whose business heavily relies on its potential to be accessed. In addition, it depends on a higher accessibility as to whether or not the region will make its industrial restructuring successful and make its economic structure adjust to the competing market. Such adaptation for competition is usually defined by the development of new emerging industry, which will in turn increase a region's spatial accessibility in a large extent.

Fourth, enhancing communication with the rest of the world other than the region will bring more advantages rising out of frequent contact across regions. The improvement of its accessibility will benefit all sub-areas, departments and enterprises concerned, making them a unified system which brings forth a sound division of laboring force for towns and cities along the rail line, and economic integrity. Meanwhile, the massive construction of transportation facilities is a guarantee for more extension of labor division in a wider space.

4.3.2.3 Quantitative Research on Accessibility

We can see above, the quantitative study of accessibility is key to evaluating the accessibility of every point on the region map, and the effect of transport infrastructure on regional development. The basic idea for a quantitative study of accessibility is that accessibility can be substituted with certain or some indicators, wherefore they are evaluated. Such indicators are: the spatial distance, the amount of time and costs required for its efforts on economic exchanges with other regions.

With regard to the choice of indicators, some scholars have done a lot of work. Among those, Vickerman (1974), Linneker and Spence (1992), Geertman and van Eck (1995), and Gutiérrez (2001) have come up with applicable evaluation indexes which consists of three indicators: weighted average travel time, economic potential, and daily accessibility.

A. **Weighted average travel time (WATT)**

WATT is a time scale that measures the amount of time needed to travel from one node city to an economic center. It is mainly determined by the location of the node city. However, the economic strength of the central areas and the effectiveness of transport facilities whereby a node gets connected with an economic center is of paramount. The less the WATT is, the better accessibility a node has, and vice versa. Therefore, for a specific city, the score of this indicator is often lower than that of the peripheral area. If it is a big economic zone of cities in a wider space, the closer a node is to its economic centre, the lower the resulting scores will be. This can be written as in the Eq. (4.13).

$$W_i = \frac{\sum_{j=1}^{n} (T_{ij} \times M_j)}{\sum_{j=1}^{n} M_j} \tag{4.13}$$

where

W_i represents the accessibility of a node city i; T_{ij} is the time needed from city i to the targeted economic center (or destination) by a certain mode of transportation. M_j is the runoff applied to evaluating flows of social economic elements involved between an economic center and the node city. As it often indicates the strength of an economic center or its radiation power and appealing for other surrounding areas, M_j can be represented by some other variables such as the total volume of GDP, the population size, or total sales volume of goods. n refers to the total number of points or cities on the line except the city i. More specific, while it seems that WATT gives more emphasis to the division of the zones, together with daily accessibility, it appears to be more visual in calculating accessibility, because it is indicated by the length of travel time and by the size of the space.

B. **Economic potential**

As one of the indicators, economic potential is mainly determined by the location of a defined city node. The bigger the figure of a node city is, the greater accessibility it has, and vice versa. There is a positive correlation between the figure of a node and its spatial interaction with the economic center and the destination, and also between the intensity of its interaction and the economic power of the center. But the figure of the node is in inverse proportion to the distance between the node and the economic center, and the time or expenses. To calculate such economic potential, the Gravity Model originating from physics is adopted, and written as:

$$P_i = \sum_{j=1}^{n} \frac{M_j}{T_{ij}^{\alpha}} \tag{4.14}$$

where

P_i indicates the potential value of the node city i, while T_{ij} and M_j means the same as in the Eq. 4.13. α, whose value is 1, is the coefficient for distance friction. This equation for economic potential can be classified as one type of "Gravity", so the results of calculation will go down with the increase of distance. Usually indicators for potential strength give more weight to regional interaction, whereas overall economic indicators focus more on the economy, whereby we find it difficult to know more about accessibility by just looking at the calculation result.

C. **Daily accessibility**

Daily accessibility denotes the degree and amount of different kinds of economic activities (including trips for business, vacations and inhabiting) taking place from the node city i to other points of places. So daily accessibility can be measured by the amount of population flow or material flows, and by the maximum volume of daily passengers as well, which is what we use in this book. So, let us say: the time range is 8 h (daily work time) in all, if it takes one four hours to travel from the node city i to his destination. The case is that daily accessibility is directly connected with the quality and the well-being of transport facilities around the area. The more travelling destinations one is allowed to reach, the greater accessibility an area is endowed with. In this sense, daily accessibility gives more emphasis to the direct effect of transport facilities.

As described above, each of the above-mentioned three indicators reflects some degree of accessibility. But it is worthwhile to notice that they are different to some extent. So it is necessary to do some comparisons so as to give more hints to the changing picture of accessibility by defining their connotations, correcting the resulted indications, and more important, by minimizing calculation error.

4.3.2.4 The Measurement of a Node City's Accessibility on the Beijing-Shanghai High-Speed Railway

Since high-speed railway seeks to shorten one's travelling time from one place to another, it is more appropriate for us to select the equation or model built on **weighted average travel time (WATT)** as a measurement to evaluate a city's accessibility. Seeing that the population scale is mainly concerned with the labor market and the consumption market, and GDP is a better indicator to express a city's attractiveness for the outside world, we use GDP to represent the weight of WATT, M_j. In China, there are classifications of cities based on population size. Correspondingly, cities in China fall into six categories: mega-cities (with a population of over 10 million), super-big cities (with a population over 5 million but less than 10 million), very big cities (with a population over 2 million but less than 5 million), big cities (with a population over 1 million, but less than 2 million), medium cities(with a population over 50 thousand but less than 1million) and small cities (with a population less than 50 thousand). And these six categories can be classified further into more sub-tiers of cities according to Chinese divisions of administrative regions and areas, whereby cities on the Beijing-Shanghai HSR are grouped and listed as in Table 4.10:

To calculate a city's accessibility, we also need to know the length of the rail-line and the speed of a transportation. This can be seen in Table 4.11

The travel time that one spends by railway or by plane can be calculated by taking the average or the mean of the total time required for the total number of runs or flights, while the travel time for expressway and national highways is calculated through dividing the length of the route by its design speed. So now we get different travel time for different means of travelling, namely the travel time for the Beijng-Shanghai HSR Express (HSR-K), the Beijing-Shanghai HSR Extra Express (HSR-T), the Beijing-Shanghai HSR motor train (HSR-D), the expressway, national highways and civil aviation.

In line with the scientific and technological features of varied transportation, expert evaluating method is adopted to give weights to W_k so that we have the equation: $\sum_{k=1}^{n} W_k = 1$. Subsequently, for different means of transportation in different cities, we make the weighting of HSR-K: HSR-T: HSR-D: expressway: national highway: civil aviation: high-speed rail = 1:2:2.5:1:0.5:3:3, and make sure that the sum of their given weights is 1 (as in Table 4.12).

Now that we are given the run time of different transportation means and the weights in Table 4.12, we can figure out the accessibility of each individual city along the line before and after the running of Beijing-Shanghai HSR. Then, in accordance with categories of city tiers in Table 4.10, the accessibility of each city is now as listed in the following table (Table 4.13):

Since the passage of the Beijing-Shanghai HSR, our concern in this book is distributed in a banded layout, where its endpoints like Beijing and Shanghai whose weighted travel time T_{ij} is definitely higher than that of Qufu, Bengbu, and Dingyuan, cities located in a central zone of the HSR. That means, the time utility of these cities resulting from its locality is very big and it will be very misleading if we compare these cities according to the absolute value of the accessibility between cities in different zones. Therefore, our comparison in this book is only centered upon the percentage of each city's accessibility since the opening of the HSR. We exclude the horizontal comparison based on the absolute value of indicators for accessibility between cities.

As illustrated in Table 4.13, with the HSR, all cities on the rail line have witnessed an increase in its accessibility. In terms of both the equation and definition of accessibility, the increasing of accessibility helps to cut down some travel time between

Table 4.10 Categories of city tiers along the Beijing–Shanghai High-speed Rail line

Province/municipality directly under the Central Government	Name of the City	Station name	Types of administrative region	Population (million)	City type	City tier
Beijing	Beijing	Beijing Station	Municipality	20.186 m	Mega-city	First
Tianjin	Tianjin	Tianjin West Station	Municipality	13.5456 m	Mega-city	First
		Tianjin South Station				
Hebei Province	Langfang	Langfang Station	Prefecture-level city	4.249 m	Super big city	third
	Cangzhou	Cangzhou West Station	Prefecture-level city	7.3482 m	Super big city	Third
Shandong Province	Dezhou	Dezhou East Station	Prefecture-level city	5.7018 m	Super big city	Third
	Jinan	Jiangxi Railway Station	Sub-provincial city	6.0408 m	Super big city	Second
	Tai'an	Tai'an Station	Prefecture-level city	5.5701 m	Super big city	Third
	Qufu	Qufu East Station	County-level city	0.638 m	Medium city	Fifth
	Tengzhou	Tengzhou East Station	County-level city	1.681 m	Big city	Fourth
	Zaozhuang	Zaozhuang Station	Prefecture-level city	3.9104 m	Very big city	Third
Anhui Province	Suzhou	Suzhou East Station	Prefecture-level city	6.4207 m	Super big city	Third
	Bengbu	Bengbu South Station	Prefecture-level	3.6223 m	Very big city	Third
	Dingyuan	Dingyuan Station	County	0.9678 m	Medium city	Sixth

(continued)

cities, but enhances a city's time and space competitiveness, which is arranged in a descending order as in (4.4)

Table 4.10 (continued)

Province/municipality directly under the Central Government	Name of the City	Station name	Types of administrative region	Population (million)	City type	City tier
	Chuzhou	Chuzhou Station	Prefecture-level	4.508 m (Dingyuan county included)	Very big city	Third
Jiangsu Province	Xuzhou	Xuzhou East Station	Prefecture-level city	9.5761 m	Super big city	Second
	Nanjing	Nanjing South Station	Sub-provincial level city	6.2977 m	Super big city	Second
	Zhenjiang	Zhenjiang South Station	Prefecture-level	2.6988 m	Very big city	Third
	Danyang	Danyang North Station	County-level city	0.8082 m	Medium cit	Fifth
	Changzhou	Changzhou North Station	Prefecture-level city	3.5982 m	Very big city	Third
	Wuxi	Wuxi East Station	Prefecture-level city	4.6565 m	very big city	Third
	Suzhou	Suzhou North Station	Prefecture level city	6.3329 m	Super big city	Second
	Kunshan	Kunshan South Station	Prefecture-level city	1.647 m	Big city	Fourth
Shanghai	Shanghai	Shanghai Hongqiao Station	Municipality	23.4746 m	Mega-city	First

4.4 The Impact of High-Speed Railways on Time and Space Competitiveness of Cities Along the Rail Line

4.4.1 Less Significant Influences of HSRs on Relatively Developed Cities

Before the operation of a high-speed railway, the accessibility of a city is varied due to its basic infrastructure and geographical location. For some first-tier and second-tier cities (like Beijing, Suzhou, Nanjing and Shanghai) with comparatively good infrastructure and regional resources, the opening of a high-speed railway does not have very significant influence on its accessibility. But, compared to other cities,

Table 4.11 The road network of Beijing-Shanghai high-speed railway

Means of transport	Name of route	The start and end point	Length of the route (kilometres)	Design speed (km/hour)
Rail road	Jing-Hu railway	Beijing-Shanghai	1,463	120, 140, 160
	Jing-Hu HSR	Beijing-Shanghai	1,318	350
Expressway	Jing-Hu expressway	Beijing-Shanghai	1,262	120
	Jing-Fu expressway	Beijing-Nanjing	988	80
	Jing-Jin-Tang expressway	Beijing-Tianjin	145	120
	Hu-Ning expressway	Nanjing-Shanghai	274	120
National highways	103 national route	Beijing-Tianjin	162	100
	104 national route	Beijing-Nanjing	1169	80
	105 national route	Beijing-Dezhou	355	80
	205 national route	Tianjin-Nanjing	1,088	100
	312 national route	Nanjing-Shanghai	380	100
Civil aviation	Beijing, Tianjin, Jinan, Xuzhou, Nanjing, Wuxi, Shanghai			

Table 4.12 Weights given to different means of transportation on Beijing-Shanghai High-speed railway

Transportation means	Weights of $W_k W_k$											
	Mix 1		Mix 2		Mix3		Mix 4		Mix 5		Mix 6	
HSR-K	0.1	0.08	0.14	0.1	–	–	–	–	0.33	0.22	–	–
HSR-T	0.2	0.15	0.29	0.2	0.33	0.22	–	–	–		–	– –
HSR-D	0.25	0.19	0.36	0.25	0.42	0.28	0.625	0.36	0.42	0.28	–	
Expressway	0.1	0.08	0.14	0.1	0.17	0.11	0.25	0.14	0.17	0.11	0.67	0.22
National highway	0.05	0.04	0.07	0.05	0.08	0.06	0.125	0.07	0.08	0.06	0.33	0.11
Civil aviation	0.3	0.23	–	–	–	–	–	0.43	–		–	–
HSR	–	0.23	0.3	–	–	0.33	–	–	–	0.33	–	0.67

Table 4.13 A comparison of city accessibility before and after the running of the Beijing-Shanghai HSR

City		GDP (billion)		Accessibility		Increase in accessibility (with a lower percentage) %
		In 2008	In 2011	Without HSR	With HSR	
First tier	Beijing	1048.803	1600.04	292.24	259.57	11.18
	Tianjin	635.438	1119.099	334.13	276.11	17.36
	Shanghai	1369.815	1919.569	246.59	226.40	8.19
Second tier	Jinan	301.740	440.629	268.22	229.66	14.38
	Suzhou (Jiangsu Province)	707.809	1050	265.21	235.73	11.12
	Nanjing	381.462	614.552	224.18	200.49	10.57
	Xuzhou	211.884	355.165	245.01	211.06	13.86
Third tier	Langfang	106.149	161.2	341.92	273.01	20.15
	Cangzhou	162.016	260	361.42	277.49	23.22
	Dezhou	140.091	195.071	329.87	268.82	18.51
	Zaozhuang	109.283	156.168	302.59	234.8	22.40
	Tai'an	151	247.5	309.77	217.77	29.70
	Suzhou (Anhui Province)	51.110	80.24	267.08	220.45	17.46
	Bengbu Chuzhou	48.639 52.011	78.024 85.049	272.59 266.90	226.99 219.20	16.73 17.87
Fourth tier	Zhenjiang	149.183	231.04	261.37	222.64	14.82
	Changzhou	226.632	358.04	270.03	234.77	13.06
	Wuxi	446.062	688.015	241.20	214.29	11.16
	Kunshan	150.026	243.225	276.94	232.45	16.06
Fifth tier	Tengzhou	47.713	72.813	279.73	225.68	19.32
	Qufu	20.053	26.576	269.38	166.91	38.04
	Danyang	42.545	72.490	241.61	153.27	36.56
Six tier	Dingyuan	5.34	10.06	232.51	182.53	21.50

since the city Wuxi (a fourth-tier city) is most economically developed and its GDP rank is among the top list, its accessibility resulting from an HSR is also not improved significantly. As can be seen from Table 4.14, the change of its rate in TSC is the lowest among the five listed cities. Also, although Tianjin ranks the third in GDP and is comparatively less developed than Beijing and Shanghai, its accessibility has been significantly improved. But when compared with the other 23 cities under investigation, Tianjin is still one of the averaged cities.

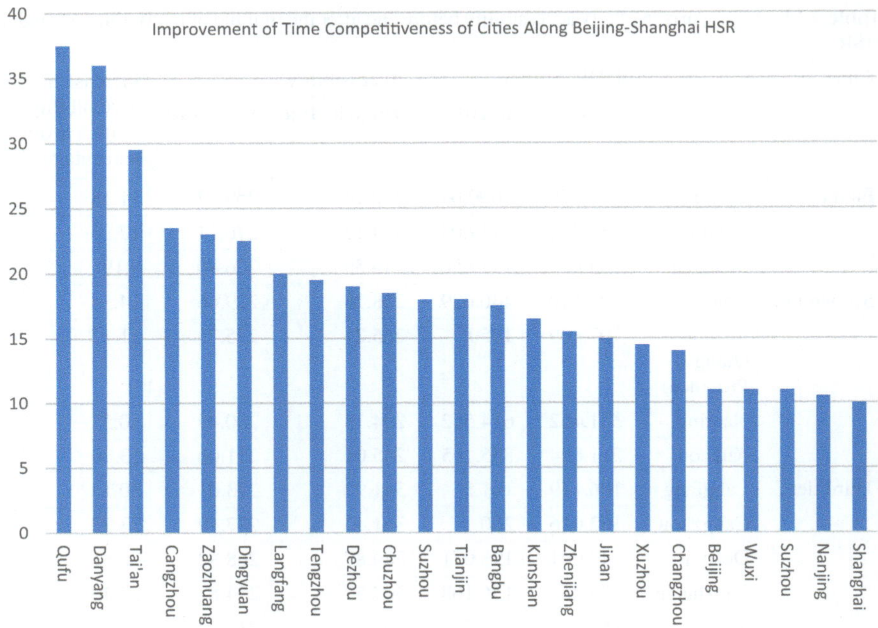

Fig. 4.4 Ranks of expansion in time and space competitiveness for cities along the Beijing-Shanghai HSR

Table 4.14 Five cities with the least significant changes in time and space competitiveness

Name of the city	Beijing	Wuxi	Suzhou	Nanjing	Shanghai
GDP in 2008 (billion; RMB)	1048.803 b	446.062b	707.809b	381.462b	1369.815b
GDP rank of the 23 cities	2	5	4	6	1
GDP rank for cities on the same scale (Numbers in the bracket are the number of cities at the same level)	2(3)	1(4)	1(4)	2(4)	1(3)
Percentage of increase in TSC	11.18	11.16	11.12	10.57	8.19

4.4.2 *Significant Influence on Economically Underdeveloped Cites Relying on Resources*

Overall, the improvement of some city's accessibility is relatively more significant. As listed in the following table (Table 4.15), for some underdeveloped cities like Qufu and Danyang (which is on the fifth level), the significance of an HSR lies more in increasing its attractions while an HSR helps to reduce a passenger's travel time. As a result of being a historical city and being the origin of Confucian culture, Qufu has become the first priority for tourists on the high-speed railway. In the first month of the Beijing-Shanghai HSR, Qufu saw an increase of visitors higher than 30% from China and abroad. Different from Qufu, Danyang has made the best use of the HSR by sparing no effort to draw in more capital and talent. Additionally, with the operation of the Beijing-Shanghai HSR, Tai'an, a third-tier city which is relatively more economically developed than Qufu and Danyang, has attracted more and more visitors owing to its unique traveling advantages and resources like Taishan Mountain, a scenic spot on the list of UNESCO world heritage.

But for Dingyuan, a small county in which one stop of the Beijing-Shanghai HSR is set, its change in accessibility is only improved to a moderate level due to its poor quality and underdeveloped economy. Before the operation of the Beijing-Shanghai HSR, road had been the basic transport (except for trains). But now, local residents in Dingyuan have one more efficient choice for travelling, which should not be measured only by the increase of its accessibility.

Further, according to Fig. 4.5, before the operation of the Beijing-Shanghai HSR, it took one hour from Beijing to Cangzhou, and also one hour from Shanghai to Danyang by inter-city rail. With the running of Jing-Hu HSR, it now takes one hour from Beijing to Cangzhou, and also one hour from Shanghai to Danyang. That means there is no observed change of time in the radius of its hour-circle. However, before the operation of this HSR, an hour may have been just enough to go from Qufu to Jinan and Xuzhou. With the HSR, an hour's time may be enough to travel from Qufu to Changzhou and Dezhou. In this sense, its extension in the radius hour-circle is much bigger than that of Beijing and Shanghai. Subsequently, the influence of the Beijing-Shanghai HSR on the time and space competitiveness of Qufu is much greater than that of Beijing and Shanghai.

Regional quality, which is represented by a city's GDP, is a key factor in time and space competitiveness (TSC). Besides, the basic transport infrastructure also plays an important part in a city's TSC. Based on the above-mentioned calculations of a city's accessibility, we hold that the TSC for cities with endowed resources may experience the most significant changes, whereas TSC for big cities, which are equipped with more elaborate transportation facilities and infrastructure, will not be affected greatly. And for economically underdeveloped cities, the impact of an HSR on its TSC is also not obvious. The relationship between the impact of HSR on a city's TSC and the degree of regional economic development is shown in Fig. 4.6:

Table 4.15 The three cities with the most significant change in accessibility

Name of the city	Qufu	Danyang	Tai'an
GDP in 2008 (billion; RMB)	20.053b	42.545b	151b
GDP rank of the 23 cities	22	21	13
GDP rank of cities on the same scale (Numbers in the bracket are the number of cities on the same level)	3(3)	2(3)	2(8)
Percentage of increase in accessibility	38.04	36.56	29.70

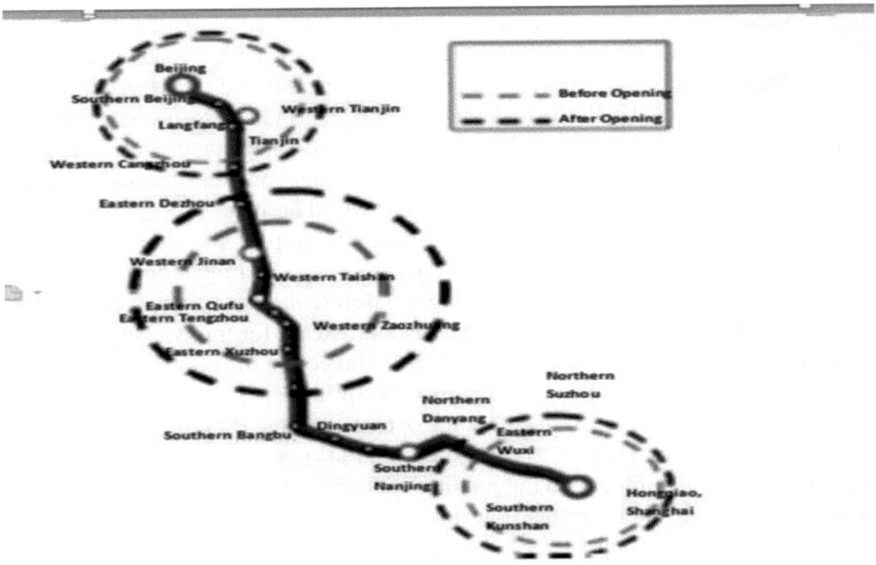

Fig. 4.5 Changes of traffic hour-circle for an hour's travel starting from Beijing, Shanghai, and Qufu

Fig. 4.6 The relationship between the effect of an HSR on a city's accessibility and its economic development

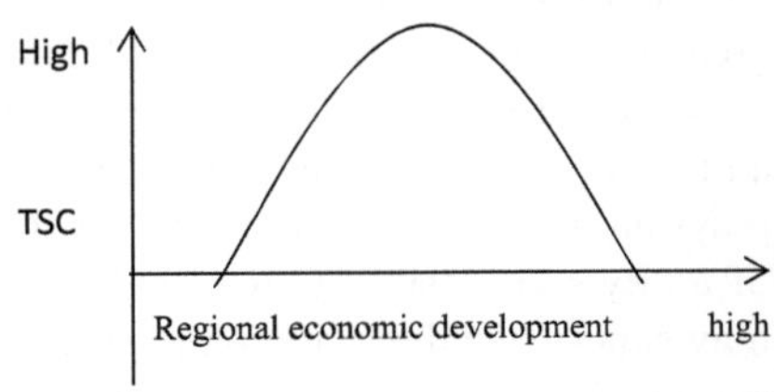

To conclude, the influence of the HSR on the regions along the rail line is profound. It does not only cut short passengers' travel time between cities on the rail line, but also stimulate people's willingness to travel around and, most importantly of all, improve greatly the accessibility of a city, making it more closely connected with other cities and regions by allowing more economic, social and cultural contacts, and an integrated spatial network and economic landscape with a strong radial force to foster regional economic development.

The regression analysis results of SPSS software are as follows:

The variables entered/removed

Model	The input variable	The variable removed	Method
1	Virtual variable D, GDP per capita, total population[1]		Input

[1] All requested variables have been entered

Model summary

Model	R	R square	Modified R square	Estimated standard error
1	0.999[1]	0.999	0.998	60.61528

[1] *Predictors* (constant), dummy variable D, GDP per capita, total population

Anova[2]

Model		Sum of squares	df	Mean square	F	Sig.
1	Regressed value	3.345E7	3	1.115E7	3034.998	0.000[1]
	Residual error	47764.751	13	3674.212		
	Total	3.350E7	16			

[1] *Predictors* (constant), dummy variable D, GDP per capita, total population
[2] *Dependent variable* Overall passenger turnover

Coefficient[1]

Model		Unstandardized coefficients		Standard coefficient	t	Sig
		β	Standard error	Trial		
1	(constant)	–10,552.119	1489.773		–7.083	0.000
	Total population	1.328	0.166	0.462	7.990	0.000
	GDP per capital	0.003	0.000	0.502	9.589	0.000
	Dummy variable D	131.978	58.936	0.047	2.239	0.043

[1] *Dependent variable* Overall passenger turnover

References

Deichmann U, 1997. Accessibility indicators in GIS. Department for Economic and Social Information and Policy Analysis, United Nations Statistics Division, 24.

Delucchi, Mark A. (1995). The Social Cost of Motor Vehicle Use in the U.S. (Can We Get the Big Numbers Roughly Right?) University of California, Davis, CA.

Geertman, S.C.M. & van Eck, J.R.R. GIS and models of accessibility potential: an application in planning. International Journal of Geographical Information Systems, 1995, 9(1): 67–80.

Goodall B, Dictionary of Human Geography. London: Penguin, 1987.

Gutiérrez, J. Location, economic potential and daily accessibility: an analysis of the accessibility impact of the high-speed line Madrid–Barcelona–French border, Journal of Transport Geography, 2001, 9(4): 229–242.

Hansen, Walter G. How Accessibility Shapes Land Use, Journal of the American Institute of Planners, 25: 2, 73–76, 1959. https://doi.org/10.1080/01944365908978307

Haynes, K. E., & Fotheringham, A. S. Gravity and Spatial Interaction Models. Beverly Hills, CA: Sage Publication, 1984.

Linneker, B. J., and N. A. Spence. "Accessibility Measures Compared in an Analysis of the Impact of the M25 London Orbital Motorway on Britain." Environment and Planning A, 1992, 24, 1137–54.

Luce, R.D. (1959) Individual Choice Behavior: A Theoretical Analysis. Wiley, New York.

Marley, A. A. J. (1965). The relation between the discard and regularity conditions for choice probabilities. Journal of Mathematical Psychology, 2, 242–253.

Marschak, J. (1960). Binary choice constraints and random utility indicators. In J. Marschak (Ed.), Economic information, decision and prediction: Selected essays (1974) (Vol. 1). Dordrecht: D. Reidel.

McFadden, Daniel. 1974. The measurement of urban travel demand. Journal of Public Economics 3(4): 303–328.

Reilly, W. J. The Law of Retail Gravitation, Knickerbocker Press, New York, NY, 1931.

Vickerman, R.W. Accessibility, attraction, and potential: a review of some concepts and their use in determining mobility. Environment and Planning A 6(6): 675–691, 1974.

Williamson, Oliver, & Masten, Scott, eds. 1999. The Economics of Transaction Costs. London: Edward Elgar Publishing, Ltd.

Chapter 5
High-Speed Railway and Regional Traffic Structure

5.1 The Mechanism of Optimizing High-Speed Railway and Transportation Structure

5.1.1 Optimization of Transportation Structure and Its Objectives

In a broad sense, optimization is to make a system as effective as possible. In a narrow sense, optimization is a way and method, that is, find the best way and method to achieve the goal from many solutions. In terms of mathematics, optimization refers to maximizing or minimizing the objective function under certain constraints. The optimization of the so-called traffic structure, therefore, not only refers to the change in the proportion of the traffic structure represented by the unequal distribution of the growth rate of various modes of transport, but also the replacement of the main transport in the entire traffic structure.

The optimization of transportation structure originated from the study of "the sharing proportion of transportation modes" in the source of the study: Chicago Area Transportation Study 1957. A Description of the Chicago Area Transportation Study. Chicago: Chicago Area Transportation Study. in the late 1950s. The core idea of it is that in order to ensure the smooth urban traffic, the utilization of various transportation modes should be properly balanced according to different occasions, purposes and times. According to previous studies in this field, to improve passenger transportation structure is to achieve the following comprehensive goals: (a) to more fully meet the transportation needs of passengers, (b) to improve the service quality and efficiency of passenger transportation system, and minimize the transportation cost; (c) to minimize resource consumption; (d) to minimize environmental pollution. To optimize the transportation structure, we should be fully aware of the advantages and disadvantages of various modes of transportation, and how to comprehensively allocate various modes of transportation, minimize the cost or maximize the efficiency,

X. Lin, *High-Speed Railways and New Structure of Socio-economic Development in China*, Research Series on the Chinese Dream and China's Development Path,
https://doi.org/10.1007/978-981-19-6387-2_5

optimize the comprehensive transportation system, and give full play to the overall functions and comprehensive economic benefits of the transportation system. Also, each mode of transportation has its own characteristics and functions, and which mode of transportation should give priority to development varies from place to place.

Traditional research on transportation structure mainly focuses on mode division model. According to the different basic units of mode selection, the mode models can be grouped into two types: the lumped model (that the traffic activity of each person is statistically processed and analyzed according to the traffic district, so as to obtain the model with the traffic district as the analysis unit) and the disaggregate model (that the data obtained from the survey is directly used to build the model with the individual who actually generates traffic activities as the unit, without statistical processing according to the traffic district, etc.). The lumped model came earlier, and its operation was relatively simple, but the calculation was not accurate. In order to make up for the deficiency of lumped model, the disaggregate model was proposed in the early 1960s. Since the 1970s, McFadden (1974) and others from the Massachusetts Institute of Technology (MIT) have made great progress in theoretical research, thus driving the team researchers like Manheim (1987), Ben-Akiva and Lerman (1985) and others in the United States to push the research of disaggregate model to the practical stage. Up to now, the research of various disaggregate models is still going on. However, there is a defect in both the plumped model and the disaggregated model: the individual interests of travelers are considered, but not the overall interests of society (Shao and Zhang 2003).

After the "four-stage method" was introduced into China, many experts began to study the mode division model. The existing research is generally aimed at the structural optimization of a single transportation mode, while the research on the comprehensive transportation structure optimization model composed of various transportation modes is still in the exploratory stage. In China, various modes of transportation are managed by different departments. When planning transportation, different sectors concerned usually care more about their own industries and take other modes of transportation as their competitors, instead of comprehensively considering the planning of the whole transportation network. For example, railways, highways, water transport and aviation all have their own industry network planning. Generally speaking, a transportation corridor can only achieve the structural optimization of one transportation mode, but not the structural optimization of the whole transportation corridor. Just because a transportation corridor is generally a group of naturally formed metropolitan areas and markets, the transportation demand will inevitably be provided by various transportation modes such as air, highway and railway, so the improvement of transportation corridor structure should comprehensively consider the advantages and an orchestra of various transportation modes. Although each mode of transportation has its own relatively independent transportation network, we can study the structure optimization of each single transportation network, but this is not the focus of transportation structure optimization. Instead, the way to optimize should be placed on how to make all modes of transportation play their own advantages, complement and compete with each other, and develop harmoniously, so as to optimize the whole transportation network.

5.1.2 Mechanism of Optimizing Transportation Structure of High-Speed Railway Driven by Demand and Technical Support

In the development history of human transportation, transport of all kinds did not come into existence all at a time, but came one after another along with the human activities, the development of science and technology and civilization. The earliest transportation mainly relied on the natural land roads and waterways. In social life, human beings renovated these natural land roads and waterways, and used simple means of transportation to realize the displacement of people and things. These simple means of transport may be vehicles driven by human beings or animals, and ships that rely on manpower and wind power. Later, with the development of economy and science and technology, new means of transportation such as railway, automobile and air transports gradually emerged. The birth and development of these transportations have a profound historical background, which are closely related to the transportation demand and the level of science and technology at a particular time. Therefore, the development of transportation in each period has its own emphasis. From the perspective of the focus and leading role of transportation in the world, the development of the transportation industry can be divided into four stages, each of which may highlight a certain transport: (1) water transportation, (2) railway, (3) road, air and pipeline transportation and (4) comprehensive transportation:

The First stage: the water transport. Water transportation in modern sense includes ocean, coastal and inland transportation. Water transportation has a very long history, and inland and coastal transportation were the earlier ones. In this stage, the spatial displacement of people and things mainly depends on ships, rafts and other floating tools, natural waterways such as rivers and lakes, and artificial canals. The development of ocean shipping is the product of human scientific and technological civilization in a certain period. Before the advent of railways, water transportation had more advantages than land transport powered by manpower and animals in terms of transportation capacity, transportation cost and convenience. The transportation demand in this period was mainly met by water transport. Therefore, most of the early industries in capitalist countries set up factories along navigable waterways, hence the development of water transportation greatly affected the industrial layout. However, in the historical period when air transportation has not yet come into existence, ocean-going transportation across oceans and seas can hardly be replaced by other types of transportation. Therefore, transportation demand and technical level determine that water transportation is the dominant mode of transport in the early stage of transportation history.

The second stage: the railroad. The first railway in the world was born in Britain in 1825. Since then, the transportation industry has entered the railway era. After the British Industrial Revolution, textile industry, mining industry and metallurgy industry developed rapidly. So, water transportation alone could no longer meet the needs of economic development which called for a new mode of transportation that can carry a large number of materials and personnel and move without geographical barriers. With the invention of steam locomotives and gasoline diesel locomotives and the development of science and technology, railway transportation has developed rapidly. Because the railway can transport passengers and goods in large quantities at high speed, it almost monopolized the land transportation at that time, which greatly changed the dominance of land transport, providing a new and powerful transportation for the development of industry and agriculture. Meanwhile, with the railway construction, the industrial layout was gradually lifted out of its dependence on water transportation, making the sources of raw material and consumer markets approachable, and the development of industry and agriculture speeded up. At that time, railway transportation was in an advantageous position in technology and economy, so European and American countries with developed industries successively entered the gold time of railway construction in the nineteenth century. Later, it expanded to Asia, Africa and South America, which made railway transportation almost in a monopoly position at this stage of development.

The third stage: the road, aviation and pipeline transportation. From 1930 to 1950s, in order to adapt to social and economic development, more diversified transportation tools were needed, as the progress of science and technology provided technical support for the successive development of highway, aviation and pipeline transportation. With the development of the automobile industry and the vigorous construction of the highway network, it competes fiercely with the railway on land, because it is flexible and things can be transported door to door, and has great advantages in short-distance passenger transportation. With the development of industry and the progress of science and technology, people have become sensitive to the value of time, which requires more efficient transportation. The great progress of aviation technology has met people's needs in this respect. With the advantage of speed, air transport not only gradually occupies an important position in passenger transportation (especially over long-distance), but also makes great progress in freight transportation. Moreover, as the demand for oil, natural gas and other energy sources is increasing, pipeline transportation that is suitable for long-distance continuous transportation appears. Although the categories of goods it transports are limited, it is developing rapidly because of its low transportation cost and convenience. All in all, it is the change of transportation demand and the development of science and technology that makes the role of these three modes of transport rise significantly at this stage.

The fourth stage: the comprehensive transportation. Since 1950s, people began to realize that in the development of transportation, the five channels of transportation, namely, railway, water transport, highway, aviation and pipeline, are mutually restricted. Therefore, it is necessary to make overall planning and systematic thinking to deal with the relationship among those different transportations. To build a modern comprehensive transportation system, one of the key points is to make a reasonable division of labor among railway, water transportation, highway, aviation and pipeline transportation, which may highlight the advantages of every transportation. At present, the expansion speed of the world transportation network has slowed down. Adjusting the layout and improving the quality of transportation have become the main trends in the comprehensive transportation stage. In a word, the demand for the optimization of transportation system and the development of high-tech such as information technology have jointly contributed to the development of comprehensive transportation system.

From the development history of transportation over the four stages, we can see that the evolution of transportation is the result of the combined action of transportation demand and technical support. Like all previous transportation, the birth and development of high-speed railway also has its unique background. On the one hand, with the emergence of the world energy crisis and the increasingly severe traffic congestion and environmental pollution caused by the rapid development of road transportation, people begin to re-examine the development model of transportation industry. As a result, a new type of transportation demand characterized by high speed, large capacity, low energy consumption and environmental protection has emerged. The technical and economic characteristics of high-speed railway are well adapted to this transportation demand, which makes the development of high-speed railway have a broad market space. On the other hand, the development and maturity of a series of related technologies such as high-speed train, ballastless track, HISR bridges and tunnels, traction power supply and safe operation control have solved the technical issues of high-speed railway, which makes its vigorous development technically feasible. Thus, it is the dual factors of new transportation demand and new technology that contributed to the birth and development of high-speed railway. It is precisely because the high-speed railway has the dual benefits of meeting the market demand and meeting the direction of transportation technology progress that the it is endowed with the dual factors of optimizing the existing transportation structure.

5.1.3 The Main Path to Optimize the Transport Structure of High-Speed Railways

The emergence of new transportation is an inevitable result of meeting certain transportation needs, and also the inevitable result of competition among varied transportations. The birth of the new transport has two impacts on the existing traffic structure. On the one hand, the new transport will inevitably bring an impact on the original transportations, since it has the technical and economic characteristics that the existing transportation does not have, meets certain transportation demands and create new ones. On the other hand, through the competition and gaming with the existing transportations, the new and the old may find their own areas suitable for giving full play to their technological and economic advantages. Then they may

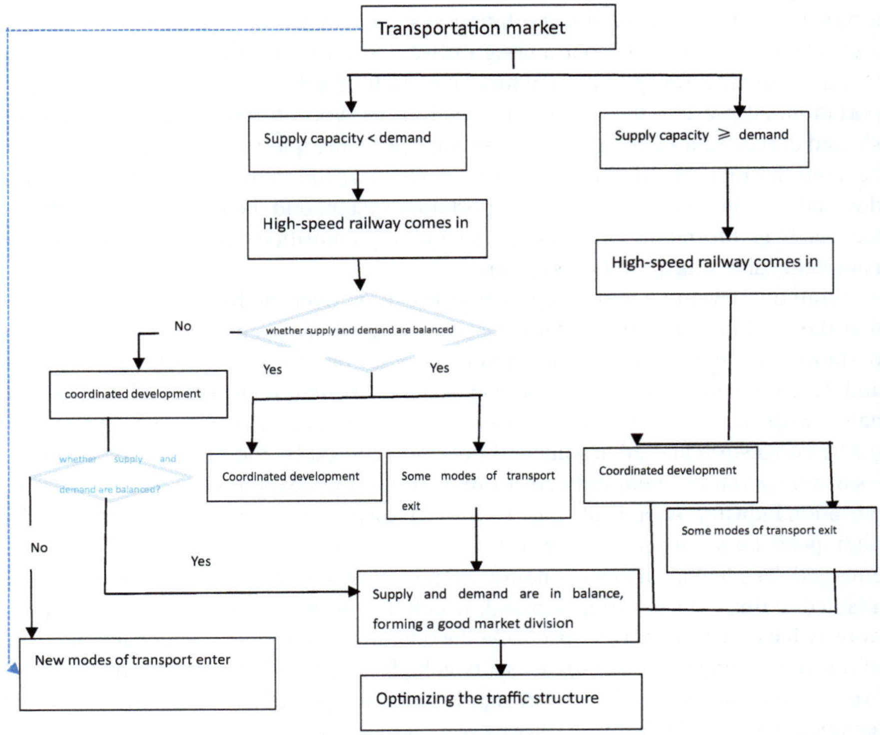

Fig. 5.1 Schematic diagram of the path of high-speed railway to improve transportation market structure

gradually withdraw from areas that they do not have comparative advantages, and finally form a market structure in which they could play full their own advantages in the competition. Then, the relationship between them changed from competition to cooperation, leading to a restructuring of the whole transportation system. (See Fig. 5.1).

5.2 Impact of High-Speed Railway on Various Modes of Traffic in the Region

As a new mode of transport, high-speed railway usually has a relatively mature transportation system in the areas where it is introduced. The network economy attribute of the transportation industry means that its smooth operation must be grounded on the effective embedding with the existing network. This implies that the high-speed railway, the new participant, should not only compete with the existing transportation network, but also collaborate with it, that is, competition and cooperation coexist.

5.2.1 Competition and Cooperation Between High-Speed Railway and Other Modes of Transportation

The modern transportation system is mainly composed of four modes of transport: highway, water transport, railway and civil aviation. The coexistence and development of these four modes of transportation is rooted in their unique technical and economic characteristics, which can meet different transportation needs. For a long time, various modes of transportation have been competing and cooperating with each other, thus forming the existing transportation structure. Competition and cooperation among various modes of transportation can be understood from a deeper level:

(1) There is competition among different modes. A certain transportation demand can often be met by more than one transport, so there is inevitable competition among different transports that can meet the demand. In addition, within a certain range of travel, the transportation services provided by these modes of transport are mutually substitutable. People can select their favorable means to travel according to their contexts and needs.
(2) There is cooperation between transportation modes. Because different modes of transportation have different technical and economic characteristics, all means of transport have their advantages and disadvantages. However, the demand in the transportation market does not necessarily correspond to the advantages of a certain transport completely, which requires the coordination between various transportation means to provide diversified transportation supplies to meet the demand.

By reviewing the development history of transportation, we can find that in a transportation market where there is not sufficient the railway transportation capacity but large passenger demand (i.e. the transportation capacity is not saturated market), where the original transportation system in the region lags behind the needs of social economic development, the transportation has become the bottleneck to impede the further development of society and economy. So this is time for the high-speed railway to come in, which may help to improve the transportation capacity of the transportation system and meet certain transportation needs. In this case, the high-speed railway may help to make up for the deficiency of the original traffic situation and build a cooperative relationship with other modes of transport.

From a practical point of view, the emergence of high-speed railway is the inevitable result of meeting certain transportation needs, and of competing among various modes of transport. The impact of high-speed railway on the original traffic system can be seen from two aspects: on the one hand, the high-speed railway brings inevitable impact on the existing modes of transportation. And because of the technical and economic characteristics of the high-speed railway that those modes of transportation do not have, it can meet and create certain transportation demands. On the other hand, because of the competition and game between the high-speed railway and the existing transportation, each mode of transport will find areas that can highlight its own technological and economic advantages, and may gradually withdraw from areas that they do not have comparative advantages. Finally, in the competition, a market structure that each can play full its own advantages will be formed, and the relationship between them will change from competition to cooperation, thus optimizing the whole transportation system.

Generally speaking, although railway, highway, water transport, aviation and other modes of transportation have been basically connected, they still cannot meet the needs of the rapid development of regional and foreign trade cooperation. In order to meet the essential requirements of efficient, fast, convenient and low-cost modern logistics, the speed of the construction, opening and operation of high-speed railway may be reasonably designed ahead of the times. Taking Guangxi high-speed railway as an example, its opening and operation will help to form an intercity high-speed railway network centered on Nanning, Liuzhou, Guilin and other large and medium-sized cities. In this way, inter-city high-speed trains can be opened in inland cities, and the construction of a modern fast railway transportation network with one axis, four vertical and four horizontal directions can be accelerated, so that the whole region can enter the era of high-speed rail economy.

In addition, there are many examples in the world from which we can see how regional transportation and economic integration is realised because of high-speed railway. For example, because of the developed high-speed transportation network between Tokyo and Yokohama in Japan (here mainly refers to high-speed railway and expressway), people don't need to consider the concept "city" when they travel. Amsterdam and The Hague, the Netherlands, are closely linked by convenient roads and railways. It can be predicted that when the regional high-speed railway network is basically built, the convenient and comfortable traffic will effectively subvert people's traditional regional concept, narrow the psychological and cultural distance of people of all ethnic groups, enhance their sense of identity and regional traffic integration, and also will be conducive to the development of regional tourism cooperation.

From the perspective of international experience, in order to integrate the relevant regions into the unified market within the China-ASEAN Free Trade Area, it is an essential hardware measure to improve the diversified transport system. The construction and development of the railway, especially the high-speed railway, and its connection with other methods are extremely important for the regional traffic integration planning. Because the demand for regional trade will inevitably lead to the integration of various modes of transportation in the region, the free flow and rational allocation of various commodities, resources, and labor in the region, as well as a unified internal market, will be realized. However, the free flow and configuration of various materials and personnel requires the transportation system to provide as much transportation capacity as possible through the connection and collocation of different modes, but with the lowest possible transportation costs. This is the fundamental driving force for the integration of regional transportation integration networking.

In a word, the high-speed rail not only integrates the existing transportation system in a region, improves the efficiency of the transportation system, but also expands the circulation of production factors and structures the basic transportation framework. By learning from the experience of Japan Shinkansen and Taiwan (China)'s high-speed railway, we can find that, on the one hand, the infrastructure and equipment demand of high-speed railway during the construction period will create a large number of employment opportunities, where the government needs to formulate investment policies to attract high-speed railway construction and equipment manufacturing enterprises to invest and build factories. However, after the opening of a high-speed railway, every mode of transportation in the region will find its own place where its own technical and economic advantages can be fully applied, and finally form a transportation system that each can maximize its own advantages in the competitive market, and finally an optimized transportation system finds its way. On the other hand, as the framework to support the regional transportation system, the high-speed railway should have a reasonable construction schedule that is coordinated with the regional economic development, thus shaping an exchange roadmap of large-scale, long-distance, large-scale and rapid circulation of production factors within the belt.

5.2.2 *Effective Passenger and Cargo Transportation Between High-Speed Railway and Existing Railway*

High-speed railways have many significant advantages over ordinary railways. For example, high-speed rail is characterized by high passenger capacity, large transportation capacity, fast travel speed, good safety, high punctuality rate, comfortable and convenient, low energy consumption, light environmental impact, and good economic returns. But it also has a certain degree of particularity, as it is mainly designed for passenger and cargo transportation. Therefore, its completion and opening will affect the passenger and freight transport of ordinary railways to a certain extent.

As shown in Fig. 5.2, after the high-speed railway is built and operated, it will divert certain passenger sources from ordinary railways by virtue of its comparative advantages over ordinary railways, thus causing unavoidable consequences for ordinary railway passenger transport. However, the degree of impact of high-speed rail on ordinary railways needs to be determined according to specific conditions. Compared with ordinary railways, high-speed rail has its obvious advantages, but it also has the disadvantages of high prices and fewer lines. In general, tourists and business people tend to take high-speed rail when they travel. Whereas for ordinary people, especially migrant workers, travel, they often choose to take fast trains. In addition, for short trips, between high-speed and ordinary trains, people tend to take ordinary trains, and they prefer high-speed trains when traveling long distances. Generally speaking, the open of a high-speed railway will inevitably have a certain impact on

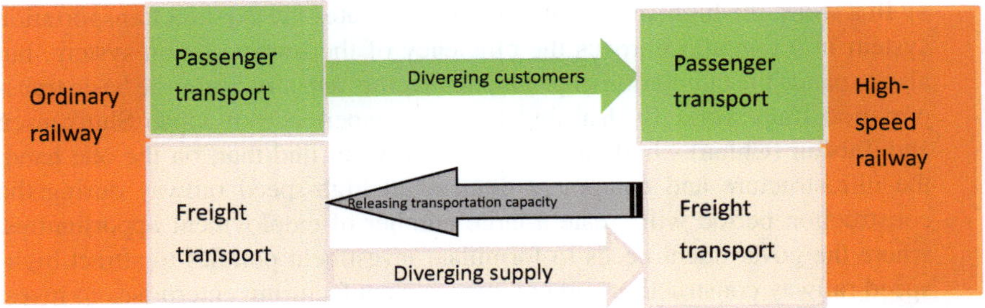

Fig. 5.2 Schematic representation of passenger and cargo distribution between high-speed rail and existing railways

the passenger transport of ordinary railways in its region, but it also lightens the passenger transport burden of ordinary railways to a certain extent by offering a new choice for people to travel.

From the freight point of view, the completion of the high-speed railway will release the freight capacity of the existing ordinary railway to some extent, and improve the freight volume of the ordinary railway. Moreover, the freight volume will be increased in some degree due to the mixed passenger and freight transportation of the high-speed railway. Therefore, the completion and operation of high-speed rail will increase the freight volume, which will greatly improve the railway freight capacity of its surrounding provinces and cities.

5.2.3 The Negative Impact of High-Speed Railway on the Highway Transport

In the competition between traditional highway and railway, with its advantages of short running time, comfortable riding environment, door-to-door transport and flexible departure time, highways serve different groups of people and share the task of carrying passenger and cargo together with traditional railway transport. However, with the development of high-speed rail and the increasing networking, it not only facilitates people's travel and accelerates urban development, but also brings unprecedented challenges to the road transport market with its advantages of high speed, large transportation capacity and high safety. The advantages of highway passenger transportation in terms of time and comfort disappeared, and it was also at a disadvantage in terms of transport capacity, environmental impact and safety factor. It can be seen from Table 5.1 that compared with road transportation, high-speed railway has obvious advantages in transportation characteristics and social cost. The completion and operation of high-speed railway may change people's choice for traveling tools, and may have a certain adverse impact on highway freight transportation.

Table 5.1 Comparison of high-speed rail transport and road transport characteristics

Items	High speed rail transport	Road transport
Road network	The route is single; Mainly connects large and medium-sized cities	The route layout of highway transport is diversified, connecting major small and medium-sized cities and towns, and gradually targeting rural areas
Management	It is managed by the central government on the whole, but operationally managed by regional government	The main business entities of highway transport are diversified, small-scale and decentralized, but the regional division is serious
Transportation characteristics	Generally, it needs other modes of transit – Have large single transportation volume, fixed fares and limited trains – Is suitable for medium and long-distance passenger and freight transportation	Smaller transportation volume, Door-to-door transportation, More frequent commuter shifts, Suitable for medium and short-distance passenger and freight transportation Flexible, easy to transfer Floating fare
Social costs	More advantages in terms of environment, accident rate, and energy dependence than roads	Less original investment, Fast capital turnover, High accident rate per unit turnover, Great environmental impact and Highly dependent on oil

From the perspective of passenger transport, according to the theory of random utility in economics, consumers are pursuing maximization of "utility" when choosing a travel tool, which means that travelers always tend to choose the method that maximizes their "benefit". The key influencing factor for the choice of travel method is the time value of the traveler. The attributes that embody the time value are: safety, economy, speed, comfort, convenience, and punctuality. From Table 5.2, it can be seen that choosing high-speed rail makes passengers' utility greater, so the opening of high-speed rail will cause certain adverse impact on road transport. At present, judging from the situational context of high-speed rail operation in China, the air and road transport of all passenger lines running high-speed rail have been comprehensively impacted. Judging from the travel performance of the majority of passengers, more and more passengers are keen to choose to travel by high-speed rail. Statistics show that since the start of Hefei-Nanjing, Hefei-Shanghai and Hefei-Wuhan EMUs, the passenger transport volume of Hefei-Nanjing highway and Hefei-Shanghai highway has dropped by 58%, and by 80% respectively. Passenger flow of the trains from Hefei to Suzhou, Wuxi, Changzhou, Kunshan and southern Jiangsu all declined to varying degrees, with an average drop of about 40%. The Hefei-Wuhan shuttle bus even stopped running in July 2009.

Table 5.2 Comparison of advantages and disadvantages of high-speed railway and road transport

Characteristics		High-speed railway	Highway
Traveler experience	Convenience	☆	★
	Safety	★	☆
	Stability	★	☆
	Comfort	★	☆
	Service	★	☆
	Punctuality	★	☆
	Time in transit	★	☆
	Operating conditions	★	☆
Market operation	Carrying capacity	★	☆
	Price	☆	★
	Market openness	☆	★
	Transportation Network	☆	★
	Sales level	☆	★
Environmental protection	Land	★	☆
	Ecological damage	★	☆
	Emission pollution	★	☆
	Energy consumption	★	☆
Comprehensive		12★5☆	5★12☆

Note ★ indicates relative advantage, ☆ indicates relative disadvantage

In terms of freight transportation, price, carrying capacity, and punctuality are the main indicators that customers are concerned about. Compared with road transport, high-speed railways have comparative advantages in these areas. Therefore, the opening of the high-speed railway will affect the cargo transportation on the highway to some extent. In addition, from the perspective of environmental protection, the relative advantages of high-speed railways are even more pronounced.

5.2.4 *Multimodal Transport by High-Speed Rail and Waterways*

As to passenger transport, the opening and operation of the high-speed rail has basically no effect on water passenger transport. In recent years, with the continuous expansion of the highway network and the improved quality of railway operations, passenger transport in the waterways has gradually turned to tourism and leisure. The purpose of the tourists to choose to take a boat may be mainly sightseeing, such as enjoying the scenery along the way and experiencing boat rides, needs that the high-speed rail may not meet in this regard. Therefore, after the opening of the high-speed rail, water passenger transport is usually not affected.

As far as freight transportation is concerned, the waterway freight industry is less affected because the goods transported by water are generally bulk goods with relatively low requirements for shipment date. However, the opening of high-speed rail is beneficial to the development of waterway cargo shipping to a certain extent. For example, Fangchenggang, one of the largest ports for goods entry and exit, has a serious backlog of goods, which has made the pressure of transportation is constantly increasing. With the substantial increase of coal and iron ore to Fangchenggang port, the stock of port yard is gradually piling up. In September, 2011, the stock has exceeded 65% of the normal inventory, but unluckily, the railway was unable to meet its demand for dredging. By contrast, with the opening of the high-speed railway, the freight transport capacity of the existing railway will be improved so that the pressure of the port will be relieved to a certain extent, thus helping the development of waterway freight.

Overall, the construction of multimodal transportation collection and distribution system with sea and railway transportation will help to improve the domestic and international transit transportation system of bulk goods and containers such as oil, coal, ore and grain.

5.2.5 *Passengers Diverted from Civil Aviation into High-Speed Rail*

Compared with civil aviation, high-speed rail has a certain relative advantage, as shown in Table 5.3, especially in terms of price and delivery. According to the related research at home and abroad and the situation of high-speed railway operation, after the high-speed railway is completed and operated, it will divert a certain number of passengers from civil aviation by virtue of its technical and economic advantages, such as large volume, high departure density, convenience and comfort, all-weather driving, relatively cheap price, etc., so that the market share of aviation will drop sharply, which will inevitably impact civil aviation transportation. For example, Zhengzhou-Xi'an high-speed railway was officially put into operation on February 6th, 2010. Every day, seven pairs of trains run from Xi'an and Zhengzhou, with the lowest fare of 240 yuan. The shortest running time of trains between Xi'an and Zhengzhou is shortened from 6 h to about 2 h. The flight time from Zhengzhou to Xi'an is about one and a half hours, but if the time to and from the airport, check-in, boarding and waiting for luggage is taken into account, it is more time-consuming than the high-speed rail. So, the influence of Zhengzhou-Xi'an high-speed railway on the existing traffic pattern has gradually emerged. The flight from Xi'an to Zhengzhou was completely grounded on March 25th, 2010, and so was the Zhengzhou-Xi'an route of by Henan Airlines. This means that the flights between Xi'an and Zhengzhou will be completely grounded.

There are also well-known examples abroad. For example, the opening of the French high-speed railway has pushed some routes of Air France to a desperate situation, forcing Air France to give up the route from Paris to Brussels. Furthermore,

Table 5.3 Comparison of advantages and disadvantages of high-speed railway and civil aviation transportation

Characteristics		High-speed rail	Civil aviation
Traveler experience	Convenience	☆	★
	Safety	☆	★
	Stability	★	☆
	Comfort	★	☆
	Service	☆	★
	Punctuality	★	☆
	Time in transit	☆	★
	Operating conditions	★	☆
Market operation	Carrying Capacity	★	☆
	Price	★	☆
	Market openness	☆	★
	Transportation Network	☆	★
	Sales level	☆	★
Environmental protection	Land	☆	★
	Ecological damage	☆	★
	Emission pollution	★	☆
	Energy consumption	★	☆
Total		8★9☆	9★8☆

Note ★ indicates relative advantage, ☆ indicates relative disadvantage on the lines with high-speed railways, the share of passengers by Air France has also dropped sharply. After the Paris-Marseille high-speed railway was put into operation, the railway market share rose from 38 to 60%. On the Paris-London line, the railway also captured 65% of the market share. In addition, from Seoul, South Korea to Busan, from Taipei, Taiwan Province, China to Kaohsiung, the completion of the high-speed railway has greatly impacted on air transport.

The completion and operation of the high-speed railway will also have some impact on the civil aviation in its area, and the degree, scope and the way of its influence depends on the specific situation of civil aviation.

First of all, high-speed railway will have a great impact on medium and short-haul flights, but its substitution effect on long-haul flights is weak. For short-and medium-haul ones, the travel time difference between high-speed railway and civil aviation is not very obvious. In effect, because the railway has the advantages of high departure frequency, little weather influence and low fare, the attractiveness of high-speed rail is greatly improved, and the crashing and substitution effect of high-speed railway on civil aviation on short-distance segments are strong. Therefore, after the high-speed railway is put into operation, it will divert some of its customers from civil aviation by virtue of its advantages in speed, price, and grid density. This will lead to fierce competition between high-speed railways and civil aviation, as well as among civil aviation companies. For long-distance trips, although the high-speed railway still has certain price and density advantages, the speed advantage of civil aviation is more apparent as the passenger's railway travel fatigue over a long distance will be increasing, so the substitution and negative impact of high-speed railway on civil aviation is weak.

Secondly, the high-speed railway may create different competitive pressures on civil aviation in different periods. According to the relevant data of foreign countries and the effect analysis of previous railway speed increases in China, in the short term, some air passengers will be diverted at the initial stage of high-speed railway construction, which will exert great market pressure on civil aviation. This is mainly due to the fact that after the high-speed railway is put into the market as a new mode of transportation, people tend to try it because of its advantages in price and speed, as well as the psychological effects of novelty and experience, which will lead to a large drop in the number of civil aviation tourists in the short term. In the medium term, if civil aviation can take active and effective measures, some customers with high loyalty could be retained, and new customers may increase, so that the market share will gradually return to a normal level. In the long run, the development of high-speed railway may result in fierce competition among various modes of transport. With the rapid growth of China's total passenger demand, if civil aviation can seize the opportunity to solve some deep-seated problems that perplex its own development, it is entirely possible to regain and expand the market share and achieve a win–win situation in the competition with the railway.

Finally, the impact of high-speed railway on civil aviation is not limited to passenger transport. The completion of the high-speed railway will not only affect the passenger transport of civil aviation, but also the civil aviation freight industry to some extent. After the completion of the regional high-speed railway, the existing lines will be mainly used for freight transportation, so the freight capacity and transportation speed of the existing lines will be greatly improved (compared with the present ones). Moreover, most of the newly-built high-speed rail lines are lines with mixed passenger and freight transport, which is bound to take away some of the cargo flow of air freight. Especially in the short and medium distance, the time advantage of air freight over railway freight is weakened, which will lead to a large number of ordinary goods turning to railways with low price and large capacity.

5.3 Case Analysis of the Influence of High-Speed Railway on Regional Traffic

In this part, to show the influence of the opening of the high-speed railway on the existing passenger traffic structure, two cases will be analyzed: (1) the high-speed railway in Taiwan Province, China, and (2) the high-speed railway in Hainan, China.

5.3.1 Case Study of High-Speed Rail in Taiwan Province

(A) **Background**

Taiwan Island, located on the continental shelf of the southeast coast of China, is the largest island in China and an important transportation hub for maritime links between China and other countries in the Pacific region. Its geographical location is very important. Situated in the subtropical zone, Taiwan is known as China's "Treasure Island" because of its mild and pleasant climate, beautiful scenery and abundant natural resources.

Taiwan Island is spindle-shaped and covers an area of about 36,000 km^2. It is long from north to south and narrow from east to west. The longest distance from north to south is about 394 km, and from east to west the widest part is about 144 km. Mountains and hills account for more than 2/3 of the total area. According to the statistics of relevant parties in Taiwan Province, as of 2010, it has a population of about 23.162 million, and the population density is about 639.99 people/km^2. From 2000 to 2010, the average annual growth rate of its population was 0.45%, and its population was mainly concentrated in the western plain.

Since 1960s, Taiwan Province has seized the opportunity of the change of international division of labor at that time. Relying on attracting foreign capital and technology, and vigorously developing labor-intensive processing and export industry, Taiwan Province has achieved economic take-off in a short time. From then on to 1980s, its economy has maintained rapid growth.

Economic growth has generated a large amount of transportation demand. In the 1970s, Taiwan Province invested a large amount of funds for the development of the transportation industry. Today, Taiwan Province has formed an integrated transportation network consisting of railways, highways, aviation, and water transportation. There are six highway systems in Taiwan Island, including expressway, roundabout highway, cross highway, longitudinal highway, coastal highway and connecting highway. By 2010, the total length of the highway in Taiwan Province was 41,383 km, with a road density of 1,143 m/km^2. The railway has formed a complete railway network around the island, with a total length of 3,915 km and a density of 108 m/km^2 by 2010.[1] There are many coastal harbors in Taiwan Province. Besides the five major ports of Kaohsiung, Keelung, Taichung, Hualien and Su'ao, there are small and medium-sized ports such as Taitung Danshui, Green Island, Magong, Donggang, Anping, Lugang and Fenggang. Civil aviation on the island is also well developed. Taiwan Province has 16 airports, including Taipei Taoyuan and Kaohsiung International, with 93 international routes. There are 50 scheduled flights from Taipei to Kaohsiung on the island every day.

By far, Taiwan's intra-island transportation mainly relies on the railway and road transportation networks that extend in all directions. Off-island convenient sea and air transport effectively supports the development of Taiwan (China)'s economy and foreign trade.

The gross national product (GNP) of the western corridor[2] of Taiwan Province accounts for 98% of the province, and its population accounts for 94% of the total population in Taiwan Province. The average population density of Taiwan Province is 750 people per square kilometer, while the average population density of Taipei and Kaohsiung in this area is 10,000 people per square kilometer and 9,000 people per square kilometer respectively. In order to enhance the competitiveness of railway, meet the growing passenger transportation demand and solve the problems of traffic congestion and environmental pollution caused by highway development, Taiwan Province made the plan of building high-speed railway in 1990s.

Taiwan High-speed Railway (referred to as Taiwan HSR for short) (see Fig. 5.3), starting from Taipei and passing through Banqiao, Taoyuan, Hsinchu, Taichung, Chiayi and Tainan to Kaohsiung, has eight stations with a total length of 345 km. Taiwan HSR runs through the west coast of Taiwan Province. It is a high-speed railway system that connects Taipei City and Kaohsiung City. The whole line was officially opened to traffic on March 2nd, 2007. Taiwan HSR adopts Japanese Shinkansen technology, with a maximum operating speed of 315 km/h. Its opening has shortened the travel time between Taipei and Kaohsiung to one and a half hours.

During the period from 1998 to 2010, the average annual growth rate of GDP in Taiwan Province was 3.37%, and the average annual growth rate of passenger transport was 2.21%. The growth in economic development and passenger demand was relatively stable (As shown in Figs. 5.4 and 5.5).

---

[1] Source: Taiwan Province Statistical Yearbook (2011).

[2] The western corridor, the center of economic development in Taiwan Province, is located in the western half of Taiwan Province Island, bordering the Taiwan Province Strait, accounting for about half of the total area of Taiwan Province Island.

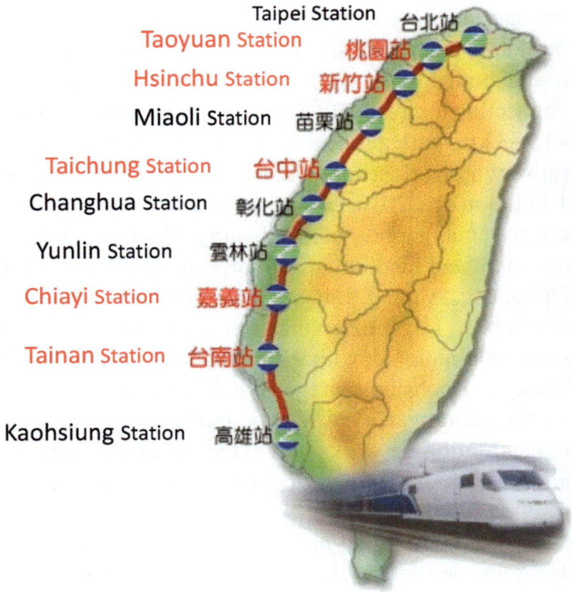

Fig. 5.3 High-speed rail lines in Taiwan Province

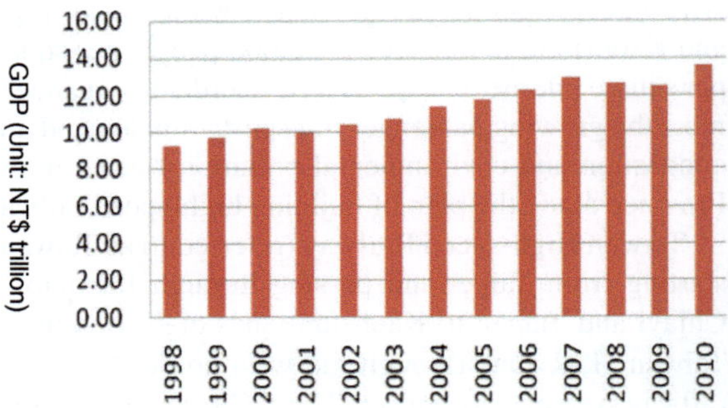

Fig. 5.4 Changes of GDP in Taiwan Province, China (1998–2010)

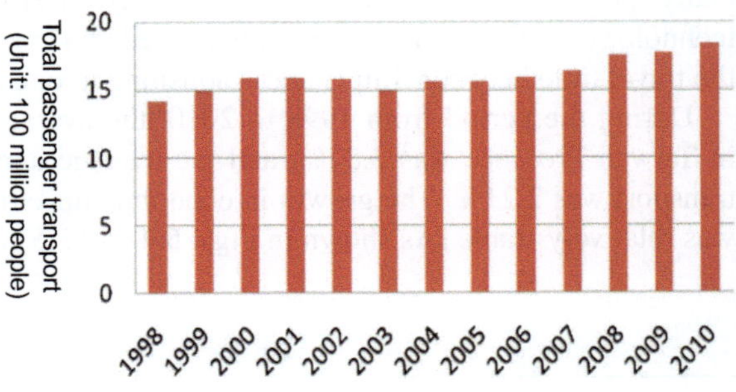

Fig. 5.5 Changes of total passenger traffic in Taiwan Province, China (1998–2010)

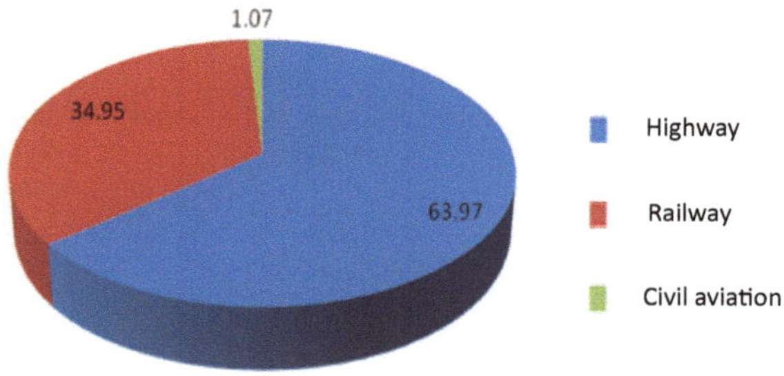

Fig. 5.6 Passenger traffic sharing rate of various modes of transport in Taiwan Province in 2006. (unit: %)

Taiwan HSR can be regarded as a new mode of transport, which enters a market structure with relatively stable economic development and basically saturated transportation market. In 2006 (the year before the opening of the Taiwan high-speed railway), the passenger transport market structure in Taiwan Province mainly consisted of roads and railways, in which the market share of passenger transport turnover of roads was as high as over 63% (see Fig. 5.6). This section will then look at the impact of the high-speed rail on its original passenger traffic pattern in Taiwan Province with regard to the total passenger traffic and structure.

(B) **The impact of Taiwan HSR**

(1) Impact of the opening of Taiwan HSR on total passenger traffic

In this section, the impact of Taiwan HSR on the total passenger traffic in Taiwan Province will be mainly focused on the total passenger traffic of railways, highways and civil aviation on the island. Because water transport accounts for a very small proportion in passenger transport, it is not included in the analysis. Due to the short opening time of Taiwan HSR, the impact of the high-speed rail on the total passenger traffic before and after its opening in 2007 will be analysed by means of comparative analysis of statistical data. The passenger volume of various modes of transport from 1998 to 2010 in Taiwan Province is shown in Table 5.4.

From the above table, we can see that the total passenger traffic between 1998 and 2006 in Taiwan Province increased from 1.421 billion passengers in 1998 to 1.585 billion passengers in 2006, with an average annual growth of 0.21 billion passengers, an average annual growth rate of 1.37%. From 2007 to 2010, the total passenger traffic in Taiwan Province increased from 1.637 billion passengers in 2007 to 1.835 billion passengers in 2010, with an average annual growth of 66 million passengers, an average annual growth rate of 3.88%. This shows that after the opening of Taiwan HSR, the total passenger traffic in Taiwan Province has increased to a certain extent. Figure 5.7 below shows the change of passenger volume of various modes of transport in Taiwan Province from 1998 to 2010.

From Fig. 5.7, we can see that after Taiwan HSR was put into operation in 2007, the total passenger volume and railway passenger traffic both increased rapidly. This is because the high-speed rail provides a brand-new mode of transport for passengers travelling to and from major cities in the western corridor in Taiwan Province, which makes people have more choices when traveling. Further, the high-speed railway features fast and convenient passenger transport services, comfortable

Table 5.4 Passenger traffic volume of various modes of transport in Taiwan Province, China (Unit: million people)

Year	Railway passenger volume	Highway passenger volume	Air passenger volume in Taiwan island	Total passenger volume
1998	233	1155	33	1421
1999	309	1149	33	1491
2000	460	1104	27	1591
2001	476	1091	25	1592
2002	500	1057	22	1579
2003	478	992	20	1490
2004	520	1019	21	1560
2005	531	1013	19	1563
2006	554	1014	17	1585
2007	603	1021	13	1637
2008	690	1054	10	1754
2009	719	1039	9	1767
2010	779	1046	10	1835

Source Adapted from *Taiwan Province Statistical Yearbook* (2011)

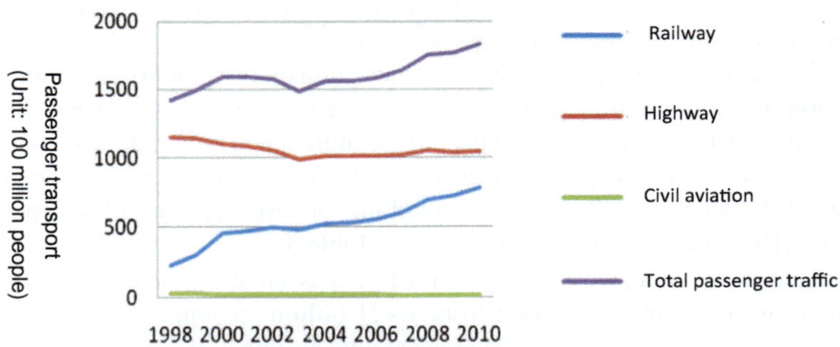

Fig. 5.7 Changes of passenger volume of different transport in Taiwan Province from 1998 to 2010. *Source* Adapted from *Taiwan Province Statistical Yearbook* (2011)

riding environment and moderate freight rate, which improves the quality of transport services and reduces the transport cost to a certain extent. These factors stimulate the growth of people's travel demand.

(B) Impact of the opening of Taiwan Province's high-speed rail on passenger transport structure

From 1998 to 2010, the market share of passenger traffic of railway, highway and aviation in Taiwan Province has changed greatly, with an increasing proportion of railway passenger traffic. However, the proportion of passenger traffic of highway and civil aviation on the island is decreasing continuously, as shown in Fig. 5.8 and Table 5.5.

Here, this section takes the data of three years before and after the opening of Taiwan Province's high-speed rail as the object, so as to analyze the changes of passenger traffic market share of various modes of transport before and after the opening of Taiwan HSR (as shown in Table 5.5). According to the data in Table 5.5, we can see that:

① With respect to railway passenger transport, in 2007, the average share of railway passenger traffic was 34.08% during the three years before the opening of Taiwan HSR (2004–2006) and 40.83% in the three years after the opening of Taiwan HSR (2008–2010).

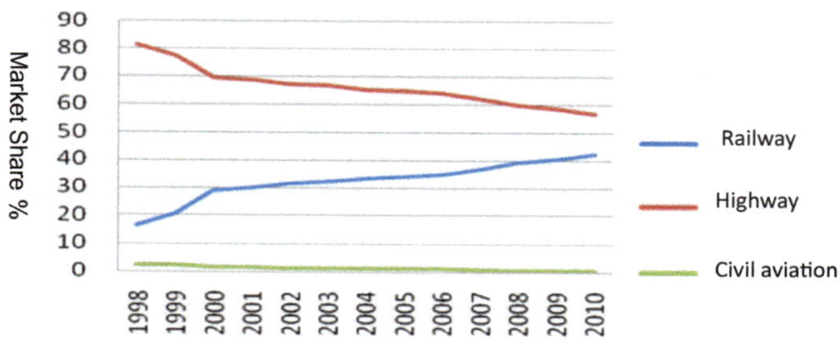

Fig. 5.8 Change in passenger traffic sharing rates for various modes of transport in Taiwan, China, 1998–2010. *Source* Adapted from *Taiwan Province Statistical Yearbook* (2011)

Table 5.5 Market share of passenger traffic of various transport in Taiwan Province (Unit: %)

Year	Railroad	Highway	Civil aviation
1998	16.40	81.28	2.32
1999	20.72	77.06	2.21
2000	28.91	69.39	1.70
2001	29.90	68.53	1.57
2002	31.67	66.94	1.39
2003	32.08	66.58	1.34
2004	33.33	65.32	1.35
2005	33.97	64.81	1.22
2006	34.95	63.97	1.07
2007	36.84	62.37	0.79
2008	39.34	60.09	0.57
2009	40.69	58.80	0.51
2010	42.45	57.00	0.54

Source Adapted from *Taiwan Statistical Yearbook* (2011)

Table 5.6 Changes in passenger structure before and after the opening of Taiwan HSR

Mode of transport	Three years before the opening of the Taiwan HSR (2004–2006)	Three years after the opening of the Taiwan HSR (2008–2010)	Average annual variation	Average annual rate (%)
	Annual average passenger Transport share (%) (A)	Annual average passenger Transport share (%) (B)	B–A	(B–A)/A
Railway	34.08	40.83	6.75	19.81
Highway	64.7	58.63	–6.07	–9.38
Civil aviation	1.21	0.54	–0.67	–55.37
Total	100	100		

② With respect to highway passenger transport, the average of the road passenger traffic share during the three years prior to the opening of the Taiwan HSR (2004–2006) was 64.7% in 2007, and 58.63% in three years after the opening of the Taiwan HSR (2008–2010).

③ With respect to civil aviation passenger transport, the average annual air passenger traffic share in the three years (2004–2006) before the opening of the Taiwan HSR in 2007 was 1.21%, and 0.54% in the three years after the Taiwan HSR was opened (2008–2010).

The changes of the above data can be concluded as follows:

① With the high-speed rail, the share of railway passenger traffic has greatly increased, with an average annual growth rate of 19.81%. However, the share of passenger traffic of civil aviation on highways and islands has dropped remarkably. The average annual decline rate of passenger traffic on highways is 9.38%, and that of civil aviation is as high as 55.37%. The impact of high-speed rail on civil aviation in Taiwan Province is the most striking. However, during the period from 2007 to 2010, the average annual growth rate of total passenger traffic in Taiwan Province was only 3.91%. Under the circumstance that the total amount of passenger transport market has not changed much, the market share of various modes of transport has changed greatly, which shows that the high-speed rail has greatly influenced the original passenger transport landscape in Taiwan Province. Through market competition, many passengers who originally chose to travel by road and by air are attracted to the high-speed rail.

② High-speed rail has different effect on highway and civil aviation passenger transport in Taiwan Province. We see its greatest impact on Taiwan civil aviation, the market share of which has decreased by 55.37% before and after the opening of high-speed rail. The reason is that each mode of transport has its own technical and economic advantages, and on this basis, it has its own advantageous distance.

From the experience of Japan, 500–750 km is the most competitive distance of high-speed rail, below which is the advantage distance of highway, and above which is the advantage of civil aviation.[3] Taiwan Province is a large island, and the transport distance on the island is just within the advantageous transportation distance of high-speed railway. Therefore, HSR has a huge influence on the civil aviation transportation in the western corridor area of Taiwan, where the HSR may attract a large number of passengers due to its advantages of all-weather, low price and short travel time.

(C) Summary

As mentioned earlier, the opening of Taiwan HSR is a new form of transport that enters a market landscape with relatively stable economic development and a saturated transportation scenario. By studying the case of high-speed rail in Taiwan, we can draw the following conclusions:

(1) The opening of high-speed rail has brought about the growth of total passenger traffic and railway passenger traffic. HSR provides a brand-new transport for passengers travelling to and from major cities in the western corridor in Taiwan Province, and more choices of traveling. Moreover, high-speed railway provides fast and convenient passenger service. Its comfortable ride environment and moderate freight charge have improved the quality of transportation services and reduced transportation costs to a certain extent. These may be the leading factors that have stimulated the growth of people's travel demand.

(2) After the opening of the high-speed railway, the passenger traffic share of the railway has greatly increased, with an average annual growth of 19.81%. Correspondingly, the share of passenger traffic of highway and civil aviation dropped proportionally (see Fig. 5.9), while the average annual growth of total passenger traffic in Taiwan Province from 2007 to 2010 was only 3.91%. Under the circumstance that the total amount of passenger transport market has not changed much, the market share of various modes of transport has changed greatly. It shows that the opening of the high-speed rail has caused great change to the original passenger traffic distribution in Taiwan Province. Due to market competition, many passengers diverted from roads and civil aviation are attracted to the high-speed rail.

In short, when the West Ring High-speed Railway in Taiwan Province was opened, the transportation market in Taiwan Province was in a period of basic saturation. After the entry of the high-speed rail, the total passenger traffic in Taiwan Province has increased to some extent. Meanwhile, the high-speed rail has a great impact on the existing passenger transport such as highway and civil aviation. The share of passenger transport by railway has been greatly increased, with an average annual increase of 19.81%, while the share of passenger transport by highway and local civil aviation has decreased by 9.38% and 55.37% respectively. With the entry of

---

[3] Source: National Gold Securities Research Institute. Analysis of the influence of high-speed railway on transportation distribution: special report of transportation service industry [R].2010:13.

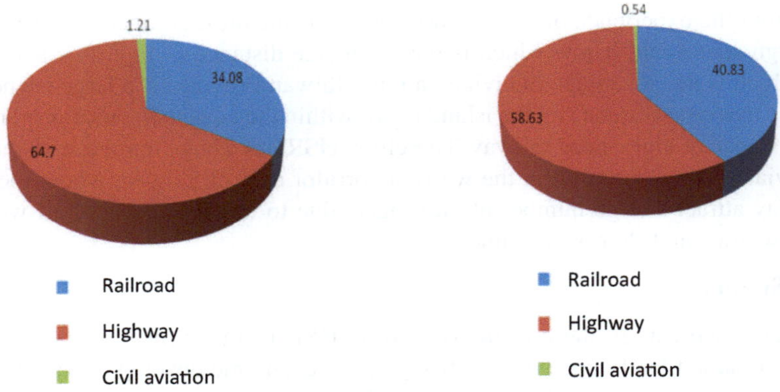

Fig. 5.9 Changes in passenger transport structure before (left) and after (right) the opening of Taiwan HSR (Unit: %). *Source* Adapted from Taiwan Province Statistical Yearbook (2011)

high-speed rail, the traffic map of passenger transportation on the island, which is dominated by railways and highways, has become increasingly changeable.

5.3.2 *Case Study of Hainan High-Speed Railway*

(A) Background

Located at the southernmost tip of China, Hainan Province has jurisdiction over 18 cities and counties and the administrative office of southwest Zhongsha Islands. Hainan has a land area of 35,400 square kilometers, and an authorized sea area of 2 million square kilometers. In 2010, the total population of Hainan was 8,960,900. Hainan Island is the main island of Hainan Province and the second largest island in China after the Taiwan Island.

On April 13th, 1988, the *Decision on the Establishment of Hainan Province* and *the Resolution on the Establishment of Hainan Special Economic Zone* was adopted at the First Session of the Seventh National People's Congress of China. On April 26th, 1988, the CPC Hainan Provincial Committee and the People's Government of Hainan Province were officially established. Since then, Hainan has become China's fifth special economic zone, thus entering a brand-new historical development period. Since the establishment of Hainan Province for more than 20 years, Hainan has made remarkable achievements in reform, opening up and economic and social development, and has become a vibrant special economic zone uplifted from a closed and undeveloped border island in the past. Since then, the economic structure with Hainan characteristics has been progressing, and the vitality of the institutional advantages of special economic zone and the economy have seen a rapid rising. In 2010, the province's GDP was 206.45 billion yuan, and the per capita GDP was 23,831 yuan.

In 2009, the State Council issued "Several Opinions of the State Council on Promoting the Construction and Development of Hainan International Tourism Island" (referred to as "Opinions"). The "Opinions" set the development goal of Hainan Island as: by 2020, the tourism service facilities, management and service level will be fully in line with the internationally accepted tourism service standards, and a world-class island leisure resort will be initially built. With the development goal, Hainan will surely have a promising prospect for economic development.

As the second largest island in China, the Qiongzhou Strait between Hainan Island and the mainland of China has an average width of 29.5 km from north to south, with a linear distance of 33.5 km at its widest point and only about 18 km at its narrowest point. The special geographical environment determines that Hainan must vigorously work on improving transportation capacity, which is the basic condition for accelerating Hainan's development. For a long time, due to historical and institutional reasons, the development of transportation in Hainan has been very slow, where port and water transportation used to be the main way to connect Hainan with the mainland of China. In 1943, the first railway in Hainan was built. But it was not until the opening of the Guangdong-Hainan Railway in 2004 that the history of no railway between Chinese mainland and Hainan Island ended. Sanya Phoenix International Airport, officially opened on July 1st, 1994, ended the history of Hainan without aviation. Up to now, with the development of Hainan's economy, after the large-scale transportation investment construction in the Ninth Five-Year Plan, the Tenth Five-Year Plan and the Eleventh Five-Year Plan, Hainan has realized a modern three-dimensional transportation system with "roads extending in all directions, trains crossing the sea, seagoing vessels wandering around the world, and planes crisscrossing clear skies".

① Highway: Highway has always been the main means of transport in Hainan, with trunk lines leading directly to ports, cities and counties, and branch lines extending to 318 towns and tourist attractions in the whole island. In 2010, the mileage of traffic was over 21,200 km, and the expressway was 660 km. Basically, a highway network running through the east, west, north and south, and covering the whole island has been formed, which takes the expressway as the aorta, the roundabout expressway and the "three vertical and four horizontal" national and provincial trunk lines as the main backbone, with the county and village roads connected to each other.

② Waterways: The domestic routes of Hainan passenger ships can reach the coastal areas and ports in the middle and lower reaches of the Yangtze River, and the international lines can reach countries and regions such as Russia, Japan, Korea, Southeast Asia, Africa and Europe. The province is planning to build a network of ports entitled "five ports in four directions" with Haikou Port in the north, Sanya Port in the south, Yangpu Port and Bazuo Port in the west and Longwan Port in the east, among which Qiongshan, Sanya, Bazuo and Yangpu are the largest. By 2010, the cargo and the passenger throughput of Hainan's major ports reached 96.1731 million tons and 11.22 million people respectively.

③ Railways: The main railways in Hainan include Hainan East Ring Railway, Hainan West Ring Railway and Guangdong-Hainan Railway. On January 7th, 2003, the Guangdong-Hainan railway passage was officially opened. The passenger train from Qiongshan to Guangzhou has been officially put into operation and connected to the national railway network. In the traffic net of Hainan Island, railway transportation develops rapidly. On April 18th, 2007, the reconstructed West Hainan Railway was officially opened to traffic, which realized the connection between the off-island railway and the on-island railway. After the transformation, the trains on the West Ring Road have been really speeded up, and the running speed has been increased from 80 km/h to 120–160 km/h. On December 30th, 2010, Hainan East Ring High-speed Railway was officially put into operation (see Fig. 5.10). At the same time, the West Ring Express Railway officially started, with a construction period of 4 years, a target speed of 200 km/h and a reserved speed of 250 km/h.[4]

④ Aviation: Hainan's aviation industry is developing very fast. Qiongshan Meilan International Airport in the north and Sanya Phoenix International Airport in the south have flights from dozens of airlines in the mainland and Hong Kong. In 2010, there were 543 civil aviation routes with a mileage of 1,069,914 km. Among them, there are 54 international routes with a mileage of 28,896 km, 489 domestic routes with a mileage of 787,618 km, and 15 regional routes with a mileage of 23,412 km. Hainan Airlines introduced two small passenger planes into the island for transportation. With departures from Qiongshan and Sanya, a convenient air bridge between Hainan and major cities at home and abroad, east, west, north and south is set up. And HNA (Hainan Airline) has become one of China's four major airlines and a leader in China's aviation industry.

Today, on the 34,000 km^2 of Hainan Island, a three-dimensional transportation network consisting of airports, railways, ports, highways, and pipelines has begun to take shape, linking Hainan Island with the mainland and the rest of the world.

As a matter of fact, the entry of Hainan's high-speed railway coincided with a period of rapid economic development, rapid growth in transport demand and an unsaturated transport market in Hainan, where the passenger market structure in 2010 was dominated by roads and railways, with a market share of over 93% of passenger traffic on roads (see Fig. 5.11).

(B) The impact of high-speed rail in Hainan on the original transport layout on the island

In the twenty-first century, relying on its unique location advantages, Hainan has made great efforts to develop regional economy and achieved remarkable results. But its economic base is relatively weak and its economic aggregate is small. As the overall level of economic and social development is still low, Hainan is still

[4] On July 7th, 2011, the feasibility study report of Hainan West Ring Railway was approved by the National Development and Reform Commission of the People's Republic of China. In April 2012, Hainan West Ring Railway Project started and the construction began in September, 2013. On December 30th, 2015, the Hainan West Ring Railway was opened to traffic.

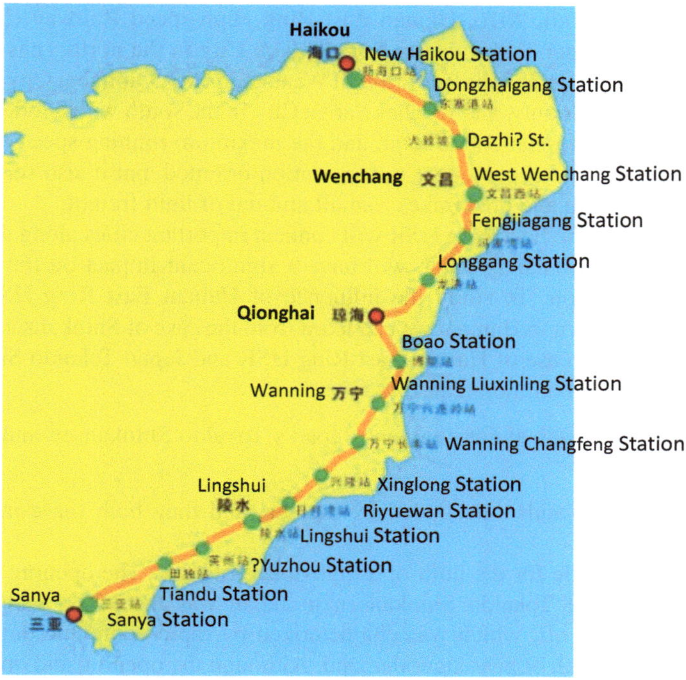

Fig. 5.10 Hainan east ring high speed rail line

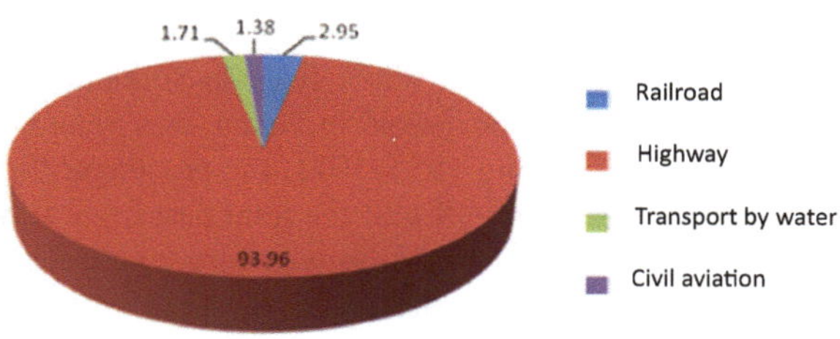

Fig. 5.11 Schematic diagram of passenger traffic share of various transportation modes in Hainan in 2010 (Unit:%). *Source Hainan Statistical Yearbook* 2011

an underdeveloped province, where many of its economic and social development indicators are below the national average. With the advancement of international tourism island planning, Hainan is expected to develop better and faster in the future.

The rapidly developing economy needs supporting transportation infrastructure. At present, Hainan Island has initially built a passenger transportation system covering civil aviation, shipping, highways and railways, but major transportation facilities such as aviation, railways and sea transportation are still lagging behind. As a result, the accessibility of Hainan Island is relatively poor, and its ability to open up overseas markets, expand economic hinterland, introduce consumer demand and expand the proportion of outflow is insufficient.

On December 30th, 2010, Hainan East Ring High-speed Railway (HSR) was officially put into operation. It starts from Haikou City in the north, runs along the eastern coastline of Hainan, passes through Wenchang City, Qionghai City, Wanning City and Lingshui County, and reaches Sanya City in the south with a total length of 308 km. There are 15 stations in total, and the maximum running speed of the line is 250 km per hour. It is passenger transportation-oriented, but it also serves as the rail transit in Haikou and undertakes a small amount of light freight.

The opening of the East Ring HSR will connect important cities along the eastern coastline of Hainan Island, which will have a significant impact on the layout of Hainan transportation. To study the influence of Hainan East Ring HSR on the existing traffic pattern in Hainan, we can learn from the case of Shinkansen in Japan. This is because the case of Hainan East Ring HSR and Japan Tokaido Shinkansen share the following common features:

(1) Similar geographical environment. Japan's Tokaido Shinkansen and Hainan's East Ring

HSR are all islands surrounded by the sea, and they both serve intra-island transportation.

(2) Similar rationale for the built of a high-speed railway. The opening and operation period of Tokaido Shinkansen in Japan was the 1960s when Japan's economy took off, which was characterized by rapid economic development and rapid growth of passenger demand. Although the opening and operation of Hainan East Ring High-speed Railway is more than 40 years later than that of Shinkansen, the economic development stage of Hainan at this time is similar to that of Japan in 1960s, a time when the demand for economy and passenger transportation has been growing rapidly, as shown in Figs. 5.12 and 5.13.

(3) Similar competitors. When Tokaido Shinkansen was opened in 1964, domestic passenger transportation in Japan was mainly undertaken by railways and highways. During 1955–1971, the passenger turnover of railway and highway accounted for 97–98% of the total passenger turnover, while the share of aviation and inland navigation was very small. The competition in the passenger transport market on the island is mainly between roads and railways.

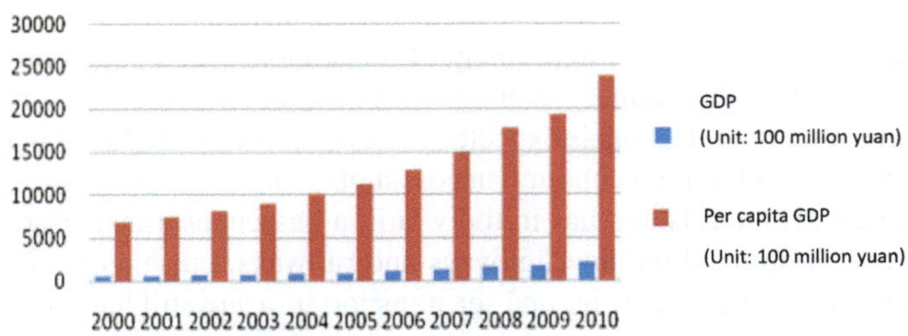

Fig. 5.12 Trend of economic development in Hainan Province (2000–2010). *Source Hainan Statistical Yearbook* 2011

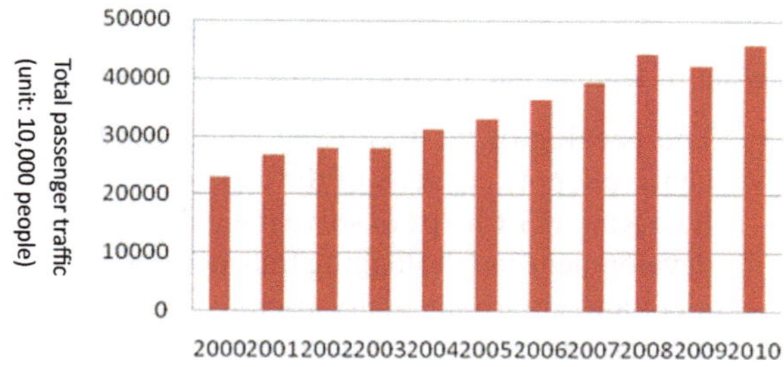

Fig. 5.13 Trend of total passenger transport in Hainan Province (2000–2010). *Source Hainan Statistical Yearbook* 2011

Based on the above analysis, we observe that the impact of the opening of Hainan East Ring HSR on the original local passenger traffic distribution is similar to that of Tokaido Shinkansen. That is, due to the rapid economic development, the passenger transportation market is not yet saturated, which offers a great opportunity for high-speed rail to enhance the supply capacity of the whole transportation system. Also, its relationship with other means of transport is mainly collaborative.

In addition, the case of Hainan East Ring HSR also has its own characteristics:

First, Hainan Island is different from the Japanese archipelago in island shape. Japan's archipelago is long and narrow, while Hainan Island is pear-shaped. So, the landscape of Hainan Island is quite different from that of Japanese archipelago, but similar to that of Taiwan Island. This means that the average transportation distance between cities in Hainan Island is shorter. The long axis of Hainan Island is about 300 km from northeast to southwest, and the short axis is about 180 km from north-west to southeast. Such distance is not suitable for civil aviation transportation, but is exactly within the advantageous market scope of highways and railways. Therefore, the passenger traffic on the island will be dominated by roads and railways.

Second, the passenger transport market is different. Tourism is the leading industry in Hainan Province. In 2011, the number of overnight visitors in Hainan Province reached 30.0134 million, an increase of 16.0% over the previous year. Among them, it received 29,198,800 domestic tourists, an increase of 15.8%, and 814,600 inbound tourists, an increase of 22.8%. The total tourism revenue of Hainan Province in 2011 was 32.404 billion yuan, an increase of 25.8% over the previous year. Among them, domestic tourism revenue was 29.947 billion yuan, up by 27.1%, and inbound tourism revenue was 2.457 billion yuan, up by 11.6%.[5] With the construction of Hainan International Tourism Island, there will be more domestic and foreign tourists. This market structure determines Hainan's transportation industry, which mainly relies on civil aviation and water transportation to the outside and developed road and railway networks to the inside. It is expected that the high-speed rail will play an increasingly important role in the future passenger transport market because of its advantages of speed, convenience, huge traffic volume, energy saving and environmental protection.

[5] Source: Statistical Bulletin of Economic and Social Development of Hainan Province in 2011.

To sum up, the impact of the opening of the East Ring HSR on the original traffic layout of Hainan Island can be summarized as follows:

(1) The opening of the East Ring High-speed Railway will improve the comprehensive supply capacity of the existing transportation system on the island. The East Ring HSR helps to solve the unbalance between Hainan's rapid economic development and the lagging transportation system, provides an efficient, convenient and environmentally-friendly new means of transport and more choices for passengers on the island. Since the opening of the East Ring HSR, the passenger turnover in Hainan Province reached 47.289 billion person-kilometers in 2011, an increase of 12.9% over 2010.[6] The East Ring High-speed Railway has accelerated the growth of total passenger transport in Hainan Province.

The East Ring HSR will cooperate and better connect with the original transportation system, which will promote the all-round development and the improvement of the efficient and convenient comprehensive transportation network in Hainan. After the opening of the East Ring HSR, the passenger turnover of Hainan Railway, the water transport, and the civil aviation increased by 4.3 times, 12.8, and 13.9% respectively in 2011, while the highway decreased by 2.1%.[7] It can be seen that the entry of the East Ring HSR has little effect on the original passenger transportation. Except for a slight decrease in the passenger turnover of highway, the passenger turnover of water transport and civil aviation has greatly increased. The opening of the East Ring HSR will transform the market structure that the internal traffic of Hainan Island depends on roads and railways, while the external traffic depends on civil aviation and water transport.

With the opening and operation of the East Ring HSR and the improvement of Hainan's transportation system, the problem that Hainan's transportation lags behind the economic development will be gradually solved. The East Ring HSR and the high-speed railways to be planned in future will play an important role in the formation of Hainan's comprehensive transportation system.

References

Ben-Akiva, M., and Lerman, S. (1985). *Discrete Choice Analysis: Theory and Application to Travel Demand*. MIT Press.

Manheim, Jarol B. (1987) Rites of Passage: The 1988 Seoul Olympics as Public Diplomacy. *The Western Political Quarterly, 43*(2): 279–295.

McFadden, Daniel. 1974. The measurement of urban travel demand. Journal of Public Economics 3(4): 303–328.

Shao Yaming, Zhang Xiuyuan, Research on Spatial Distribution Method of Railway Express Passenger Transport in China, Proceedings of Academic Research on Reform and Development of Railway Transport Enterprises, 2003.

[6] See Footnote 5.

[7] See Footnote 5.

Part III
Performance of the Regional Basic Functions of High-Speed Railways

Chapter 6
High-Speed Railway and Regional Economic Aggregate

As a transportation infrastructure investment, high-speed railway is not only an investment that can affect economic growth in a short period of time through demand pull and capital accumulation, but also a quasi-public goods of "externality" that can indirectly have a long-term impact on economic growth. In terms of the macroeconomic investment multiplier theory, as an element of the national economy, high-speed railway investment not only brings a direct increase in total output, but also may have an impact on capital accumulation through the multiplier effect, increasing investment several times the aggregate demand, thereby augmenting the national income and economic growth. In addition, according to the modern endogenous economic growth model, high-speed rail is a public product of infrastructure provided by the central government, which has an overflow effect on both local and private capital and will eventually promote economic growth.

As a typical contributor to economic growth, high-speed rail has a profound impact on the industrial distribution in various places by coordinating the flow of investment, labor force and other factors of production. It eventually changes the economic structure of cities alongside it, and the patterns of economic growth in various regions. The construction and operation of high-speed railways will greatly shorten the time and space distance, effectively promoting the exchange of people and goods between regions, strengthening economic ties between cities and regions along the line. In addition to these tacit benefits, high-speed railways can also help foster market economy and improve the efficient allocation of resources and professional collaborations, which will translate into the development of economies of scale. Consequently, the overall efficiency of the local economy will increase, with improved cross-region traffic conditions and a better business environment that appeals to foreign investors.

High-speed rail construction allows higher accessibility of cities. The flow of people and funds among cities will be frequent and the opportunities for economic cooperation will be greatly enhanced. It will also bring tremendous economic vitality to the less developed areas along the line, promoting their economic development

X. Lin, *High-Speed Railways and New Structure of Socio-economic Development in China*, Research Series on the Chinese Dream and China's Development Path,
https://doi.org/10.1007/978-981-19-6387-2_6

and integration, and give a vital and powerful impetus to the economic growth along high-speed railway lines.

The World Bank (1994) classified infrastructure into economic and social infrastructure. High-speed railway, post communications and telecommunications, energy supply and other economic infrastructure, which are directly involved in the production process and play a crucial role in helping improve social productivity and speeding up economic growth, are viewed as material capitals. Therefore, as an important infrastructure, the construction and operation of high-speed railway have a fundamental influence on the regional economy along the line.

6.1 A Theory of the Regional Economic Impact of High-Speed Rails

In the process of economic globalization, the economies of various countries are integrating in the world, especially in Europe, North America and the ASEAN region. The formation of an open, easily accessible mesh economy in these regions places fierce competition in most parts of the economy. While the result of mutual fierce competition is bound to benefit some areas, it damages the rest of participants globally. This is also true to a nation. Thus, to a large extent, high-speed railway is an important guarantee for enhancing regional economic competitiveness and coping with globalization, and the balanced development of the region.

In recent years, scholars have been paying more and more attention to assessing a variety of regional impacts brought by major high-speed railway projects, including the impacts on regional productivity growth, industrial development planning, employment and regional development. On the regional level, these impacts can be either positive or negative, and often involve crucial policies such as those associated with sustainable development. Although it is not difficult for decision makers to collect and analyze relevant empirical evidence, quantifying how high-speed railway projects influence economy and welfare remains an important part of high-speed railway planning and decision-making. In fact, regional development is not only the product of a rational allocation of production factors such as labor and capital, but also requires a well-organized transport network. Building high-speed railways will help increase the efficiency of the production factors in the region.

Opinions diverge as to whether the supply of high-speed railways will generate benefits or bring detrimental impacts to economic development. Briefly summarized, the investment effect of high-speed railway on economic development is shown in Table 6.1.

Temporary effect is caused directly by construction and indirectly by other related economic activities. In addition, the crowding-out effects (especially large-scale investments) of high-speed railways are often overlooked. Because high-speed railway construction must have strong financial support, governments have to issue

Table 6.1 The investment effect of high-speed railway

Effect type	The demander	The supplier
Temporary effect	Construction effect: crowding effect	–
Long-term effect (constructive)	Maintenance and operating costs	Impact on productivity and new activity areas

government bonds to solve this problem. This may lead to a higher debt burden, thereby reducing the construction investment in other regions.

Long-term effect is the outcome of operation and maintenance once construction is finished. Here we mainly discuss one special long-term effect: the project effect. It refers to the long-term, indirect change in the private sector's revenue, employment or investment during the construction of high-speed railways. This effect is induced by the improvement or increase of high-speed railways, and the opportunities they bring.

Due to the diversity and complexity of regional economic impact, theories and evaluation methods from different perspectives are various. Few theories or methods can cover and predict all the impacts in all aspects. However, different methods can be integrated into a concept map, which can be used as a starting point to understand its diversity. In this chapter, through summarizing previous research and doing original work based on this research, we designed a set of standardized and applicable evaluation methods for practical operation to identify and quantify the regional economic impact of high-speed railways.

Impacts of transportation investments have been there since the very beginning of projects, with the introduction of labor, capital, materials and intermediary services, all of which will normally stimulate regional and local economies. However, while these multiplier effects above have important policy implications (for example, as part of many stimulus packages), they are generally short-term in nature, contributing very little in project cycles and to project benefits.

In general, a large number of economic impacts are accumulated in project operations. The most common method used in transportation project reviews is based on transportation market and involves the estimation of direct transportation costs, travelers' or the user's benefits gained, safety improvement and environmental benefits throughout the evaluation period.

However, the evaluation methods used within the transport sector are based on two assumptions: the market is in perfect competition condition, and the output of goods and services is excluded from the influence of external factors. In practice, these two assumptions are not valid. Instead, transportation may have a varied impact on an enterprise, as it depends on (1) where companies invest and grow their business, (2) how they purchase their production materials or resources from different places, (3) how they grow and expand the market reach of their products and services, and (4) how they learn from their rivals and establish their own market position.

Transportation services also affect every family and individuals in different aspects such as: where they choose to live, how they attain their life necessities and services, how they gain and expand thier career opportunities, how they learn from other competitors, and how they build their own social circle. The effects of monopolistic competition and various external factors, especially the effects of agglomeration, are ubiquitous in almost all of these processes. These effects are accompanied by the diversification and expansion of markets, and the increase in the suppliers with special market positions.

The UK DfT (2006) method proposes a comprehensive theoretical framework for these effects. The work started with a series of case studies. These studies concluded and confirmed that the traditional high-speed railway cost–benefit analysis ignores the impact of many transportation, and identified the following four potential additional effects: (1) increased competition resulting from improved traffic conditions, (2) the increased output from unbalanced competitive markets, (3) the financial benefits from increasing labor supply, and (4) agglomeration economy. The improvement of high-speed railway conditions reduces transportation costs and transit time, thereby influencing production activities and household consumption, leading to the redistribution of sub-economic circles and regions. Here, we will look at the interrelationship between transportation and interregional trade. The standard model for interregional trade is illustrated in Fig. 6.1.

As shown in Fig. 6.1, products will be exported from Region 1 to Region 2 when the transportation cost is less than the equilibrium price (p2 − p1 shown in Fig. 6.1). This is different from the pattern with the absence of trade: in Region 1, excessive production (area A) appears due to the increase in suppliers; in Region 2, there is a production deficiency (area B) caused by the growth in demand. Therefore, based on this model, both regions benefit from trade. Improvements to high-speed railway projects have led to lower transportation costs and consequently higher transport capacity. Trade pushes the equilibrium price of Region 1 up but at the same time pulls the equilibrium of Region 2 down. As a result, in Region 2, consumers will benefit from the improvement of high-speed railway projects, while it damages the interests of suppliers. In Region 2, the situation is diametrically different. As for regional employment, Region 1 benefits from the improvement of HSR projects while Region 2 suffers losses.

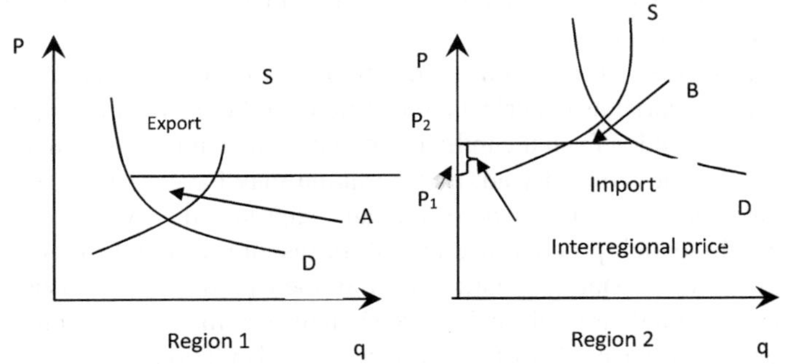

Fig. 6.1 Supply and demand in two regions

The model discussed above is a partial equilibrium model, which is used to analyze the situation where there is only one product in the market. However, when it comes to the effects of high-speed railway changes, the general equilibrium model is more complicated but more proximate to reality.

Figure 6.2 reveals the main effects of the improvements in transport infrastructure on multiple sectors (Lin and Chen 2006). However, it is difficult to tell which effects can be completely attributed to such improvements. The intermediate transitional effect is also extremely complicated. In addition, in some regions, employment will be adversely affected by the increase in the degree of competition, which will inevitably require compensation. As a result, the price of the related products will decrease and consumers can allocate more budgets to other products, some of which may be produced in these regions.

In the long run, improvements in high-speed railways will certainly lead to some changes in social production, which are mainly reflected in productivity. They will also cause the restructuring of capital and population, that is, the restructuring effect.

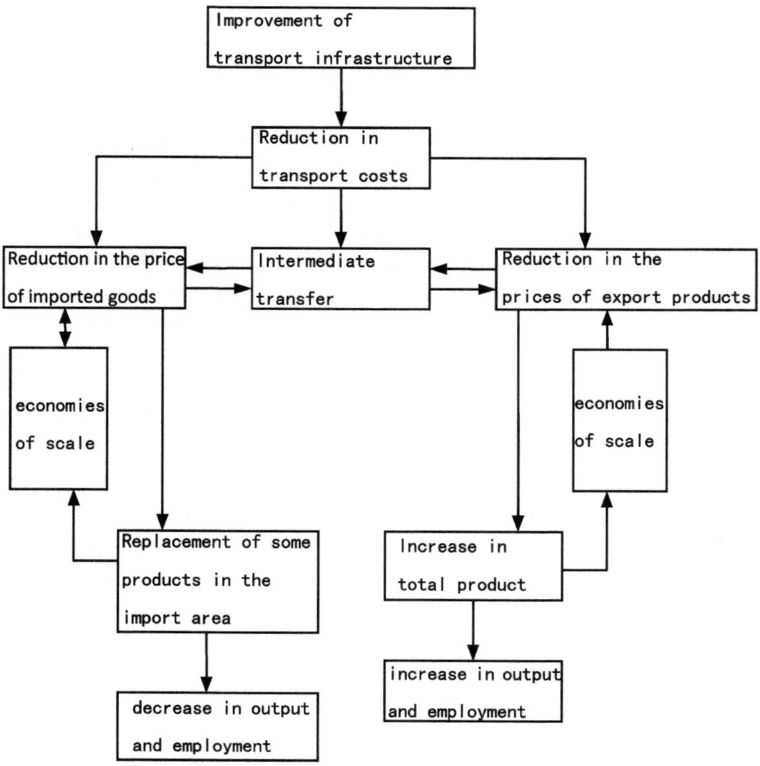

Fig. 6.2 Effect of improved high-speed railway conditions

6.1.1 Impact on Regional Productivity

In the process of social reproduction, in addition to production factors such as labor and private capital, high-speed railways also play an important role as an investment. The enhancement of transport services in a certain area means that the economy in this area can make full use of these production factors with higher production efficiency. Better high-speed rail conditions ensure the same productivity with a lower input of labor and capital. Especially for producers, economies of scale are likely to form when improved transport conditions reduces the workload of the whole industry, or when more firms collaborate to occupy more markets. As the scale of economy grows, so does its return, leading to a lower average cost of production and growth in social welfare.

We can analyze the contribution of improved high-speed railways on a manufacturer scale and a regional scale. On a manufacturer scale, it is measured by the reduction of transport cost, whereas on a regional scale, regional production function is introduced and the involvement of other production factors is taken into account.

6.1.1.1 Analysis of Firm

Microanalysis helps to track individual behavioral decisions and responses caused by new high-speed railway projects. There is much controversy about the influence of high-speed railway projects on regional economic development.

NEA, in the Netherlands, studied the economic costs in the province of Noord Brabant due to inadequate high-speed rail facilities in 1990. They studied the very congested S20, a region where manufacturers were overburdened with transport costs. As it turned out, the shortcomings of high-speed railway project affected the productivity of transport enterprises, but had a relatively small impact on other manufacturers. In this case, the impact of inadequate high-speed rail projects on regional productivity is not significant. But this may have a bigger impact on the profitability of individual vendors (Lin 2006).

NEA has also studied the situation in the province of Zuid-Holland. It is ex post factor research on improved high-speed railways. As the capital of the province of Zuid-Holland, Hague is adjacent to the capital of the Netherlands, so in the 1980s, an industrial park near Hague naturally had a better connection with the national highway system. Manufacturers in the region may save between 2 and 10% of the time. Clearly, manufacturers outside the park could save less time on average than those in the park. However, the research found that the manufacturers in the park rarely made full use of the opportunities that would make them benefit in practice. For example, manufacturers did not reschedule travel plans, which made the time saved by improvements to high-speed rail projects wasted by the increased negligence of the manufacturers. In this case, the reduction in transport costs resulting from improvements in high-speed rail projects and the related productivity gains were not significantly linked.

The limitation of the above case studies is that it takes the increase of manufacturer productivity directly caused by the improvement of transport facilities as its main concern, but overlooks the indirect effect on other manufacturers. In addition, the role of a particular type of high-speed rail is the only concern in the above studies. Therefore, it is more meaningful to combine the case study method with the method of establishing model, put all elements concerned into consideration and make them explicit by setting up a total production function.

6.1.1.2 Impact on Production Functions

When there are different types of high-speed railways in the region ***r*** and sector ***i***, the general form of the production function is as follows:

$$Q_{ir} = f_{ir}(L_{ir}, K_{ir}; A_{ir}, \ldots, N_{ir}) \tag{6.1}$$

Among them, Q_{ir} is the function value of region r and sector i; L_{ir} is the employment of regional r, sector i; K_{ir} is the private capital of regional r, sector i; $A_{ir}, \ldots, N_{ir}$ are regional r infrastructure projects of different types. These infrastructure projects mainly refer to transport, communications, energy, water resources, education, health services, etc.

In the case of high-speed rail projects, it is difficult to include the entire transport network in the production function, but it can be included by dividing the spatial scope of high-speed rail projects (such as within regions, between regions and even between countries) accordingly. In addition, the impact of high-speed rail projects may exceed the boundaries of a given area. For example, in some areas they may not own their own airports, but they may still benefit from airports in nearby areas. The concept "accessibility" of high-speed railways can be used in the production function.

Worldwide research on the contribution of railways to productivity has been uninterrupted. For example, in a collection of empirical research papers on economic growth done by the United States Department of Transportation (US DoT), most researchers did a comprehensive observation in several time ranges and several cross-sectional areas. Overall, these papers point to a small positive relationship between public investment and productivity. Output-investment coefficients ranged from 0.03 to 0.39 (the highest value was from Aschauer (1989)). Compared with the results of time series analysis, the correlation is greatly weakened. The disagreements in the studies indicate that research in this field is very difficult. The most well-known research is that of Aschauer's, who compared economic growth rates and transport infrastructure investments across the United States. He came up with the conclusion that transport investment is the main cause of economic growth. His conclusion was once widely cited in the United States and other countries to support road construction projects. Other analysts, however, argue that transportation investment is the result of economic growth, not the cause (because it is natural that wealthy states tend to have more money and spend more on transportation, and such is true in many other areas).

6.1.2 Impact on the Reconfiguration of Regional Production Activities

As noted above, high-speed rail projects can increase regional productivity. The increase in productivity may lead to the expansion or redistribution of factors of production in the region, such as the scope of capital and labor, or rather, the redistribution of economic activities. In studies on the effect of relocation, the level of employment and capital are usually a concern of the location theory. In this theory, the relative price of labor, private capital and high-speed rail play a role with a range of other location factors. These factors mainly include urbanization economy, sector structure, labor quality, market accessibility and special regional policies.

To analyze the influence of high-speed railways on employment and private capital distribution, the following four approaches can be adopted: (1) the influence of a high-speed railway on accessibility can be modeled by the influence of a high-speed railway project; (2) the influence of a high-speed railway project on the marginal cost of transportation is modeled, which can be calculated by a linear programming transport model; (3) A regional economic model helps to get the investment for a high-speed railway project connected with the private investment directly; (4) The role of high-speed railway can be seen by doing surveys on entrepreneurs (the importance of high-speed railway project is more important than other location factors).

6.1.2.1 The Term Accessibility and Its Impact on the Regional Layout

The study of accessibility originates from Johan Heinrich von Thunnen's agricultural location theory[1] and Alfred Weber's theory of the location of industries.[2] It is a very important and basic concept in the study of transportation and cities. With

[1] Johan Heinrich von Thunnen (1783–1850): His agricultural location theory is a product of the special social and economic background of Germany (Prussia) in the nineteenth century. At the beginning of the nineteenth century, Prussia carried out the reform of the agricultural system, banning all subordination relations attached to the owners of the land. All citizens can own movable property and can freely divide and trade it. Farmers have legally become free farmers and can independently control their own farms. The reform of the agricultural system abolished many privileges of the aristocracy, but also promoted the aristocracy to become a large landowner and an independent agricultural entrepreneur. At the same time, a large number of peasants, who have obtained personal freedom, have become agricultural laborers, who can freely sell their labor force. As a result, an agricultural enterprise-style operation composed of agricultural entrepreneurs and agricultural laborers emerged.

[2] Alfred Weber (1868–1958), German economist, sociologist and cultural theorist. He was Max Weber's brother. His work ***Theory of the Location of Industries***, published in 1909, founded the theory of industrial location and deeply influenced the development of modern economic geography.

the increasing attention of scholars in the field of urban planning and transportation geography, the connotation of accessibility is deepening, and the scope of its research is expanding, which covers not only the evaluation of the accessibility of infrastructure such as railways, highways and aviation, but also the related research of urban planning, transportation geography and marketing.

Accessibility refers to the ease of getting from one place to another. It includes the starting point, the ending point and the connection between them. It is determined by the mobility of people, the nature of land use and the transportation system. Increasing the accessibility of a region or location means that the accessibility and connectivity of the region is improved, resulting in shorter transport distance, shorter traveling time, lower freight costs and greater convenience, and therefore can be measured by spatial distance, travel time, and transportation costs. For the convenience of this book, the term accessibility is defined as the size of opportunity, the easiness and cost to reach the destination from a given region or node via transportation facilities.

The improvement of high-speed railway projects reduces transit time, reduces transport costs, improves accessibility of markets and inputs, and thereby affects the redistribution of labor and capital. A region accessibility variable Z is defined as:

$$ACC_r(Z) = \sum_i Z_r \cdot f(c_{r\prime r}) \tag{6.2}$$

In this formula, ACC_r (Z) *denotes accessibility,* $(c_{r\prime r})$ *represents the travel cost index between region r' and r, and* $f(c_{r\prime r})$ is a function that decreases as their distance increases. Variable Z refers to employment, production, inputs, and so on.

Botham (1983) represents the concept of location employment and accessibility in the following equation:

$$\Delta E_r = \alpha_1 ED_r + \alpha_2 W_r + \alpha_3 LAPE_r + \alpha_4 ACC_r(Z)$$

where: ED_r indicates employment density; W_r indicates the rate of wages; $LAPE_r$ indicates the labor stock index; Z represents the variables mentioned above; ΔE indicates the difference of labor transfer between regions; $\alpha_1,\ \alpha_2,\ \alpha_3,\ \alpha_4$ were the coefficients respectively.

Using the above formula, Botham analyzed and calculated the situation of 28 regions in the UK from 1961 to 1966. This period was right before the construction of the British National Highway System. This formula was used to simulate the impact of highway construction on the regional employment distribution. The construction of expressways reduced the cost of transportation and makes some areas easy to enter. The influence of expressway construction on labor transfer is calculated by this formula. The conclusion shows that the construction of expressway system has little influence on the regional labor transfer. Dodgson (1974) used the same analytical method to analyze the impact of the M62 highway in the UK. Bruinsma et al. (2008) have done the same in the Netherlands and come to the same conclusion.

According to Li and his colleagues (2005), the main features of accessibility are as follows: (1) Accessibility is a concept of space, which reflects the degree of difficulty in communication between space entities in overcoming distance barriers. Therefore, it is closely related to the concepts of location, spatial interaction and spatial scale. In the sense of space, accessibility expresses the closeness between spatial entities. Second, it has time significance. Space entities interact with each other or approach the center of economic activity mainly through the traffic system. Time is the most basic impedance factor in travel. The cost of traffic depends to a large extent on the cost of transit time, so time units are usually used to measure spatial distance. Third, accessibility has social and economic value. Higher levels of accessibility are associated with a higher quality of life and satisfaction, attractiveness, and economic development. Lastly, the starting point, ending point and transportation system are the three essential elements of accessibility. Accessibility measures the relationship between a specific starting point and an ending point. A form of a transportation system is used as the connection tool between the start and end points.

Most of the studies on accessibility have found that accessibility is positively correlated with total employment. In Fig. 6.2, it means that the difference between sectors benefiting from lower transport costs in these areas and sectors that have suffered as a result is absolutely positive. Thus, specifically in some sectors, increased accessibility may have a negative impact on employment.

6.1.2.2 Impact on Regional Employment Through the Marginal Cost of Transportation

The above discussion about the concept accessibility is very similar to the universal gravitational model in that the force matrix is formed in interactions in space, and variables are entwined via gravitational force. The linear programming model, a substitute for the transport model, can describe the mutual forces of the spatial matrices and allows many zero cases (Nijkamp and Reggiani, 1992). When the aggregate supply and demand are fixed, we can apply this model to find the minimum total transportation costs in these regions. The aggregate supply and demand in each region are viewed as dual variables to represent the marginal transport cost of the input and output materials and products.

Some scholars from other countries have found that high-speed railway investments bring changes in transportation costs in the studies of industrial distribution model. At the same time, the dual variables, the aggregate demand and supply, determine the locations of industrial complexes. According to the model, other factors that influence the distribution of industry are: labor cost, land value, previous investment, and variables of agglomeration. This model, an outcome covering approximately 3,000 counties and up to 100 different sectors, was originally designed to study the situation in the United States. According to this model, in the first two years, investment in roads and railroads in rural counties in the United States directly increased employment in these areas. However, there were also structural side effects of such investments. After the fourth year, railway investment had a slight negative influence on employment in these areas. This well explains the lower part of Fig. 6.2: The improvement of railway conditions may lead to an impediment to regional development.

Amano and Nakagawa (1990) studied the influence of the Tokaido Shinkansen on numerous cities along the route. They compared the population growth of the two station cities and the other four neighboring cities without stations from 1960 to 1985. It should be noted that the sample in their study was small and the sizes of cities concerned were also different. One of the conclusions drawn from the survey is that the average annual population growth rate of station cities (1.7%) is higher than that of the surrounding cities (1.3%). This may be due to the smaller sample. The ratio of new residents (1:5.3) between the two types of cities proved this assumption. The data on the two types of city density and information on the overall economic trend may be more instructive, but they are not included in their research.

Hirota's (1984) study of the Shinkansen high-speed railway in Japan showed that when the Sanyo line was completed in 1975, the number of passengers from other counties in all station cities along the line increased significantly. For example, there were 7.5 million passengers in the station city Fukuoka along Hakata Line, increasing by 93.5%; in Hiroshima City, the number of passengers reached 7.2 million passengers, with a 52.3% growth. Nakamura and Ueda (1989) studied how the Joetsu and Tohoku Shinkansen Lines affected the regional population from 1975 to 1985. They compared the population growth rate of 10 counties with the national figure. Six of these counties had one or more Shinkansen stations, while the rest of four counties did not have Shinkansen stations. They classified these counties according to whether high-speed railways traversed through them or not. The benefits of the high-speed railway include, to a large extent, the following: half of the HSR counties had a higher growth rate than the national average, but the population growth rates of all the non-HSR counties were lower than the average. Brotchie et al. (1991) continued Hirota's research and found that the number of hotels and guest rooms in Hakata (the terminal of the Sanyo Line) doubled between 1972 and 1974. The number of hotels increased from 20 to 40, and the number of hotel rooms jumped from 2,069 to 5320. The study shows that after several years of operation of the Tokaido high-speed railway, the population growth rate of station cities was 22% higher than that of non-station cities. However, it is not obvious whether the stations actually contributed to population growth or inflow.

Nakamura and Ueda (1989) also conducted further analysis. They divided the 10 counties into 735 areas and found that the population growth was very significant in the vicinity of Shinkansen stations (often adjacent to prefecture capitals or other regional centers, and where there were highways passing by). They further divided these regions into 104 Daily Life Circles, where the household economic activities were located. They concluded that populations in these circle areas increased in general, and there were 19 more populous circles and 14 less populous ones. In addition, Shinkansen stations showed a strong correlation to highways and revenues, and it is interesting to note that of the daily living circles that had an access to freeways but no station, 16 of them experienced a population growth, whereas only three circles saw a decline; of the daily living circles that had stations and highway access, the more densely populated ones (17) outnumbered those with a population decline (10). The study shows that the Shinkansen station was largely responsible for population growth, and highways amplified this effect. However, because almost no cases proved this conclusion, which may also be affected by the way that the living circles were divided, the logic of this study was not reliable. The economic growth trend of Shinkansen station cities or their urban belts was still not obvious.

6.1.3 Economic Effects of Regional Agglomeration

The theories of interactions between high-speed railways and regional economy include but are not limited to new economic geography and the related research on endogenous growth models. All these theories show that minor changes in accessibility have huge impacts on productivity and growth in industrial agglomeration, and these impacts may be greater than the direct benefits calculated by the evaluation methods in traditional industries. Since the very first beginning of the twenty-first century, it has been found in developed countries that the empirical results of these agglomeration effects had statistic implications. With better accessibility to transportation, agglomeration economies increase potential efficiency through inter-enterprise and urban activities in ways such as: expanding the variety of products available, sharing a larger, more extensive and more flexible labor market, capital and raw material input, promoting the exchange of well-known entrepreneurs and technical experts, creating more opportunities for learning and competition to achieve the transfer of technology and innovation. Initially, this benefit may only be reflected in a small number of companies within a particular region. However, other companies and individuals in this region will also enjoy this efficiency as they deepen their understanding of new technologies and raise their awareness of innovation.

In areas of improved traffic, cross-city externalities are shown in the increase in the number of enterprises with tight connections. Microeconomic theories believe this type of agglomeration brings immeasurable benefits to all companies located in the same region. For a long time, people have been aware that the average income in large cities is significantly higher than small cities. This can be explained by the positive externalities resulting from the agglomeration of economic activities. The production advantages of companies are reflected in their relationship with suppliers (to reduce the purchase price), with the labor force (to increase labor productivity), and with obtaining information (to improve technology). Therefore, when a company is located in an agglomeration area, it may increase the output of other companies through the above connections.

6.1.4 *Quantitative Method for Regional Economic Effects of High-Speed Railway*

Regional economists and econometricians have done a lot of research on regional economic impacts, most of which were described by the percentage of productivity or GDP. Spatial impact assessment (SIA), which was based on a series of socio-economic data related to regional development, was used in Germany, to rate the accessibility and the degree of stagnancy of the structural economy in each region. The results were applied to the German federal transport planning. The method used by Japan Railway Construction, Transport and Technology Agency combined the regional economic Computable General Equilibrium (CGE) model[3] with the model for population migration.

In China, the most important regional economic indicators are GDP and employment. The earliest research paper on large-scale construction projects and their economic impacts in China was mainly conducted by Zheng et al. (1994). Zheng Youjing gave a preliminary theoretical analysis on the regional economic meanings of construction and proposed an index system method and an input–output method. They also did a comprehensive feasibility analysis of the social, economic, and natural factors related to the mega-projects. Li (1998) demonstrated the regional economic benefits of the South-to-North Water Diversion Project from the perspectives of water supply, economic aggregate growth, and industrial development. Li et al. (2003) put forward the elementary concepts of mega-investment projects and their regional and macroeconomic influence. The authors also analyzed the similarities and differences between regional and macroeconomic impact assessment and national economic evaluation, and developed an index system and methods to measure the overall economic contributions of large-scale construction projects. From the perspective of the index system, Li and his colleagues proposed a number of gross indexes such as added value, structural indicators like coefficients of influence, sensitivity and employment; from the analytical angle, they built a macroeconomic measurement model, an input–output model, a system dynamics model and other models. They also introduced several methods including expert assessment. Li and Xu (2004) analyzed the regional and macro-economic impacts of the South-to-North Water Diversion Project on northern water-receiving areas in terms of regional development, population, employment, poverty reduction, economic development, public revenues and regional development gaps. Methods like location quotients and CGE models were applied to quantify the regional economic impact on northern water-receiving areas, and build an evaluation index system for the economic involvement of this national project (World Bank, 1994).

---

[3] Computable general equilibrium (CGE) models are widely used by governmental organizations and academic institutions to analyze the economy wide effects of events such as climate change, tax policies, and immigration. A CGE model consists of (a) equations describing model variables and (b) a database (usually very detailed) consistent with the model equations. The equations tend to be neo-classical in spirit, often assuming cost-minimizing behavior by producers, average-cost pricing, and household demands based on optimizing behavior. [Wikipedia].

In fact, there are many other ways to analyze the role of high-speed rail projects that are not discussed here.

The relationship between high-speed railway projects and land use purposes is notable. Geographically speaking, as a response to existing high-speed rail projects, public or private activities will gather in a certain area. It is also possible that the high-speed rail project is only one way to solve existing problems, such as traffic congestion. Therefore, a detailed cause and effect analysis is required to evaluate the impact of high-speed railway projects on economic activities. Undoubtedly, high-speed railways can also lead to diminishing marginal productivity. Therefore, when high-speed railways in a region are already well-established, it is not imperative to have more such construction. But, in countries or regions with relatively slow high-speed railway development, high-speed railway project investment may play a significant role in regional economic promotion.

Better high-speed rail projects are not a sufficient condition for regional development. Many other intermediates will also play a role. However, related research is still relatively small in number. To further study some certain high-speed railway projects, a better approach is to probe into the actual users and potential users of high-speed railways on the micro-level, so the drawbacks of models can be largely circumvented. Improved high-speed railway condition will produce both distribution effects and output effects. If the high-speed railway improvement rate is the same in all regions, the distribution effect is not obvious. When the targeted regions are strictly confined, it is very possible to overestimate the output effects but underestimate the redistribution effects, since output effects are usually the result of redistribution on a large scale.

In addition, we need to pay attention to the following issues: Firstly, the model of high-speed railway projects is more inclined to be applied at the enterprise level. However, it is also important to notice its application at the household level, because when families choose to migrate, high-speed railways are usually taken into account. In the long run, high-speed railways will also affect companies' locations. Secondly, most models, depicted in the form of formulas, are tools to study the impact of high-speed rail projects. In these models, high-speed railways exist as an exogenous variable, but this is not appropriate in the modeling process. As mentioned earlier, the construction of high-speed railways does not necessarily guide the economic activities of private sectors; instead, it is more likely that the private sector guides the construction of high-speed railways. In addition, emphasis should be put on the dynamics of high-speed railways. The reason is that the long-term effects of high-speed railways on supply and demand are different, so these models are also applicable to production and region analysis.

6.2 The Relationship Between High-Speed Railway and Regional Economic Aggregate

At the moment high-speed rail projects are being planned, local governments along the routes have already taken corresponding adjustments to make full use of the high-speed rails' advantages. These responsive measures include optimizing the urban space layout, positioning stations, and drawing land requisition and demolition plans, establishing investment and financing platforms, and making industrial restructuring schemes. With the inauguration of high-speed rail projects, local capital flow becomes linked to local policies. As a result, the financing and investments in supporting facilities, and the capital flow induced by better transport infrastructures, will finally be converted into new local investment in fixed assets, exerting a multiplier effect on the regional economy. After HSR operations, the consequent capital and labor flow, and an accelerated exchange of technology, will prompt the optimization of economic structure in various regions. High-speed railway operations will also help create the economic growth points for cities along the track, and increase total economic output, as shown in Fig. 6.3.

After the initiation of high-speed railway projects, some high-speed railway investments directly make up the GDP of the region along the line. Supporting facility investments from local governments will also increase the GDP of the region along the lines. As the private capital of a high-speed railway construction project moves on, new industrial chains will gradually form, further pulling up the GDP of high-speed railway cities. The procurement of various products and equipment for high-speed railway construction will directly generate effective demand with the multiplier effect. Meanwhile, the demand brought about by the high-speed rail projects usually involves cutting-edge technology, which stimulates the upgrade of technical standards for related products and equipment, improves enterprises' technological standards, enhances their competitiveness, and ultimately accelerates local economic growth. For example, the operation of the Beijing-Shanghai high-speed railway directly prompts flow of private and foreign capital, exchange of intelligence, and redistribution of production factors and demands. However, these effects on different cities vary. In urban growth points, known as pillar industry clusters, the re-distribution of production factors will result in growing outputs, but due to the outflow of production factors, the situations in areas outside these growth points may be diametrically different. It is also worth mentioning that the stratification and cross-regional movements of consumption, together with consumption agglomeration, may impede service industries of second- and third-tier cities and hence lead to the slowdown of local economic growth.

In short, the construction and operation of high-speed railways are highly relevant to the GDP growth, because it can exert a considerable influence on the flow of capital, employment, technology, and other factors of production along the lines. In practice, the interactions between high-speed railways and other factors are very complicated and intertwined. In this study, the effects of high-speed railway on aggregate economy

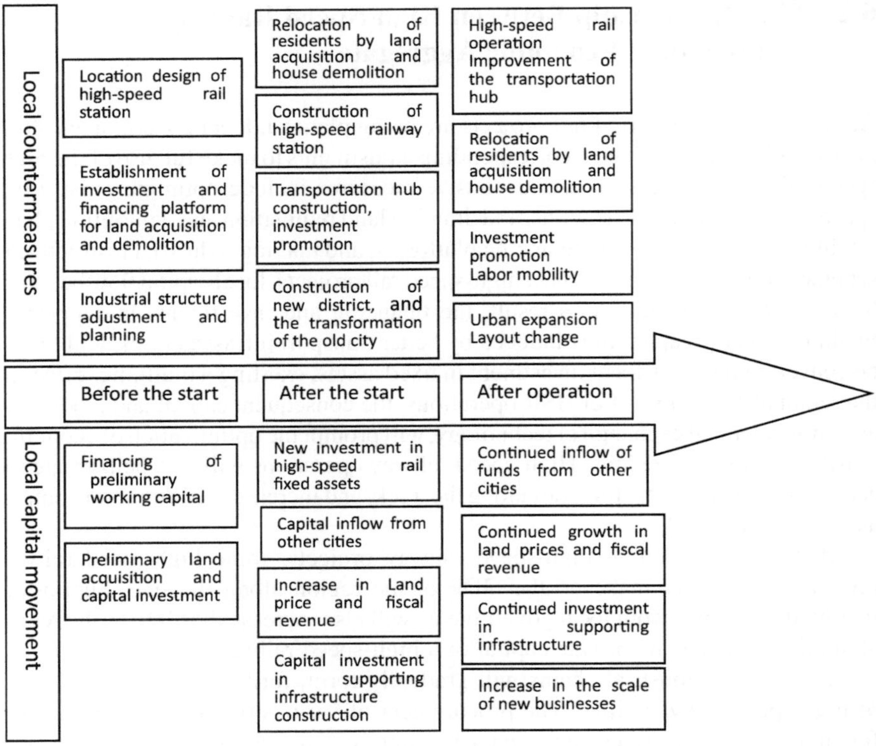

Fig. 6.3 Impact of high-speed rail project investment on local economic policies and capital movement

and industrial structure are viewed as the effects of such interactions, in which high-speed rail (as an important propeller) plays a key role, as shown in Fig. 6.4.

6.3 The Impact of High-Speed Rail on Trade—Taking the Xiang-Gui Line and Its Extension Line as an Example

The construction of high-speed rail has a major role in the development of trade. Guangxi is a case in point. Thanks to its unique geographical advantages, Guangxi has made full use of its unique geographical position and national preferential policies to gradually develop foreign trade into its economic pillar. In particular, the establishment of the China-ASEAN Free Trade Area on January 1st, 2010 effectively accelerates the fully open-up of Guangxi, and its foreign trade, especially its bilateral trade cooperation with other ASEAN countries.

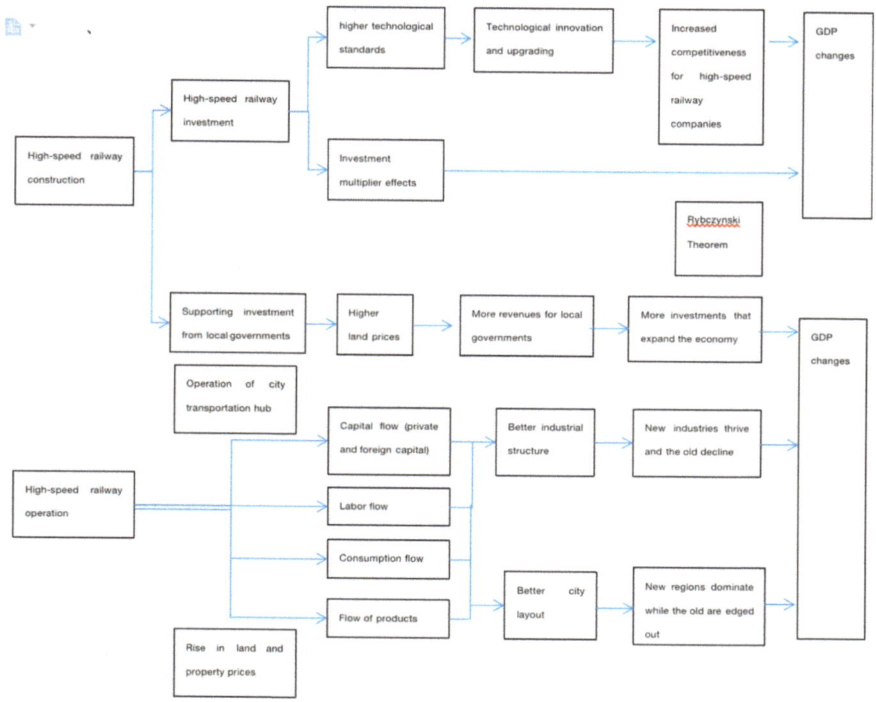

Fig. 6.4 Impact of high-speed rail project on local economic aggregate

In 2010, the bilateral trade between Guangxi and other ASEAN countries reached US$6.53 billion, of which its bilateral trade with Vietnam reached US$5.13 billion. And Vietnam has become the largest foreign trading partner of Guangxi Province for nine consecutive years. In 2010, Guangxi's largest foreign trade city was Chongzuo City, whose import and export volume reached US$3.34 billion, accounting for 21.1% of Guangxi's total foreign trade volume. As Hunan-Guangxi High-speed Railway Line runs through three central cities in Guangxi, traverses Chongzuo and extends to the Pingxiang City, which is on the border between Guangxi and Vietnam, the operation of this high-speed railway artery will undoubtedly provide firm transportation support for the exchange of goods and commodities between Guangxi and other ASEAN countries.

Guangxi Beibu Gulf Economic Zone is composed of the four cities: Nanning City, Beihai City, Qinzhou City, and Fangchenggang City. Among them, Beihai, Qinzhou, and Fangchenggang are located at the southern coastline of Guangxi and the northern shore of Beibu Gulf. The three cities are centered around Qinzhou City, forming a triangular layout. Beihai and Fangchenggang are on the east and west of Qinzhou City respectively. The map of the Beibu Gulf Economic Zone is shown in Fig. 6.5.

Beibu Gulf Economic Zone has natural location advantages and an outstanding strategic position. Situated at the junction of the South China Economic Circle,

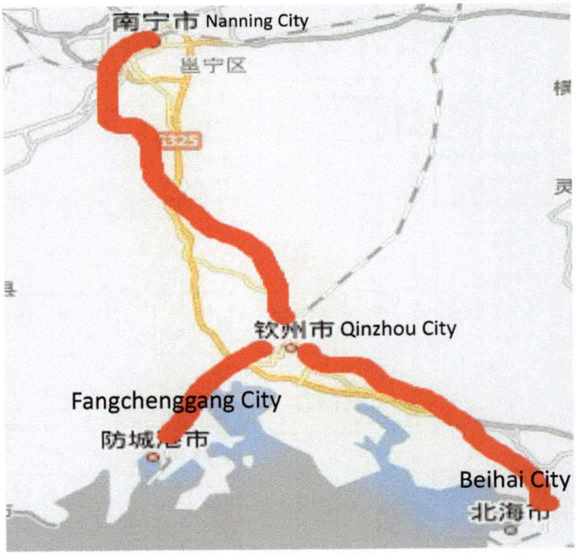

Fig. 6.5 Map of Beibu Gulf Economic Zone

the Southwest China Economic Circle and the ASEAN Economic Circle, Guangxi Zhuang Autonomous Region enjoys the only coastal area in China's Great Western Development program, and both sea and land borders between China and other ASEAN countries.

According to the strategic positioning of Guangxi Beibu Gulf Economic Area approved by the National Development and Reform Commission (NDRC),[4] it is a key area for China's western development and opening cooperation with ASEAN, and is of great significance for the country to implement the overall strategy for regional development and an opening strategy of mutual benefit and win–win outcome. Therefore, it is necessary to build the Guangxi Beibu Gulf Economic Zone into a base for logistics, trade and business, and centers for processing, manufacturing and information exchange. This economic zone will become a strategic highland with high openness, powerful influence, economic prosperity, social harmony and good ecology. High-speed railway lines along the Guangxi coastline include Nanning-North Qinzhou, North Qinzhou-Beihai, North Qinzhou-Fangchenggang, and lines connecting these cities with Laibin and Chongzuo. These high-speed rail lines meet in Qinzhou, which helps to get different node cities in the Beibu Gulf Economic Zone connected with each other as a network. Consequently, Guangxi Coastal High-speed Railways will undoubtedly change the way railway passengers and freight are

[4] The National Development and Reform Commission (NDRC) is in charge of China's macroeconomic planning. It is responsible for formulating and implementing strategies for national economic and social development and coordinating major economic operations. The NDRC submits a plan for national economic and social development to the National People's Congress each year on behalf of the State Council. For more information about the main functions of the NDRC, refer to http://en.ndrc.gov.cn/, and http://en.ndrc.gov.cn/mfndrc/.

delivered in related port areas, and provide strong infrastructure for the formation and development of foreign trade in the Guangxi Beibu Gulf Economic Zone.

6.4 A Comprehensive Analysis from the Perspective of Production Function: A Case Study of Guangxi High-Speed Rail Network

Relying on a modern transportation network with high-speed railway as its backbone and Guangxi's location advantages, it is helpful to give full play to the radiating and driving role of central cities. intensively develop counties and small towns, optimize the layout of cities and towns, improve the urban system, accelerate the cultivation and formation of urban agglomerations and town belts with orderly structure, complementary functions, overall optimization and joint construction and sharing, and help them become a new growth pole for leapfrog development within the economic belt.

The Beibu Gulf urban agglomeration with high-speed railway as its support and Nanning as its core is a priority area for industrialization and urbanization. To promote the development of ports, transportation, logistics, urban construction and tourism, it is necessary to accelerate the networking of high-speed railway construction and give full play to the leading role of the Beibu Gulf Economic Zone. Above all, it is especially necessary to speed up the expansion of the throughput capacity of Beibu Gulf Port, improve the infrastructure and logistics facilities such as the high-speed railway that connects the collection and distribution system for the ports with the surrounding areas, and build an vital coastal hub port and an important transportation hub connecting multiple regions in China as soon as possible. At the same time, strengthening the spatial economic ties of Nanning, Chongzuo, Beihai, Qinzhou and Fangchenggang will help the cities to highlight more complementary functions and industrial division of labor, making the urbanization rate in this region exceed 55% and build a national key urban agglomeration (See Fig. 6.6).

On July 8, 2014, the State Council officially approved the "Development Plan of Pearl River-Xijiang River Economic Belt" by the letter of the State [2014] No.87. The planning covers Guangzhou, Foshan, Zhaoqing and Yunfu in Guangdong Province, and Nanning, Liuzhou, Wuzhou, Guigang, Baise, Laibin and Chongzuo in Guangxi Zhuang Autonomous Region, with an area of 165,000 square kilometers and a permanent population of 52.28 million at the end of 2013 (See Fig. 6.7 for Xijiang River Economic Belt).

The urbanization area of the mainstream of Xijiang River in Guangxi covers the town groups in central and southeast Guangxi, which is the key area for advancing industrialization and urbanization. Therefore, in order to effectively reduce logistics costs, provide strong support for industrial development, upgrading, agglomeration and urbanization, and realize the overall rapid development of the Xijiang River Economic Belt, it is imperative to speed up the development of the Xijiang golden waterway and the central cities along the river. It will also be necessary to construct an

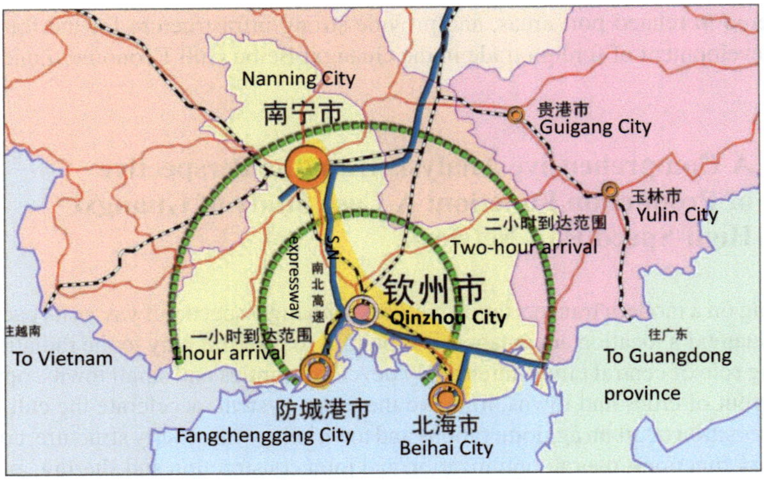

Fig. 6.6 Guangxi Beibu Gulf urban agglomeration with high-speed railways as the backbone

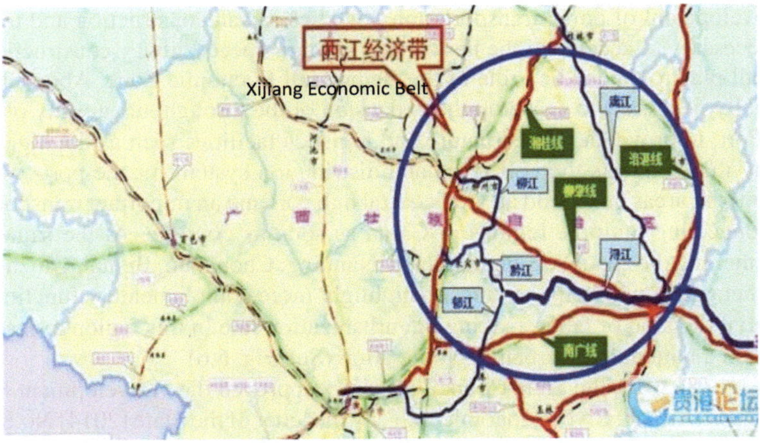

Fig. 6.7 Xijiang River Economic Belt Relying on "Iron-Water" multimodal transportation

efficient and smooth integrated transportation network with high-speed railway as the backbone, and form a transportation hub for China's important inland river shipping and a multi-modal transportation high-speed railway. On the other hand, it is important to take the central cities of Liuzhou, Laibin, Guilin, Wuzhou, Yulin, Guigang and Hezhou as the nodes in the belt, with the high-speed rail network as the framework to improve the spatial organization. Building high-speed rail economic corridors in Liuzhou-Laibin, Guigang-Wuzhou, Yulin-Wuzhou, Yulin-Guigang, Hezhou-Wuzhou and other places, will help build the clustering of cities with strong regional influence.

At the same time, the rising influence of Guilin, Baise, Hechi, Chongzuo, Hezhou and other central cities, helps to promote the development of counties, towns, and high-speed railways. This will help form the industrial and population concentration in the zone, and the distinctive town clusters of northern Guangxi, Youjiang River Valley Corridor, Guizhou-Guangxi (Qian-Gui) Corridor, Southwest Guangxi and Northeast Guangxi.

As the above analysis shows, Guangxi high-speed railway will speed up urbanization while promoting the economic development and industrial upgrading of the areas along the line. First of all, the construction of high-speed railways has strengthened the central position of node cities in a transportation network and an urban system network, thus accelerating the urbanization process of node cities. It is predicted that there will be obvious differences between cities (such as Nanning, Liuzhou, Guilin, etc.) in regions where high-speed railways will traverse in the next decade or so, to several decades in the future. This can be mainly manifested in the increase in the size of urban population, the rapid expansion of urban space, and the rapid growth of the total urban economy.

Secondly, the construction of high-speed railways will help to build unique cities and towns with their own characteristics along the line. For example, Guilin, a city with tourism as its main featured industry, may reduce its spatial–temporal distance with other cities at various stations through the construction and operation of a high-speed railway, thus greatly shortening the travel time between cities and allowing people to travel more conveniently. Moreover, the construction and opening of high-speed railways will bring more tourists to Guilin which will have a positive impact on its related industries, such as the hotel industry, food specialties, information consulting, shopping malls, exhibitions, post and telecommunications, entertainment industry and so on will also undergo rapid changes.

Thirdly, the opening of Guangxi's high-speed railway can accelerate the city integration effect in the areas along the line. Considering that once the regional passenger transport network with high-speed railway as the backbone is basically formed, the fast passenger transport channels between large and medium-sized cities such as Nanning, Liuzhou and Guilin will be further improved, thus further enhancing the "city integration" and "one-city" effects of the regional economy. Moreover, the completion and improvement of the high-speed rail network will greatly facilitate the business travel and play of ordinary people in these cities. Such large-scale and high-density frequent flow requires the completion and operation of large-scale high-speed rail in Guangxi, which will also have a significant impact on Guangxi's economic region and industrial structure in the future.

The current large investment in railway construction has injected a strong impetus to the rapid economic development of Guangxi, which has brought about the improvement of the technical level of high-speed rails and related industries. The opening of high-speed railway will also drive the flow of people, information, capital and technology to Guangxi. Evidence has it that the contribution rate of technological progress to output growth in Guangxi's high-speed rail cities is relatively low, but after the ground of high-speed rail is broken, the contribution rate of technological progress to output growth has increased to a certain extent than before. Although the contribution rate of technological progress to output growth in Laibin and Chongzuo is still negative, it also shows an increasing trend. Moreover, the construction and opening of Guangxi's high-speed rail will help improve the technological level of cities and the contribution rate of technological progress to output growth. The high-speed rail will also accelerate the development of high-tech services closely related to modern manufacturing, help establish investment and management consulting, software development, service outsourcing, and exhibition services. Other industries will develop such as intermediary services such as legal arbitration, accounting and taxation, consulting and evaluation, engineering design and the financial industry. The high-speed railway will encourage the "introduction of gold into Guangxi", and domestic and foreign banks, securities, insurance, futures, trust, venture capital funds and other financial institutions to set up branches and back-office services in Guangxi. Furthermore, high-speed railways will greatly promote the development of tourism by giving play to the potential of tourism resources, improve the overall development level of tourism, and build a strong tourism area. The construction of Guangxi high-speed railways has effectively accelerated the transformation and upgrade of traditional industries, played an active role in cultivating high-tech enterprises and supported the development of strategic emerging industries, led to the emergence of a number of new enterprises, and cultivated many new growth points for the transformation. Also, their full operations will lay a solid foundation for cities' sustainable development, and the transformation of Guangxi's economic growth mode.

Before and after the opening of Guangxi's high-speed railway, it seems that labor input has little effect on output growth, which indicates that labor input has little effect on output growth at this stage. However, with the further construction and opening of Guangxi's high-speed railway, it will play a positive role in effectively promoting the transformation and upgrading of traditional industries, cultivating high-tech enterprise groups and supporting the development of strategic emerging industries, bringing more and more high-quality and high-tech talents to Guangxi. This will give a strong impetus to the contribution of labor input to economic growth and bring the contribution rate of labor input to economic growth to a new level.

At present, as the driving force for the growth of railway transportation capacity, such abundant capital investment has pulled the growth of railway transportation capacity, and the rapid economic development in Guangxi. Due to the driving effect of high-speed rail on changing the mode of economic growth, the transformation and upgrading of economic structure, and on attracting talent, Guangxi's economic development is changing from extensive to intensive realizing the optimization of industry and economic structure.

References

Amano K, Nakagawa D. Study on urbanization impacts by new stations of high-speed railway[C]. Conference of Korean Transportation Association, Dejeon City. 1990.

Aschauer, D.A. (1989) Is Public Expenditure Productive? Journal of Monetary Economics, 23, 177–200. https://doi.org/10.1016/0304-3932(89)90047-0

Botham R (1983) The road programme and regional development: The problem of the counterfactual. In: Button KJ, Gillingwater D (eds) *Transport, location and spatial policy*. Gower, Aldershot, pp 23–56.

Brotchie, J., Batty, M., Hall, P., and Newton, P., eds. Cities of the 21st Century: New Technologies and Spatial Systems [M]. New York: Longman Cheshire. 1991

Bruinsma F, Pels E, Priemus H, Pietveld P, van Wee B (2008) Railway development. Impact on urban dynamics. Physica, Amsterdam.

DfT (2006). Transport, Wider Economic Benefits, and Impacts on GDP. London: UK Department for Transport (2006). See http://www.dft.gov.uk/pgr/economics/rdg/webia/transportwidereconomicbenefi3137

Dodgson, J. S. (1974). Motorway investment, industrial transport costs, and sub-regional growth: A case study of the M62. *Regional Studies, 8*(1), 75–91. https://doi.org/10.1080/09595237400185061

Hirota, R.. Present situation and effects of the Shinkansen[R]. International Seminar on High-speed Trains, Paris, 1984 (11):15.

Li Jingwen. The influence of Beijing-Shanghai high-speed railway construction on the economic development in the areas along the line [J]. *China Railway*, 1998, 10: 44–50.

Li, P., Li, W., Guo, S. Analysis of the Regional and Macroeconomic Impacts of Mega Investment Projects, *Journal of Quantitative & Technological Economics*, 2003(02): 5–10.

Li Pinghua, Lu Yuqi. Metropolitan accessibility: literature review and research progress in the west [J]. *Urban Problems*, 2005, (01):69-74.

Li, S., Xu, X. *South Water to North and Chinese Development*. Beijing: Economic Science Press, 2004.

Lin Xiaoyan, Chen Youxiao. Quantitative Evaluation of Infrastructure Investment Effect[M]. Beijing: Tsinghua University Publishing House; Beijing Jiaotong University Press, 2006.

Lin Xiaoyin. Privatization Reform and Market Financing of Railways [M], Economic Science Press, Beijing, 2006.

Nakamura H, Ueda T. The Impacts of Shinkansin on Regional Development[J]. *Proceedings of WC*, 1989, 3:95-109.

Nijkamp, P., and Reggiani, A. (1992). Impacts of multiple-period lags in dynamic logit model. *Geographical Analysis*, 24(2): 159–173.

The World Bank. *World Development Report, 1994: Infrastructure for development*, Oxford University Press, New York, 1994. Available at: The World Bank annual report 1994

Zheng, Y., Bao, L., Numov, S. J. Prospects for Economic and Technological Cooperation between China and Canada—An Analysis of Canada's Development Potential in Jiang and Zhejiang Provinces, *Journal of Quantitative & Technological Economics*, 1994(1): 71–75.

Chapter 7
High-Speed Rail Economic Belt and the Improvement of Regional Economic Structure

7.1 The Development of High-Speed Railway and New Economic Belt

7.1.1 The Background and Significance of High-Speed Rail Economic Belt

As a major artery of the national economy, an important national infrastructure and a popular means of transport, railway plays an important part in regional economic and social development. With the development of China's economy, a high-speed and large-scale passenger transport system becomes a must, which could be satisfied by building a high-speed railway network. According to the plan of the former Ministry of Railways, China will have completed a high-speed rail network including dedicated passenger lines and inter-city passenger transportation systems by the end of 2020, with operating mileage exceeding 16,000 km. By then, China will become the country with the longest operation mileage of high-speed rail in the world.

As the Chinese saying goes, "if the traffic is not smooth, all trades will be unlikely to be prosperous". Traffic is the prerequisite for prosperity. The underdeveloped infrastructure is the bottleneck that hinders the current regional economic development. Especially for the development in remote areas of China, the undeveloped railway network and poor traffic conditions may be one of the most important reasons for poverty. However, with the rapid development of China's regional economy, the railways built in the past may not be able to meet the needs of commercial and fast transportation in modern society. Therefore, the construction of modern high-speed railway is not only a general call, but also brings a new concept, "high-speed economic belt".

High-speed rail will become an important link to promote economic development along the route. With the expansion of the overall regional economic development in China, the interdependence of economic development in different regions has

X. Lin, *High-Speed Railways and New Structure of Socio-economic Development in China*, Research Series on the Chinese Dream and China's Development Path,
https://doi.org/10.1007/978-981-19-6387-2_7

increased, and an exchange with large-scale, long-distance, large-volume and fast-flowing production factors as its characteristics has begun to take shape. Presumably, the high-speed rail transport will play an extremely important role in the economic development.

High-speed railway will become the foundation for developing modern industrial system. Historically, different types of transport network have determined different industrial divisions and formed different types of industrial systems. The modern industrial system is a production system oriented by innovative services, which relies on talents as its main resource. And the application-directed innovation takes materials as a carrier of technology. The transportation system brings more credits for the allocation of talents and material resources. High-speed rails, which can move frequently in a wider space, is an economical method for excessive space disposition. Therefore, the allocation of talents and small quantities of material resources will be more effective, which will stimulate the innovation-oriented and culture-oriented industrial development, thus enabling the modern industrial system to evolve under such transportation systems.

High-speed railways will change the way the economy develops. As it is offering fast, comfortable ride on a fixed schedule, and good communication affordances, and is almost free from the change of weather, it becomes easier for travellers to balance their time for travel and work. The coming of high-speed rail has thus transformed people's cognition about work and their working styles, and made production more economical and life more convenient. For this reason, the production, consumption and industrial structure along the high-speed railroad tend to have lower energy consumption, lower pollution, but more added-value. The development of tourism industry along the line will further stimulate the culture and leisure industry, and make the rational allocation of cultural resources possible in a larger space, thus changing the way of economic development.

High-speed railway will bring about changes of the transportation system in areas along the line. With the regional networking of high-speed railways, 1–2 h of work and life circle is changing the concept of life and work of future residents. Subsequently, people may rely more and more on high-speed rail as a means of transport, which may change its property of a physical transport network into a more economic one. While the utility increases, the construction and operation costs and ticket prices will be declining. These are bound to become the basis for the further development of high-speed rail, serving more users and changing people's work and life style.

High-speed railway may bring into existence a "high-speed railway economic circle". The upgrading of a regional high-speed railway network will help to open up the railway network connecting the surrounding areas, and making a place part of a high-speed railway economic circle, like a binder, which connects the region with its neighboring provinces and countries.

Based on the above-mentioned practical significance, regional governments at all levels should strive to make economic and social development strategies under the guide of high-speed railway traffic circle, breaking through the inherent positions and traditional thinking of development strategies confined to a township, a county, a region, and a province. Plus, the local government should take the scientific outlook on development as a guide, practice innovative thinking, plan ahead, and act quickly so as to make the best of the historic opportunities brought by the construction of high-speed railways. Besides, it is necessary for research institutions and relevant persons-in-charge to think deep about the realistic impact of high-speed railway construction on regional economic and social development, and provide suggestions for local government decision-making. Similarly, it is crucial that shareholders and managers of varied industries and businesses take the initiative to think and respond actively, and fully grasp the infinite business opportunities brought about by the construction progress and the completion of the high-speed railways.

7.1.2 The Prerequisite Guarantee for the Formation of High-Speed Rail Economic Belt

The Twelfth Five-Year Plan (12th FYP)[1] is a crucial period for the scientific development of high-speed railways in China. During this period, the construction task is very arduous. In order to realize the planning goals and objectives, it is important to seize the development opportunity to implement the construction plan in a scientific and orderly manner, and promote the sustainable development of high-speed railway by strengthening the leadership and supervision and providing the guarantee of railway funds. Specific suggestions for developing high-speed rail in China are as follows.

First, it is important to do reasonable and necessary adjustment to the timetable for some construction projects. To carry out the 12th FYP for the development of regional high-speed rails, governments of all levels should adhere to the principle of "highlighting key areas, balancing and coordinating various departments to carry out scheduled projects chronologically" and combine the conditions of constructions with regional real needs. At the same time, we should scientifically grasp the pace of development and seek the unity of quality and efficiency. All parties concerned should actively link up with relevant national ministries and commissions, make full use of the advantages of the regional cooperation mechanism, and clarify the different positioning of the central government, local governments and enterprises in the development of high-speed railways. Government authorities should actively fulfill their duties at all levels and encourage the market, enterprises and the lower government departments to collaborate on proper resource allocations for the construction of regional high-speed railways.

---

[1] China began implementing five-year plans in 1953 in order to align its economic development with top policy goals and to communicate this directive throughout the government bureaucracy. Five-year plans can be perceived as roadmaps for regulators and officials at all levels, who take responsibility to their implementation.

The 12th Five-Year Plan (FYP) adopted by the Chinese government in March 2011 devotes considerable attention to restructuring the Chinese economy by encouraging domestic consumption, developing the service sector, shifting to higher value-added manufacturing, conserving energy, and protecting the environment.

Second, project management should be improved. It is essential that all parties concerned see quality and safety as the first priority, and strengthen the management of the construction projects, funds, companies (especially the HSR companies with joint investment or cooperative construction projects), and attach great importance to the preliminary work for newly-commenced project to make it run on a feasible construction plan, a reliable scale and an investment portfolio.

Third, project funds must be guaranteed. It is crucial that the Ministry of Railways collaborate with local government and other sectors in the process of high-speed railway construction, bearing in mind the characteristics of regional economic development, carrying out the diversified investment policy research and mechanism practice in high-speed railway construction, and the leading role of government financial support, guidance, leverage, etc. Plus, Local governments should actively strive for the allocations and other fiscal supports from the central government, establish a system of joint investments in high-speed railway construction at the level of autonomous region and city to encourage regional high-speed railway investment groups or companies to play their roles in raising funds for high-speed railway construction, and encourage the injection of high-quality assets to enhance their financing and solvency. Besides, various effective measures should be taken to give financial institutions policy support so that the credit guarantee capacity of financing platforms for high-speed railways can be enhanced. At the same time, enterprises should be allowed to have good opportunities to obtain high-speed rail construction funds through various means, and private capital should be encouraged to get the entrance to high-speed railway construction projects. This will ensure the full implementation of high-speed railway construction plans.

Fourth, it is very important to create a good environment for HSR construction programs. Focus should be put on the research and development of high-speed railway construction and market-oriented transport, and developing a diversified high-speed railway construction development model, as well as effective corporate governance and operation mechanism taking corporate system as the main stakeholder, and then leading high-speed railway construction and transportation to the market. All departments at different levels should vigorously support the construction of high-speed railways and seek the understanding and cooperation of the public by increasing the publicity and creating a good construction environment for high-speed railway engineering projects.

Fifth, coordination should be enhanced. It is important to strengthen the connection between high-speed rail project planning and overall land use planning and forest land protection and utilization planning. Therefore, in the process of high-speed rail planning and construction, ecologically sensitive areas such as nature reserves, forest parks, and key ecological public-welfare forest areas should be protected and un-destructed as much as possible. Also, agricultural land and forest should be developed under the guidance of the national law and regulations in order that Chinese forest resources may be fully protected and rationally used. Further, the budget for high-speed railway projects should be outlined in detail to ensure that the use of forest land does not violate the law, and that the approval process for the application, inspection, and funding should be well-organized and justified to avoid unauthorized forest land use.

Last but not least, the integrated traffic and transport system should be constructed comprehensively. The construction of regional high-speed railways breaks the market division of passengers in aviation, railways and highways. Thanks to its speed and convenience, airlines are targeting at high-end travellers of medium and long distance. Railways are mainly for ordinary travellers of medium and long distance because of their low speed but much more acceptable price, while highways rely on its flexibility for short-distance transportation. Therefore, it is necessary for the regional transport departments to actively cooperate and plan the overall construction and operation of regional high-speed rails.

7.2 Features of and Underlying Causes for High-Speed Rail Economic Belts

7.2.1 Characteristics of the High-Speed Rail Economic Belt

As a typical transportation economic belt, high-speed rail Economic belt has the general characteristics of an economic belt and other unique features.

First, high-speed railways are generally constructed for better connectivity between more economically developed cities. A typical example of this is the Beijing-Shanghai High-speed Rail Economic Belt and the Beijing-Tianjin Intercity High-speed Rail Economic Belt. Before the completion of the high-speed rail, the transport network of the Beijing-Shanghai Economic Belt and the Beijing-Tianjin Economic Belt were comparatively advantageous, as their infrastructure and resource endowments are very rich and no more large investment is required. What is needed is to get the high-speed rail "embedded" into the original transportation economic belt, making it turn progressively from "balance to imbalance and then back to balance". The breaking of the existing balance, thus, may bring impetus for the economic development of the belt, and reorganize its industrial structure and economic growth modes.

Second, compared with other transport economic belts (regular railways and highways), high-speed rail is more efficient, flexible, fast, and cross-regional, therefore having a more prompt and profound impact on the regional layouts. The completion and operation of high-speed railways greatly facilitate the transportation, and the economic development along the line, as well as expand the size of the economic belt.

Third, high-speed rail Economic belt can be seen as a socio-economic organism with unique characteristics. The traffic economic belt is the result of long-term interaction between transport corridors and economic activities. In the process of its development, towns and cities, industries large and small, and societal and cultural activities benefit from the technological innovation of the transport system, so their influences tend to agglomerate and spread, which will lead to significant changes in its structure, overall functions and economic strength. The inborn high-tech characteristics of the high-speed rail will be beneficial to urbanization, industrial structure upgrading, and transformation of economic growth mode.

Fourth, the high-speed rail takes passenger transport as its main stakeholders. As a new means of transport, it may change people's travel habits and preferences. In terms of economics, the distance between interpersonal communications in a time unit is directly proportional to the people's educational level in a certain region. When people can communicate with the outside world more effectively and frequently, they feel compelled to learn more about the outside world to a large extent, thus broadening their horizons and improving their overall quality. In addition, after the opening of passenger lines, the related management and technical talents are required to support the maintenance. Therefore, such talents will also greatly promote the development potential of regions. Besides, the highspeed rail will promote the development of the tertiary industry, as the passenger flow will gain uprising increase correspondingly due to the greatly shortened travel time, which undoubtedly brings more profits to the catering industry, commerce, tourism and service industry.

Fifth, the high-speed railway is essentially an outcome of high-tech development, which is inseparable from a series of high-tech fields such as electronics, information, materials, aviation, and environmental protection. The construction of high-speed railway not only makes the railway system more knowledge-intensive, but also boost the development of related high-tech industries, driving China's industries to a high-end, sophisticated and cutting-edge direction.

Finally, high-speed rail Economic belt is an evolving economic system. Its formation is a result of the coordinated development of transport and economy, which changes dynamically with the interaction of transportation and economic activities. Besides, the high-speed rail Economic belt is an open, non-equilibrium and non-linear system composed of a variety of elements such as industry, population, towns and cities, information, and transport infrastructure. Its existence relies heavily on the mutual affinity of internal elements and the frequent exchange of materials, energy and information with external systems, and keeps growing, with more complex structure, diverse functions and broader boundaries.

7.2.2 The Formation and Development Model of High-Speed Rail Economic Belt

High-speed rail economic belt is essentially a spatial regional complex and a socio-economic organism, which means that it can be regarded as a specific stage in the development of the regional economic system.

High-speed rail economic belt gradually evolves along with the development of industrialization and transportation. During its development, population, industry, towns, information and other elements often agglomerate or disperse along traffic lines on a large scale. The construction and introduction of traffic lines or new transport technologies breaks the existing balanced and homogeneous development structure of regional economic systems. One major reason is that the increase in spatial accessibility along traffic lines leads to the agglomeration and distribution of production factors smoothly and rapidly, and the gathering of regional population, industry, towns and information. Such polarization effects make the central areas superior in location and transportation, and allow them to continuously attract talents and capital of the neighboring areas through backward linkage effects. As a result, these areas will eventually grow into the economic centers along the lines. As the economic power of the core cities increases, the areas that have economic relations with them will also expand. In addition, in this process, the imbalance of the high-speed rail economic belt will also be rising, leading to an expanded belt. When the agglomeration reaches a certain level, the diffusion effect gradually plays a leading role. The economic center continuously transmits its industrial and technical elements to the neighboring regions in the form of gradient and hierarchical diffusion, and displacement. The improvements in the functions of integrated transportation channel in turn push a more balanced development of the economic belt. When less developed regions in cities are input more production factors (such as industry, technology and capital) and grows to a certain extent, the entire high-speed rail economic belt will reach its maturity and gradually moves from the earlier low-level equilibrium to an advanced equilibrium. After that, two trends may appear: First, the belt may be integrated into the regional economic system with a generally improved overall level through its connection with other high-speed rail economic belts. Second, due to the relatively obvious poor transport technology or industrial structure in the belt, the production factors will gather in the neighboring high-speed rail economic belts, leading to its gradual shrinking or even dying. Along with the growth of high-speed rail economic belt are the dynamic changes such as its borders, size of core areas, economic strength, industrial structure level, urban population, and the influence of its city centers, as well as the flow of goods and displacement of core metropolises. This may give us a more visual and clear evidence to define the life cycle and the evolution of high-speed rail economic belt. According to the above evolution mechanism, the life cycle of a high-speed rail economic belt can be roughly divided into four stages: the initial stage, the take-off and rising period, the expansion and maturation period, and finally, the converging and declining period. From the perspective of time and space, the evolving stages of high-speed rail economic belts can be illustrated as in Fig. 7.1 below.

Stage One: The Initial Stage

At the early stage of regional economic development, the transportation modes connecting various transportation places are relatively primitive or single, and the central places are isolated, scattered and in a state of no hierarchical system due to the large and small number of passengers and freight exchanges. The regional economic system is in a low-level equilibrium stage, and the spatial structure is also characterized by a closed cycle in a small area. With the construction and introduction of new transport, mineral and agricultural resources along the lines can be

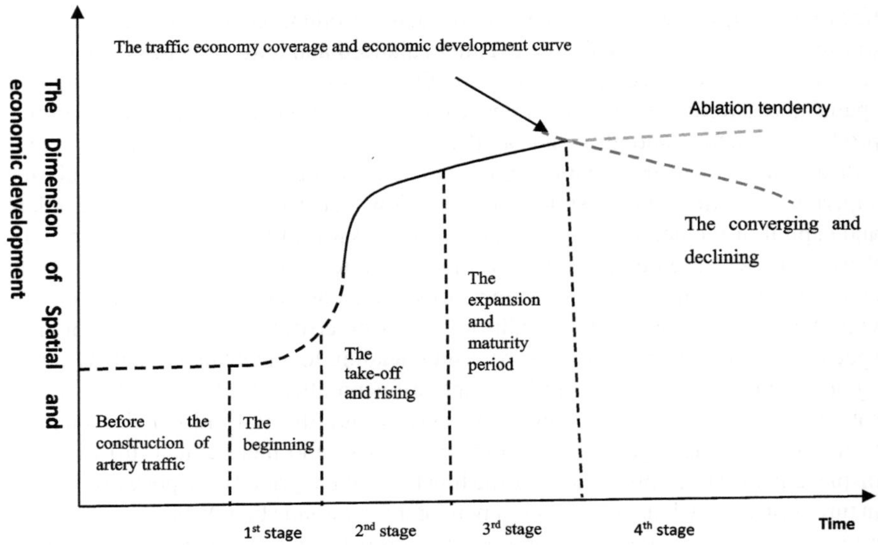

Fig. 7.1 The lifecycle of a high-speed rail economic belt

quickly exploited, therefore facilitating the local economic development. Centers with superior locations (such as areas near water or at road and railway junctions) become the harbor of modern industrial and commercial industries, gathering or processing agricultural products and mineral resources. These locations have gradually become economic centers in the region, greatly contributing to the growth of the local economy. Meanwhile, city locations, population and productivity in the region are redistributed. Therefore, the regional imbalance will gradually shape the initial form of high-speed rail economy. The salient features in this stage are as follows: Firstly, ① The mode of transportation is relatively simple; ② Logistics along the route is mainly based on internal and external exchange, with less exchange of goods between towns along the route, single structure of goods, and raw material goods as the main export goods; ③ The level of industrial structure along the route is relatively low, with some primitive processing industries such as agriculture, food, textile and mining as the main ones; ④ The main center mostly plays a role in the distribution and exchanges of passengers and goods along the route, and its economic, social and cultural functions are limited to a certain space. The conceptual model in this stage is shown in Fig. 7.2.

Stage Two: The Take-Off and Rising

With the improvement of means of transport and transportation capacity, it is more convenient for cities and towns in the economic belt to communicate with the outside world. At the same time, the external market supply and demand information is gradually obtained and valued by the regions along the route. Relying on the traffic line, based on the connection of production links, and characterized by meeting

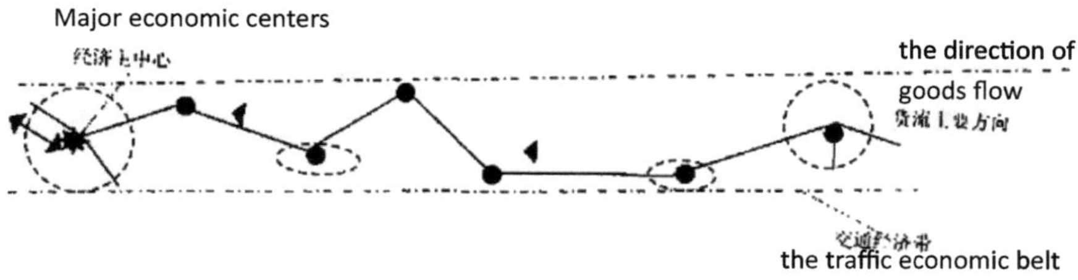

Fig. 7.2 The development model of the high-speed rail economic belt in the initial stage

internal needs, the connection between cities and towns is increasingly closer, the industrial chain and regional division of labor within the region are gradually formed, the industrial structure along the line is generally improved, the number and strength of cities and towns are rapidly enhanced, and the population further converges to the areas along the line. The location conditions in the economic zone are superior. The cities that are the first to accept emerging technologies and the first to upgrade their industrial sectors gradually develop into new economic centers, which may be formed after further polarization of the economic centers in the first stage (such as Shanghai, the economic center of Shanghai-Nanjing-Hangzhou high-speed rail economic belt), or it may be formed after the economic strength of other cities exceeds that of the previous economic centers. That is, it is possible that the main economic center will shift at this stage (for example, Shenyang in Harbin-Dalian high-speed rail economic belt has replaced Dalian as the new dominant economic center). Around the growth pole and leading departments, the economic strength of the economic center and its polarization effect is further enhanced, and the core–edge dual spatial structure is increasingly obvious.

Compared with the previous stage, the take-off and rising period has the following remarkable characteristics: ① the transport means or transportation capacity is obviously enhanced, the number of the branch lines crossing the main axis is constantly increasing, and the sphere of influence of the economic belt is obviously expanded; ② Logistics along the route pays equal attention to both internal and external exchanges. The passenger and freight exchanges between towns along the route are obviously enhanced, and the structure of goods is becoming increasingly complex. Among the exported goods, the proportion of manufactured goods is obviously on the rise, and the imported goods are still dominated by manufactured goods. ③ The industrial structure level along the route has generally improved, and the raw material deep processing industry has become the leading industry; ④ The position of the dominant economic center is enhanced, and its sphere of influence is obviously expanded, although its influence on other regions is mainly in a state of gradient diffusion. See Fig. 7.3 for a spatial model at this stage.

Stage Three: The Expansion and Maturation Period

With the further strengthening of traffic capacity and transport, comprehensive transport corridor with large capacity and high speed has gradually formed, which makes

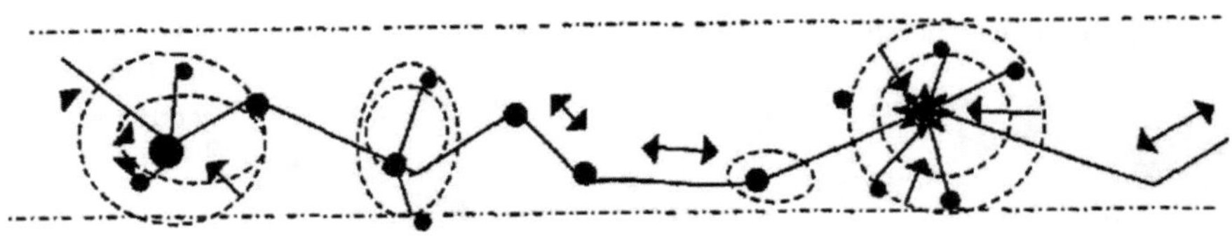

Fig. 7.3 The development model of the high-speed rail economic belt at the take-off and rising stage

the exchange of passengers, goods and information inside and outside the region more convenient. The process of industrial upgrading and diffusing along the route has been greatly accelerated, the scale of towns has been rapidly expanded, the division of urban functions and links within the economic belt have become clearer, and satellite cities in major cities have been emerging. The diffusing effect leads to an obvious process of suburbanization of big cities, creating some well-connected urban districts in some regions along the line, and the scope of the economic belt continues to expand. The main economic center which functions as a trade, finance and information center and as an incubator is becoming more and more important and powerful. It exerts influence on farther towns along traffic routes by means of hierarchical diffusion.

Compared with the previous stage, the maturity and expansion period has the following remarkable characteristics: ① The capacity of transport is further enhanced, the comprehensive transportation corridor is formed, and the influence of the economic belt is increasing; ② The directions and types of logistics along the route are more diversified, including internal and external exchanges, internal communication, and communications between central cities and satellite cities. Among the exported goods, the proportion of industrial manufactured goods has increased significantly, and the imported goods are mainly manufactured goods and raw materials. ③ The upgrading of industrial structure along the route is accelerating, and the tertiary industry has become the leading industry. Due to the limitation of local resource endowment, the number of enterprises using external resources in the economic belt is gradually increasing; ④ The position of the main economic center is enhanced, and the diffusing influence of the main center on other regions is still going on. The status of other central cities has also been correspondingly enhanced, and the suburbanization has led to the convergence of urban boundaries. See Fig. 7.4 for the spatial model at this stage.

Stage Four: The Integration and Declining Period

Two trends stand out in this stage: the integration and the declining. Integration means that with the more powerful comprehensive transport corridor, the agglomeration and spreading effects of high-speed rail economic belt are accompanied by each other, the economic hinterlands of central cities cross each other, urban boundaries are disappearing, urban continuous areas with the comprehensive transport corridor as the main trunk and a balanced network composed of interlock connections among large, medium and small towns are formed within the economic belt. In the meantime,

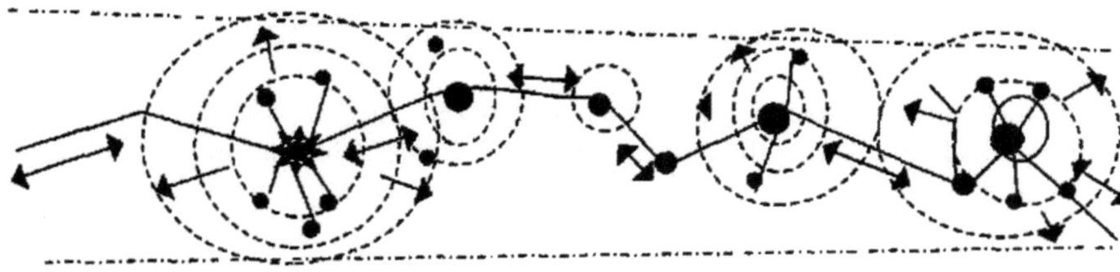

Fig. 7.4 The development model of high-speed rail economic belt at the maturity and expansion stage

with the progress of adjacent high-speed rail economic belts, there is a trend of mutual connection and integration among major economic belts. When the links between high-speed rail economic belts become increasingly closer and more accessible, the homogeneous community in a larger area will form. The possible integration at this stage is an outcome from the homogeneity of circulation and economic growth caused by the highly upgraded transport construction in the region.

The going downhill or the declining can be understood as this: due to the lagging-behind transport technology, the comprehensive transportation capacity is getting dysfunctional. And, because of location conditions, regional development policies and other factors, the strength of the economic belt is shrinking, urban construction and industrial upgrading have also been slowing down, and the gap with other neighboring high-speed rail economic belts is widening, which leads to the gradual transfer of regional economic development factors such as population, capital and information to other economic belts, and thus the dissolved economic belt.

7.2.3 The Formation of Highspeed Rail Economic Belt: A Case Study of Liu-Nan Line and Its Extension Line in Guangxi

The role of the transportation economic belt is mainly reflected in two aspects, namely the polarization effect and the spreading effect. The most striking difference between Guangxi High-speed Rail Economic Belt and other economic belts can be summarized as follows:

Guangxi high-speed railway economic belt has experienced a process of "growing from scratch", as the construction of high-speed railway is generally selected in areas with relatively high economic development level (for example, Nanning and Liuzhou), where the transport network is relatively well-built, and where there is already an economic belt that has begun to take shape. In a sense, it is to embed the newly-built high-speed railway into the existing economic belt, which will undergo a development process of balance-imbalance-balance. This process is mutually interactional in which not only the influence of high-speed railway on economic belt will be growing, but also the interaction between the high-speed rail and the existing means of transport in the economic belt.

The formation of Guangxi high-speed rail economic belt finds it basis on large and medium-sized cities along the route (Nanning, Liuzhou, Guilin, Laibin, Chongzuo, etc.), with high-speed rail entrances and hubs as the growth point. Before analyzing the formation of economic belt, it is necessary to know the types of growth points. As noted earlier in this book, before the completion of the high-speed railway, Guangxi have already had a relatively improved transport network and a preliminary economic belt. The construction of the high-speed railway broke the original balanced development. In terms of the order, there are two kinds of growth points: before a high-speed rail is built, large and medium-sized cities along the route (Nanning, Liuzhou, Guilin, Laibin, Chongzuo) are the main growth points; after the completion of the high-speed railway, the cities where the entrance and exit of the high-speed railway and the hub are located become new potential growth points.

The effects of high-speed rail construction on different growth points are as follows: (1) it promotes the development of the original central point and the extension of the central city or region to other places; (2) it brings new growth points, and the line hub becomes a new economic growth point, which exceeds the original growth point, but they are complementing each other. At first, the economic development level of the original growth point is better than that of the high-speed rail, and then the original growth point drives the development of the new growth point. And then, they are mutually beneficial. Eventually, with the growth of passenger flow along the line, the level of development of new growth points will see a striking increase, exceeding the original growth points, and driving the development of the original growth points. However, while the construction of high-speed rail brings new growth points, it may impede the development of the original growth points and is detrimental to the original economic belt.

According to the coverage of core cities in Guangxi, the Guangxi high-speed economic belt probably comprises of: the southern part of Nanning (Long'an County, Binyang County, etc.), Laibin City, some counties under the jurisdiction of Liuzhou City (Liujiang County, Luzhai County), some counties of Guilin City (Yongfu County, Lingui County, Lingchuan County, Yangshuo County, Gongcheng Yao Autonomous County, Longsheng Autonomous County, Xing'an County, Quanzhou County), and Chongzuo City (the capital city of Chongzuo City).

After the completion and opening of the high-speed rail connecting the Xiang-Gui line[2] in the north to Pingxiang in the south, a high-speed rail economic belt

[2] Hunan-Guangxi Railway is the longest railway trunk line in Guangxi. Starting from Liuzhou, Litang and Nanning South Railway Station, they are connected with Guizhou-Guangxi, Jiaoliu, Lizhan, Nanfang, Nankun, Nanguang, Guiguang, Jingguang, Hengchaji, Huaishaoheng and Luozhan railways respectively.

The new line, Xiang-Gui Express Railway, is a railway built in the wave of capacity expansion and transformation of a large number of old railways in China. It is a Class I double-line electrified express railway of the State Railway, with a designed train speed of 200 km/h (reserved speed-up conditions). In December 2013, the whole line was completed and opened to traffic. People mistakenly call it the so-called Hunan-Guangxi high-speed railway.

will be formed with the high-speed rail transportation passages as the main axis and cities along the line as the support, with a wider coverage, which covers 14 prefecture-level administrative regions in Guangxi, including Nanning, Liuzhou, Guilin, Wuzhou, Beihai, Fangchenggang, Qinzhou, Guigang, Yulin, Baise, Hezhou, Hechi, Laibin and Chongzuo, with a total area of 236,700. At the end of 2010, the total population of the whole region was 46,026,600, of which the minority population was 17,110,500, accounting for 37.18% of the total population. In 2010, the GDP of the whole region was 1,171.435 billion yuan, which was nearly 0.3 percentage points lower than the national average (1,521.174 billion yuan), and there was much room for improvement in economic strength and industrial development.

In spatial economy, there are details about defining the boundary of expressway economic belt, whereby a time-based and freight-based boundary model is constructed. Here, this book will adopt the model to look at the impact of Guangxi high-speed rail economic belt.

The influence area of Guangxi high-speed railway is centered on the entrance which has a certain radius length. The radius measurement, therefore, is the key to define the boundary of high-speed railway economic belt. The radius is determined by many factors. However, as far as high-speed railway transportation is concerned, freight is mainly used as the basis for route selection, and passenger transportation is mainly used as the basis for passenger route selection. These two factors can explain the radius of influence.

As shown in Fig. 7.5, Wuzhou-Liuzhou and Wuzhou-Chongzuo are non-high-speed rail transportation routes in Guangxi. Wuzhou may have transportation communication with Chongzuo through Wuzhou-Liuzhou-Chongzuo, where the high-speed railway plays an indirect role. It may also communicate with Chongzuo directly through the Wuzhou-Chongzuo highway, which has nothing to do with Liuzhou. Therefore, Liuzhou-Chongzuo high-speed railway has no influence on Wuzhou. When there are two entrances and exits adjacent to a high-speed rail connected by different traffic routes, we can use freight expenses and time as decisive variables to measure the influence radius of a certain entrance and exit.

Fig. 7.5 A model of the sphere of influence of Guangxi high-speed rail (Liuzhou-Chongzuo) on the areas along the route

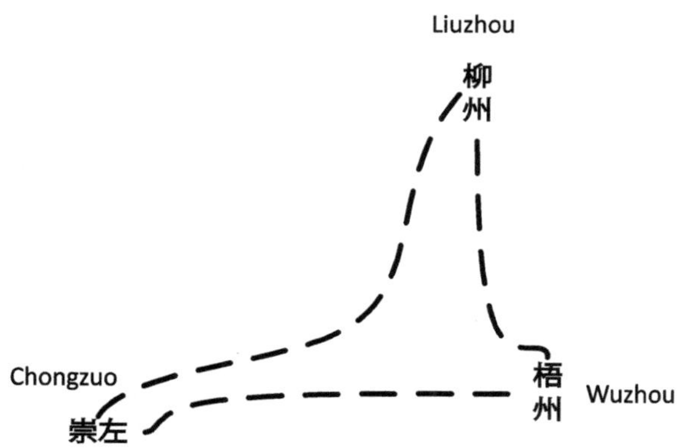

Let l be the length of traffic line, then the lengths of the routes of Wuzhou-Liuzhou, Wuzhou-Chongzuo, Liuzhou-Chongzuo can be represented respectively as $l_{\text{Wuzhou-Liuzhou}}$, $l_{\text{Wuzhou-Chongzuo}}$, $l_{\text{Liuzhou-Chongzuo}}$. Let Wuzhou be the average freight transportation cost per unit, then the transportation cost from Wuzhou to Liuzhou, from Wuzhou to Chongzuo, and from Liuzhou to Chongzuo are respectively: $C_{\text{Wuzhou-Liuzhou}}$, $C_{\text{Wuzhou-Chongzuo}}$, $C_{\text{Liuzhou-Chongzuo}}$. Let V be the average transportation speed, then the average transportation speed from Wuzhou to Liuzhou, from Wuzhou to Chongzuo, from Liuzhou to Chongzuo are $V_{\text{Wuzhou-Liuzhou}}$, $V_{\text{Wuzhou-Chongzuo}}$, $V_{\text{Liuzhou-Chongzuo}}$. Let R_{Liuzhou} be the high-speed railway entrance's influence radius of Liuzhou, then the radius of influence determined by freight is R_C, and the radius of influence determined by time is R_t. Let's assume that the two routes Wuzhou-Liuzhou and Wuzhou-Chongzuo are the path from Wuzhou to Liuzhou and Chongzuo which has the most favorable freight and the shortest travel time, then the model for determining the radius of influence of high-speed rail entrances and exits based on freight and time is:

$$opt\, R_A = \text{MAX}(R_C, R_t)$$
$$s.t.\begin{cases} C_{\text{Wuchou-Liuzhou}}R_C + C_{\text{Liuzhou-Chongzuo}}l_{\text{Liuzhou-Chongzuo}} \le C_{\text{Wuzhou-Chongzuo}}l_{\text{Wuzhou-Chongzuo}} \\ \dfrac{R_t}{V_{\text{Wuzhou-Liuzhou}}} + \dfrac{l_{\text{Liuzhou-Chongzuo}}}{V_{\text{liuzhou-Chongzuo}}} \le \dfrac{l_{\text{Liuzhou-Chongzuo}}}{V_{\text{Wuzhou-Chongzuo}}} \end{cases} \tag{7.1}$$

$$opt\, R_A = \text{MAX}(R_C, R_t)$$
$$s.t.\begin{cases} R_C \le \dfrac{C_{\text{Wuzhou-Chongzuo}}l_{\text{Wuzhou-Chongzuo}}}{C_{\text{Wuzhou-Liuzhou}}} - \dfrac{C_{\text{Liuzhou-Chongzuo}}l_{\text{Liuzhou-Chongzuo}}}{C_{\text{Wuzhou-Liuzhou}}} \\ R_t \le \dfrac{l_{\text{Wuzhou-Chongzuo}}V_{\text{Wuzhou-Liuzhou}}}{V_{\text{Liuzhou-Chongzuo}}} - \dfrac{l_{\text{Liuzhou-Chongzuo}}V_{\text{Wuzhou-Liuzhou}}}{V_{\text{Liuzhou-Chongzuo}}} \end{cases} \tag{7.2}$$

Equation (7.1) measures the influence radius R_C of Liuzhou determined by transportation cost in terms of the minimum transportation cost; and the (7.2) measures the influence radius R_t of Liuzhou determined by transportation cost in terms of minimum transportation time. The target function is to select the larger one from the freight influence radius R_C and the time influence radius R_t as the radius R_A of influence of Liuzhou.

If the influence radius is determined by the two adjacent entrances and exits on the Guangxi High-speed Railway, the radius can be infinite in theory. However, in reality, the traffic network between the entrances and exits on the Guangxi HSR has certain development limits, and some influence radius can be calculated from some main traffic lines (Liuzhou-Chongzuo). The calculation of the influence radius of different entry and exit points on the high-speed railway may form a stripping-shape of the sphere of Guangxi high-speed railway below (Fig. 7.6).

Assuming that the influence radii of n entrances and exits are measured, the average radius of Guangxi high-speed rail Economic belt is:

$$\overline{R} = \frac{1}{n}\sum_{i=1}^{n} R_i \tag{7.3}$$

Fig. 7.6 A strip of areas where a high-speed railway may exert its influence

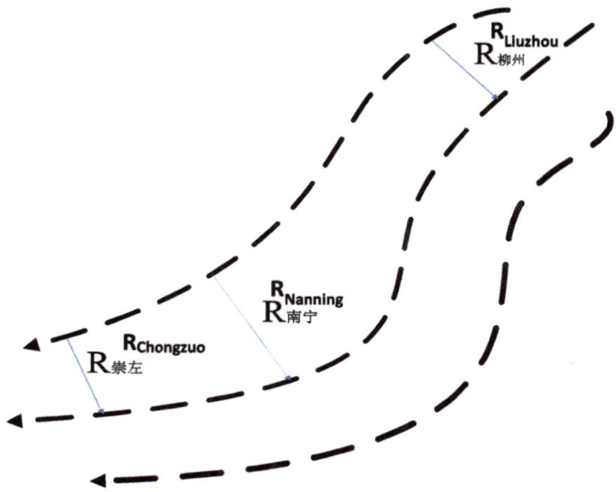

The influence radius of an entrance and exit of the Guangxi high-speed railway calculated by Eq. 7.3 depends on the condition of the traffic branch line between the other entrance and exit and the two entrances. Therefore, along the positive and negative directions, two different sequences of inlet and outlet influence radius should be calculated, which may be $(R_{11},\ R_{12},\ \ldots,\ R_{1\mathrm{n}})$ and $(R_{21},\ R_{22},\ \ldots,\ R_{2\mathrm{n}})$. Since the high-speed railway is generally closed, and the development of the traffic branch lines on both sides of the same entrance and exit is quite different, the influence radius on both sides will be different. Considering the influence radius on the other side of the high-speed railway, we may have two sequences of influence radius like this: $(R_{31},\ R_{32},\ \ldots,\ R_{3n})$ and $(R_{41},\ R_{42},\ \ldots,\ R_{4n})$.

In this way, four sequences of influence radius can be obtained through careful calculation. The precise average influence radius of high-speed railway is shown as Eq. 7.4:

$$\overline{R} = \frac{1}{4n}\sum_{i=1}^{4}\sum_{j=1}^{4} R_{ij} \tag{7.4}$$

It needs to be understood that the influence radii of the entrances and exits of Guangxi high-speed railway are different because of the development of the surrounding traffic network and the natural conditions, which shapes the stripping area. This makes it difficult to systematically and accurately define the high-speed rail economic belt. On account of the fact that the influence radius of high-speed railways measured in different directions (here refer to different entrances and exits) and different traffic branches on both sides is different, we should make choices based on our research needs to examine the actual impact of a high-speed rail on the regional economy. With the expansion of the space of the high-speed rail economic belt, the influence radius of the entrance and exit is also extending, and the average radius of the formed high-speed rail economic belt will also be growing.

An area, through which high-speed rail lines pass, usually has already had a certain level of development, especially the city where the starting point of a HSR line is located. In addition, the transportation lines passing through the region are also relatively more advanced. That is to say, the construction of a high-speed railway will integrate existing resources, and promote economic cooperation between regions along the line and accelerate the formation and development of the economic belt. The following is an analysis of the formation and development mechanism of the economic belt of Guangxi high-speed railway. Like other transport economic belts, the Guangxi high-speed rail economic belt has its own life cycle of formation and development, which includes four stages: start-up, growing, stabilizing, and maturation and reconstruction. The main development characteristics and industry types at each stage are as follows: The first stage is dominated by high-tech industrial parks along the route, so the new industry is the main feature. In the second stage, with the improvement of high-tech parks, subsidiary industries and facilities, the service industries such as trade, logistics, catering and exhibition start to move. In the third stage, tourism and other industries in the tertiary industry started to grow. The fourth stage is the maturity and reconstruction period when some new lines enter around or along the route.

Stage ONE: Start-up (before the completion of the high-speed rail)

(1) Some areas along the Guangxi high-speed rail economic belt have been economically better off. Among them, cities like Nanning (the provincial capital city), Liuzhou, Guilin, which are rich in resources, are the main ones. The interaction between regions in this economic zone is mainly reflected in the agglomeration of these central cities on the surrounding areas. Due to the favorable geographical advantages and conditions, the central city headed by Nanning harbors the manpower, capital and resources of the nearby areas for its own development, which does not have a diffusing effect on the surrounding areas. This may cause a large gap in the level of economic development among regions within the economic belt.

(2) The industrial structure of each region in the economic belt will be optimized and upgraded according to its own advantages or endowments. For example, Guilin, which is rich in tourism resources, may take tourism as the priority for its economic development, and so its industrial structure is relatively one-dimensional. In the more developed cities, Nanning, which is the central city of the economic belt, may give more weight to the tertiary industry. Other cities in the economic zone may be devoted to advantageous industries and diversify their industries with their own characteristics. These development choices may be grounded on the characteristics of the regions on the high-speed rail line.

(3) In the initial period, there are fewer spatial connections between cities in each economic zone. Despite the relatively improved traffic conditions along the high-speed railway, the frequency of urban or regional interactions in the initial stage is still very low. And the spatial connection is mainly based on the simple circulation of people and goods, which will not affect the industrial structure of the economic belt itself.

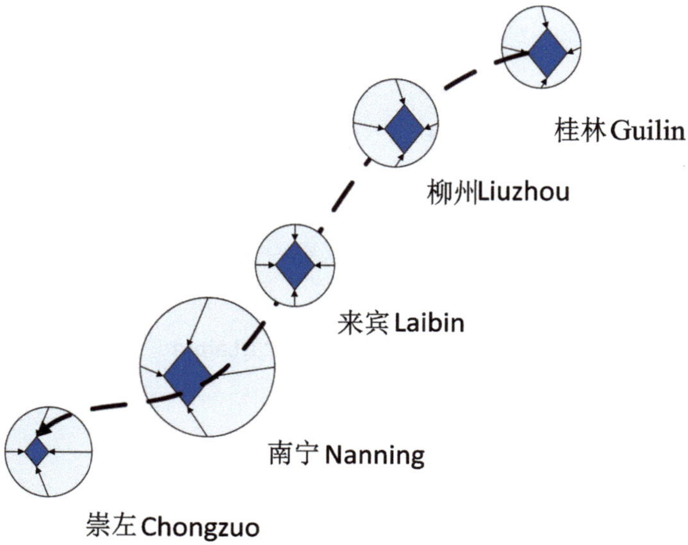

Fig. 7.7 Guangxi high-speed rail economic belt in the start-up period

(4) There is some gap between the railway network and the social development of the economic belt in the region. First, the scale of the railway network is insufficient and the level of technical equipment is low. The network density in the region is 136 km/10,000 km^2, ranking 20th in China. Among them, the ration of the railway double-track and electrification are 14 and 24%, respectively, far lower than 41 and 47% in China. Second, there are few main corridors in the area, and the transportation capacity of the main entrances and exits is very insufficient. For example, the transport capacity between the port and the railway can only meet 60% of the market demand, because of limited rail network on the Xiang-Gui line, Qian-Gui line and Nan-Kun line (Fig. 7.7).

Stage TWO: Growth (a time when there is no stable passenger flow after the completion of the high-speed rail)

(1) The completion of the high-speed railway means a growing transport capacity. As time goes on, the agglomeration effect of central cities (such as Nanning in Fig. 7.8) or regions in the original economic belt will be more obvious, as the accessibility between regions along the line is greatly improved, and more and more good resources are crowded around the central cities of the economic belt. In addition, while gathering in the central city, it also began to spread to the surrounding areas. The industrial structure is gradually diverging in the economic belt, and so advantageous industries will form in different regions. Generally, the central cities focus on developing the tertiary industry, while the secondary industry will shift outward, which, for example, moves to smaller cities or regions near Nanning, and the primary industry will be transferred

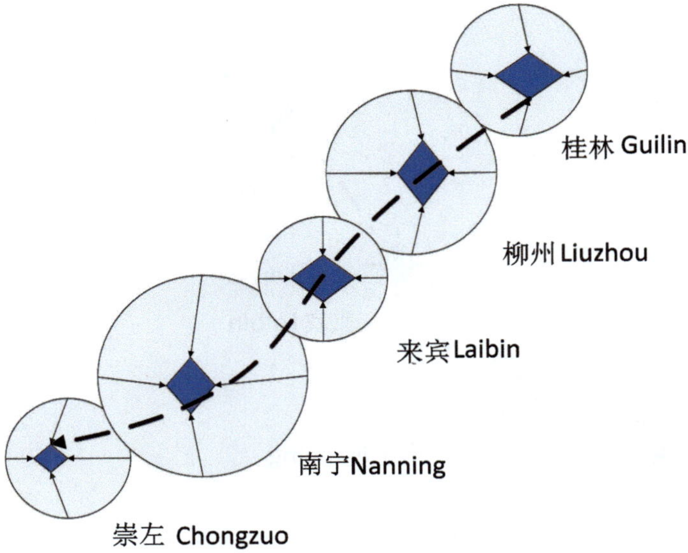

Fig. 7.8 The Guangxi high-speed rail economic belt during the growing period

to the marginal areas of the economic belt. By then, the division of industrial structure within the economic belt has initially taken shape, and the number of major towns and population along the line has rapidly increased.

(2) High-speed railway shortens the spatial distance between cities in the belt and strengthens the exchanges within the economic belt, especially the exchange of knowledge and technology with talents as the carrier. High-speed railway itself is a high-tech industry, which will drive the development of high-tech, creating new economic growth points for areas along the route. The construction of knowledge-intensive industrial parks and high-tech industrial parks (for example, Liuzhou, etc.), in particular, will promote the upgrading of industrial structure along the route.

(3) With the high-speed rail, the space distance is shortened, and the scope of economic radiation is also expanding. This not only provides a broader market space for each region, so that it prioritizes on developing its own advantageous industries, strengthening the cooperation and complementary advantages between regions, which makes the industrial division between cities or regions in the belt clearer. In the meantime, the relatively closed concept of "circle" in the existing economic belt is weakening.

(4) High-speed rail is very attractive to the surrounding passenger traffic. The high speed, convenience and comfort on the Guangxi high-speed rail will have a heavy impact on and may re-divide the passenger transport market in the region. For one thing, high-speed rail brings new passenger flows, and drives the rise and development of knowledge-intensive and technology-intensive industries along the route, as well as the massive flow of talents. With the construction

of high-speed rail, tourism resources along the line (such as Guilin) will be activated, which will also become a major source of tourists for high-speed rail. For another, the opening of Guangxi high-speed railway, because of its obvious comparative advantages, will also attract some travelers from other modes of transport, especially air tourists.

Stage Three: Toward stability (A time when there is stable passenger flow)

(1) The agglomeration and diffusion effect in the economic zone alternate with each other. Compared with the second stage, Guangxi high-speed rail economic belt has created more significant diffusion effect, which is spreading in width and depth. Besides, as the focal point of the economic belt, Nanning (the most important central city) has a stronger impact. In addition to the traditional trade exchanges, the high-speed rail economic belt will be more embodied in technical exchanges and cooperation. At the same time, as the cities and regions within the belt have basically completed the regional division of industrial structure, the economic gap is narrowing down, but the size of cities and towns keep growing rapidly.

(2) The development of Guangxi high-speed rail economic belt driven by talents and technology is gradually maturing. The allocation of related industries in the belt will be complete one after another, while the secondary and tertiary industries will be the main industries in the high-speed rail economic belt. Moreover, the infrastructure construction and supporting services around the economic belt will also be improved in this stage.

(3) The division of functions and connections of some cities in Guangxi high-speed rail economic belt will be clearer. Satellite cities of some central cities are constantly emerging. The diffusion effect makes the suburbanization of the capital city like Nanning more widespread. Meanwhile, urban continuous areas within the economic belt will also begin to appear. The size of the economic belt keeps growing. In this stage, the concentration of the population in the surrounding areas to Nanning and other central cities will gradually decrease. With the industrial transfer, the population will gradually shift to the satellite cities or suburbs of Nanning, which also reduces the population pressure of central cities like Nanning. Nanning's role will change to service-oriented. As the financial, information and trade center in the economic zone, Nanning will be more professional and powerful in providing services. Because the rapid economic development in the economic belt will require Nanning to take on this role, that is, as the "heart" of the economic belt. At the same time, the further clarification of industrial division within the economic belt will strengthen the service function of Nanning.

The positioning of the high-speed railway in the transportation market within the economic belt is basically fixed. Varied modes of transport within the economic zone have completed the redistribution of the market, obtained their own stable customers, brought their own advantages into play, and formed a good relationship of cooperation and cooperation within the economic zone.

Stage Four: Maturity and Reconstruction (with new routes coming into the economic belt from the neighborhood)

In this stage of Guangxi high-speed rail economic belt, the original urban or regional boundaries will become increasingly blurred, and a continuous urban area with high-speed railway as the main road will be formed. In the continuous urban areas or the high-speed rail economic belt, cities of different sizes are intertwined, in which the industrial structure is a well-proportioned patchwork and there is a clear division of labor among different regions. The economic development of each city or region in the economic belt has its own advantages and characteristics. With a narrowing economic gap, the effect of agglomeration will gradually decrease, and the population will be evenly distributed in the whole economic belt. At the same time, due to the growth of the economic belt, the transportation capacity of the high-speed rail line will not be able to meet its development needs, and new modes of transport are needed, which will break the internal balance of the present economic belt. Consequently, the resources in the economic belt will be re-integrated, bringing new vitality to the development of the economic belt. And then, the economic belt will go through the initial stage, development stage and stabilization stage again. By then, Guangxi is on its way to have a new economic belt based on the present high-speed rail economic belt.

Under the background that the world economic center is shifting from the Atlantic to the Pacific Rim, the Asia–Pacific economic circle will become the most active region in the world economy in the twenty-first century. Guangxi is in the core position of Asia–Pacific economic circle, and it has natural geographical advantages (Because it is bordering Beibu Gulf, close to Hong Kong, Macao and Southeast Asia, it has the natural advantages to develop import and export trade and to build a coastal industrial base). In order to catch such trend of economic development and make full use of Guangxi's natural location advantages, high-speed rails may play a key role in the process of boosting Guangxi into the core of Asia–Pacific economic circle. Meanwhile, the construction of Guangxi high-speed railway can also transform Guangxi's location advantages into obvious economic comparative advantages, as it is located in the docking area of southwest China and southeast Asia economic circles. In the big market for more than 700 million people, Guangxi has obvious location advantages and economic comparative advantages. However, over the years, the main reason why such advantages have not been translated into economic achievements is the lack of support from efficient transportation network as external thrust. Therefore, the construction of high-speed railway and the formation of its economic belt will help to make up for this defect. The construction and operation of Guangxi high-speed railway will change the traffic conditions inside and outside its jurisdiction, enhance its accessibility and circulation between cities inside and outside the region, and make the material and talent exchange in the whole region more convenient. In addition, it also saves a lot of time. Accordingly, Guangxi should grasp the preferential policies granted by the Chinese government in recent ten years, so that tourism, logistics and foreign trade relying on high-speed rail may gradually become the economic pillar and growth point of Guangxi.

Table 7.1 Ticket prices for high-speed railways in operation

	Mileage (km)	Second-class fare (RMB)	Fare (yuan/km)
Beijing-Tianjin intercity railway	120	55	0.46
Wuhan-Guangzhou high-speed railway	1069	465	0.43
Beijing-Shanghai high-speed railway	1318	555	0.42
Beijing-Guangzhou high-speed railway	2298	862	0.38

In short, the high-speed rail economic belt is the result of industrial transfer from developed areas to underdeveloped areas. Therefore, the new "high-speed rail economic belt" is more likely to appear along the high-speed rail connecting developed and underdeveloped areas. According to the above case analysis, it is found that the construction and operation of high-speed railway can help the formation of economic belt in the following aspects:

First, time shortens. According to the current speed of Wuhan-Guangzhou high-speed railway, the distance between Wuhan and Guangzhou is 1,069 km. If the speed of the high-speed railway is 300 km/h, the direct travel time is about 3 h and 30 min. If the multi-stop shift is selected, the travel time is about 3 h and 50 min, and the speed is 280 km/h.

Second, the cost of transportation is reduced. For a closer look, refer to Table 7.1 for the high-speed rail fare. Basically, it should not be less than 0.45–0.5 yuan/km, while the benchmark air fare is 0.75 yuan/km, but it is allowed to rise by 25%. Considering the distance gap, on the whole, the high-speed rail fare is equivalent to about 50–60% of the total air fare. Or rather, considering the discount of air ticket itself, the air ticket price in China is almost 1.5 times that of high-speed rail (equivalent to that of Shinkansen in Japan). Therefore, the opening of high-speed railway will greatly reduce people's travel cost. In addition, at the same distance, the comfort of high-speed rail is also very competitive.

Third, high-speed rail promotes the dissemination of technical knowledge. High-speed rails connect almost all economically developed cities, which are the rallying point of knowledge and technology. The construction of high-speed rail network greatly enlarges the sphere of influence of these cities, thus driving the upgrading of industrial structure and economic development of cities along the line.

Forth, the development of high-speed rail may give birth to the business model of "Pre-sales and Post-production", which seems more feasible. To be specific, this model implies that economically developed regions focus on marketing and R&D, while economically underdeveloped regions take the responsibility for production activities.

Fifth, high-speed rail can bring development opportunities for featured tourist cities. The increase of passenger flow will also bring new development opportunities for cities with unique (for example, tourism) resources along the route.

Sixth, high-speed rail helps to release freight capacity. With the completion of the high-speed railway, passengers and cargos has been separated on the lines that have met necessary conditions. According to relevant data, the suspension of one passenger train will release the freight capacity of two freight trains. As far as a single railway is concerned, the separation of passenger and freight can improve the freight capacity by about 80–200%.

Lastly, freight costs are reduced. With the gradual release of the existing railway transportation capacity after the completion of the high-speed railway network, the railway will replace some long-distance highway freight transport with its advantage of low-cost. This will reduce the logistics cost along the high-speed rail, which is conducive to industrial transfer.

Generally speaking, the completion of high-speed rail will dilute the peak of population and economic density in the original economic belt. The attraction radii of cities along the route will gradually increase, which will help to narrow the gap of economic development level between regions or cities along the route, and better allocate resources within the economic belt.

7.3 High Speed Railway and Optimization of Regional Structure

High-speed railway construction can stimulate domestic demand, increase employment and promote the upgrading of industrial structure, and also the development of related industries such as machinery, metallurgy, construction and synthetic materials. High-speed railway is a new passenger railway system on the base of modern comprehensive science and technology, which will promote the research and development of new materials and information industries, and the technology and product quality of related industries to a higher level. High-speed railway also brings a large number of people, capital and information flows to the areas along the way, which provides rich resources of production factors for the optimization and upgrading of industrial structure.

Industrial structure refers to the proportion of the various sectors of the national economy, including the total industrial output value, added value, fixed assets investment, capital investment and employment. The working mechanism of high-speed railway on regional industrial structure is shown in Fig. 7.9.

Generally speaking, high-speed railway, as a passenger railway, has its own characteristics in its influence on major industries in the economic belt. The competitive advantages of high-speed rail are mainly manifested in four aspects: rapidness, punctuality, comfort and safety. The speed of high-speed railway is 250 km/h, which is more than three times faster than that of highway and nearly twice that of ordinary railway. For example, the Nanning-Guangzhou high-speed railway reduced the time from Nanning to Guangzhou to less than 3 h, saving about 11 h compared with the freight time of the current railway. In terms of convenience, compared with aviation,

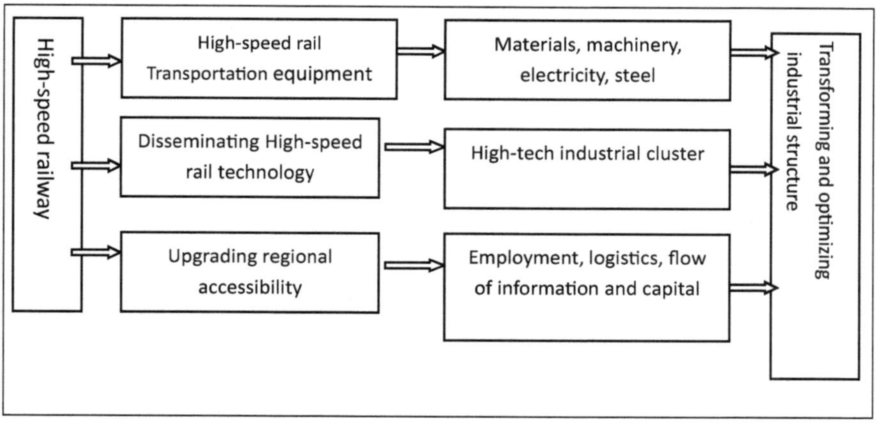

Fig. 7.9 The working mechanism of high-speed rail on industrial structure

the location of the high-speed railway station is closer to the urban area. In terms of punctuality, civil aviation and highway transport are more likely to be affected by weather factors. In addition, ordinary trains on existing lines are severely restricted by scheduling, which are the shortcomings that high-speed trains can make up for, because high-speed trains start and arrive on time. Compared with other modes of transport, high-speed rail has fully demonstrated its advantages in comfort and cost performance. At a moderate price, passengers can enjoy the comfortable and quiet environment and in-place service in the carriage in a shorter travel time. The above advantages of high-speed rail passenger transport have a greater impact on other industries in the economic belt, boosting related industries and economic growth.

There will be a couple of major industries affected in the economic belt. The opening of the high-speed rail will promote the development of tourism along the route by influencing tourists' choice of transport. In addition, the advantages of speed and comfort will affect the real estate industry in the economic zone. Due to the increasing convenience of people-to-people communications or flows of people, the price of real estate industry in the economic zone will tend to converge to a similar level. In addition, as far as Guangxi high-speed railway is concerned, the mixed mode of passenger and freight is the main mode of Guangxi high-speed railway transportation. So, with the high-speed rail freight, the impact of the high-speed rail on the industries in the economic belt will also be directly reflected in the logistics industry.

7.3.1 High-Speed Railway and Regional Tourism: A Case Study of Guangxi High-Speed Railway

Located on the southwest edge of China, Guangxi is a western province with an early start in tourism, with rich and diverse tourism resources. Its tourism industry has basically formed four major tourist areas with different characteristics: northern Guangxi, southern Guangxi, eastern Guangxi and western Guangxi. Its popularity at home and abroad has been significantly improved. In 2009, the number of inbound tourists in Guangxi was 2,908,500 ranking tenth in the country, second in the western provinces and second only in Yunnan province. In 2010, it received 1,007,300 domestic tourists and 252,300 inbound tourists, including 1,008,400 compatriots from Hong Kong, Macao and Taiwan and 1,413,800 foreigners. Among foreign visitors, Vietnamese is the most, followed by Malaysia, the United States, France and Japan. Foreign tourists mainly came from some developed countries in Europe and the United States and Southeast Asian countries.[3] In 2011, tourism revenue in Guangxi exceeded 100 billion yuan, with total tourism revenue of 127.78 billion yuan, accounting for nearly 11% of GDP.

Tourism resources are the prerequisite for developing tourism. Guangxi high-speed railway covers most areas of Guangxi, where tourism resources are very rich, with unique charm in natural landscapes, ethnic customs, coastal scenery, border expeditions, red tourist attractions[4] and so on. According to the "Opinions on the Integrated Development and Construction of Key Tourism Resources in Guangxi Zhuang Autonomous Region",[5] the overall framework of tourism development in Guangxi is "one tourism leader, two golden tourism belts, two major tourism distribution centers, five tourism characteristics, six tourism brands, eight tourism districts and ten top-quality tourist routes".

Specifically, (1) give full play to the leading role of Guilin, and improve the two tourist belts of Guilin-Hezhou-Liuzhou-Hechi-Laibin-Nanning-Qinzhou, Fangchenggang (Dongxing)-Beihai and Wuzhou-Yulin-Guigang-Nanning-Baise-Chongzuo. (2) Build and improve two major tourist distribution centers in Guilin and Nanning; (3) Highlight five major tourism features: natural landscape, coastal leisure, border resorts, ethnic customs and red tourism; (4) Make great efforts to build six major tourism brands: Guilin Landscape, Detian Waterfall, Beihai Silver Beach, Baise Tiankeng, ethnic customs and red tour; (5) Comprehensively promote

---

[3] According to *China Statistical Yearbook 2011* and *Guangxi Statistical Yearbook 2011*.

[4] *The Outline of the National Red Tourism Development Plan 2004–2010* emphasizes that red tourism refers to the thematic tourism activities organized to receive tourists to carry out Memorial study and visit, taking the memorial sites and symbols formed by the Communist Party of China during the revolutionary war as the carrier, and taking the revolutionary history, deeds and spirit it carries as its connotation. Developing red tourism is an important measure to implement the spirit of the Fourth Plenary Session of the Sixteenth Central Committee of the Communist Party of China and to continuously improve the ability of building socialist advanced culture.

[5] *Opinions on Integrated Development and Construction of Key Tourism Resources in Guangxi Zhuang Autonomous Region* [No. 112 Official Document in 2005].

the planning and construction of eight major tourist areas, such as Grand Guilin Landscape Culture Tourist Area, Nanning Zhuangxiang Greentown Exhibition Business Tourist Area, Detian Transnational Waterfall Tourist Area, Beibu Gulf Rim Coastal Transnational Tourist Area, Baise Grande Tiankeng Group Tourist Area, Dayaoshan Ecological Folk Culture Tourist Area, Eastern Guangxi Religious Historical Culture Tourist Area and the Left and Right River Red Tourist Area[6]; (6) Integrate and build high quality tour for Guilin landscape, coastal leisure tour around Beibu Gulf, border tour of Detian transnational waterfall, customs tour of "Liu Sanjie", exploration tour of Dashiwei Tiankeng Group, religious history and culture tour of Guidong (eastern Guangxi), footprint tour of Deng Xiaoping, footprint tour of Sun Yat-sen's Northern Expedition, ethnic customs tour of Zhuang, Yao, Miao, Dong and Mulao, and cross-border tour of China and Vietnam.

Under the guidance of relevant government policies, the pace of tourism development in Guangxi is gradually accelerating. By 2010, Guangxi had 2 5A scenic spots, 59 4A scenic spots and 44 3A scenic spots. At present, the high-quality tourist attractions that have been launched include the Great Guilin Landscape Cultural Leisure Resort with Guilin as the center, the Nanning Business Exhibition Green Capital Cultural Tourism Zone, the Binhai Tourism Zone with Beihai Silver Beach as the main part, the Dashiwei Tiankeng Group Tourism Zone in Leye, the Detian Cross-border Waterfall Tourism Zone, the historical and cultural tourism zone of religious sites in Guidong, the eco-cultural tourism zone of Hezhou Shanshui Ancient Town, the Zhuang Yao Miao Dong ethnic customs eco-tourism zone with Liuzhou as the center, the "Three Saints" tourism zone of Laibin and Jinxiu, and the southern Border Customs Tourist Area with Pingxiang and Jingxi as the Focus. Among them, Lijiang River in Guilin, Xishan in Guiping and Huashan in Ningming are listed as national scenic spots, and Silver Beach in Beihai as a national tourist resort.[7]

The scale and structure of tourism demand have a certain influence on the competitiveness of tourism, and tourism demand is related to the level of per capita disposable income in a region. That is to say, the higher the per capita disposable income, the greater the demand for tourism. However, Guangxi is an economically underdeveloped region, where its tourism demand is still very low and at the level of massive sightseeing, which is not conducive to the development of personalized and high-level tourism products and the whole tourism industry chain. Therefore, it is not enough for Guangxi's tourism industry to rely solely on the regional tourism market. Instead, it should expand to other parts of China and the international market to win more tourists.

---

[6] As one of the revolutionary bases led by the Communist Party of China during the Agrarian Revolutionary War, the Left and Right Rivers area is located in the southwest of Guangxi Zhuang Autonomous Region, with Baise as the center, including the Zuojiang River Basin and the Youjiang River Basin. The Zuojiang River is located in the southwest of Nanning City and the Youjiang River is located in the northwest of Nanning City. The two rivers flow in a "V" shape to the west of Nanning and converge to form the Yongjiang River.

[7] Current development in Guangxi, accessed through http://www.gxtj.gov.cn/read.asp?Typid=45 (on March 20, 2009).

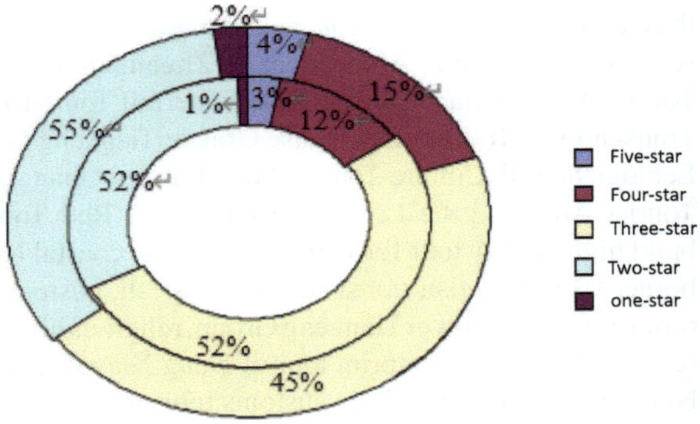

Fig. 7.10 2010 Star hotels in Guangxi and China

Tourism is a highly dependent industry, which not only relies on the unique natural tourism resources, but also has direct or indirect relations with many industries in the national economy. As far as Guangxi is concerned, the industries closely related to tourism include catering and accommodation, industries of cultural creativity, and the convention and exhibition industry.

In 2010, there were 425 tourist star hotels in Guangxi, including 13 five-star hotels, 52 four-star hotels, 221 three-star hotels, 136 two-star hotels and 3 one-star hotels. As shown in Fig. 7.10, among the star-rated tourist hotels in Guangxi, three-star and two-star budget hotels account for 84% of the total, and the national similar data is 78%. This suggests that the customers received by tourist hotels in Guangxi are mainly mid-end customers, and the development of high-star hotels is insufficient. Moreover, there are no five-star hotels in Wuzhou, Guigang, Fangchenggang and many other cities.

Tourism is inseparable from "food", which plays an important role in the tourism industry. Since the first Guangxi Ethnic and Local Cuisine Competition in 2006, the "Guicai" or Guangxi cuisine has been developing from scratch. In 2009, Guangxi cuisine was officially recognized. In 2010, Guangxi Ethnic and Local Cuisine Competition was renamed Guangxi Cuisine Tasting Competition.[8] According to the data in Table 7.2, the turnover of catering industry in Guangxi in 2010 was 2.036 billion yuan, less than 1/5 of the national average, indicating that the overall development of catering industry in Guangxi is low.

Guangxi has unique advantages in developing flower industry. First, Guangxi has superior climatic conditions, with different types of climates from subtropical to tropical, and the warm climate and abundant rainfall. Second, there are very rich varieties of flowers, with more than 7,200 plants, ranking third in China. Third, Guangxi has relatively rich land and cheap labor resources, which has certain advantages compared with developed domestic provinces and cities. However, due to the fact that the information and market system of the flower industry in Guangxi have not yet

[8] Yesterday, Today and Tomorrow of Guangxi Cuisine. Accessed on March 20, 2010. http://ngzb.gxnews.com.cn/html/2010.12/15/content_473466.htm.

Table 7.2 Accommodation and catering industry in Guangxi and China in 2010

	Guangxi	National	National average
Turnover of accommodation industry (RMB 100 million)	47.19	2797.80	90.25
Turnover of catering industry (RMB 100 million)	20.36	3195.10	103.07
Number of legal person enterprises in accommodation industry (individual)	389	15,713	507
Number of corporate enterprises in catering industry (individual)	192	21,595	697

been formed, the flower production and management activities lack market information guidance, which leads to blindness (Zhu, 2007). In addition, flower consumption belongs to a higher level of consumption, which is related to the degree of economic development and people's consumption beliefs. Guangxi may attract more tourists from other places, especially from more developed areas, by holding some flower fairs, which will greatly broaden the market and bring more industrial benefits.

Convention and exhibition industry is a new service industry and a bridge for trade, science and technology and cultural exchanges. Exhibition industry is also an open industry, which can promote the development of related industries such as transportation, accommodation, catering, communication, travelling and shopping, and in particular, tourism. In 2004, the China-ASEAN Expo (CAEXPO), sponsored by the economic and trade authorities of China and 10 ASEAN countries and hosted by the People's Government of Guangxi Zhuang Autonomous Region, settled in Nanning. And Nanning has become a permanent venue, which is an important milestone in the development of Guangxi exhibition industry. In addition, China (Guilin) International Tourism Expo, China (Yulin) Small and Medium-sized Enterprises Business Opportunity Expo, China-ASEAN Business and Investment Summit, Guilin International Tourism Forum, China-ASEAN Automobile Fair and other exhibitions have gained strong popularity and influence both at home and abroad.[9] Although the exhibition market in Guangxi has seen expanse, compared with the developed provinces, the exhibition industry in Guangxi is still in its infancy. It is necessary to improve the exhibition quality and expand its influence by attracting more tourists and exhibitors.

Different from traditional natural landscape tourism, cultural and creative tour is a development model of tourism with culture as the core, creativity as the means, technology as the support and market as the guide, which is very beneficial to creating diversified tourism product carriers and forming industrial connections (Wang 2010). Creative tourism industry is of great significance to enrich the tourism product system, optimize the structure of traditional tourism industry and promote the development of tourism economy.

In recent years, Guangxi has integrated many of its advantageous resources such as ethnic cultural resources and tourism resources to accelerate its cultural tourism

---

[9] Opinions of the People's Government of Guangxi Zhuang Autonomous Region on Accelerating the Development of Guangxi Convention and Exhibition Industry-Gui Zhengfa [2010] No. 65.

industry. Impression–Liu Sanjie, a large-scale live performance that began to be staged in Lijiang River in 2003, is the first "landscape live performance" with a brand-new concept in China, which has achieved a win–win situation in terms of income and social benefits. In addition, Nanning International Folk Song Art Festival, a comprehensive large-scale activity integrating culture, tourism, economy and trade, has achieved considerable economic benefits. With its popularity at home and abroad, Guangxi is made known to the world increasingly.

In Guangxi, the cultural and creative tourism industry has a good momentum of development, but there is still a significant gap compared with developed regions in terms of industrial scale and structure, management and market competition, which is characterized by insufficient development of cultural resources and a relatively small scale. Therefore, there are many potential cultural and creative tourism resources available, such as Danu Festival of Yao nationality, Lusheng Festival of Miao nationality, Fireworks Festival of Dong nationality and marine cultural resources of Beibu Gulf.

As most of the existing tourism resources in Guangxi are covered by high-speed railway, the opening and operation of high-speed railway in Guangxi will have a great impact on tourism and related industries in the economic zone. Specifically, the development of high-speed rail in Guangxi may result in the following changes:

(1) Changing the spatial pattern of tourist market. The construction of high-speed railways will change the traffic structure in Guangxi, involving more than 46 million people in the whole region. For example, Nanning-centered city circle makes Guangxi a "123-h city economic circle". Namely, the "one-hour urban economic circle" in Guangxi Beibu Gulf Economic Zone (Nanning, Beihai, Qinzhou and Fangchenggang), the "two-hour urban economic circle" in major cities in the region and the "three-hour urban economic circle" in neighboring capital cities such as Changsha, Kunming, Guangzhou and Haikou. This change will greatly reduce the perceived distance of the area where the high-speed rail passes, and transform the transportation distribution. Whether it is inbound tourism from Southeast Asian countries or inter-provincial tourism from Guangdong, Hunan, Guizhou and other places, the travel time will be greatly shortened. Due to the improvement of accessibility of tourist attractions in Guangxi, and the abundant tourist resources in Guangxi, on the one hand, it will divert some tourists from neighboring Hunan, Xiamen, Hainan and Lijiang; on the other hand, more tourists will be attracted to Guangxi.

(2) Changing tourists' consumption habits. As the high-speed railway greatly shortens the travel time, passengers can move long distances in a short time. The reduction of travel time brought by high-speed rail may lead to an increasing number of one-day and two-day visitors, and boost the weekend tour market. Meanwhile, tourist destinations will change. Tourists will tend to prefer those larger, well-known, distinctive and accessible tourist areas, and gradually lose interest in relatively small tourist areas with lower service quality, which may speed up the polarization between tourist attractions, and highlight the importance of traffic conditions to such polarization (Zhang et al. 2011). Relevant data

show that in the Spring Festival of the first year after the Wuhan-Guangzhou high-speed railway was opened to traffic, Wuhan received 351 tourist groups in a time of only seven or eight days, with a year-on-year increase of 160% and a 15.5% increase in tourism revenue. Also, after the Beijing-Tianjin high-speed railway was opened to traffic, Tianjin's tourism revenue increased by 35%.

(3) Promoting investment and competitiveness in tourism. After the completion of the high-speed railway network, there will be more active business investment opportunities, which creates favorable conditions for tourism, hotels, restaurants and other enterprises in developed areas to invest in Guangxi. Due to the remoteness and inconvenient transportation in some areas of Guangxi, many tourism resources have not been fully developed. Many other scenic spots are not well-known except Guilin, which is well-known in China. For example, although Baise, Pingxiang, Hechi and other places in western Guangxi are endowed with abundant tourism resources and ethnic customs, their tourism service facilities and tourist products are underdeveloped and lack of characteristics due to insufficient capital investment. As time goes on, the tourism market in the more developed areas is increasingly saturated. It will be a wise choice for investors to come to Guangxi to look for investment opportunities. This is possible and realistic with the completion of the high-speed rail, which will bring more tourism investment opportunities for investors and is conducive to the great stride development of Guangxi tourism.

(4) It is helpful to realize the complementary advantages of tourism resources. The complementarity of tourism resource refers to the complementary and interdependent relations between tourism resources resulting from their differences in terms of category, scale, shape, cultural connotation, time series and location (Zhang and Xue 2010). Cooperation among regional tourism economies may greatly integrate tourism resources, thus generating greater economic benefits. In the 12th Five-Year Plan of tourism development made by Guangxi Zhuang Autonomous Region, it is pointed out that we should focus on building two tourism development zones, namely, the east–west, the north–south zone; That is, Golden Waterway Tourist Belt along the Xijiang River and the North–South Golden Tourist Belt of "Guilin–Liuzhou–Laibin–Nanning–Beihai (Qinzhou, Fangchenggang)".[10] The construction of Guangxi's high-speed railway is conducive to regional tourism cooperation within Guangxi and between Guangxi and its neighboring provinces. For example, the links of Liuzhou-Nanning HSR, Jingchengjiang-Nanning HSR and Nanning-Qinzhou HSR have realized the joint from North to South, and Nanning-Kunming HSR and Nanning-Guangzhou HSR have realized the connection between East and West, whereby the structure of coordinated development of regional tourism is formed. Nanning-Guangzhou and Nanning-Kunming high-speed railways also help to connect Yunnan, Guangxi and Guangdong. They passed through

[10] The 12th Five-Year Plan for Tourism Development in Guangxi Zhuang Autonomous Region (issued jointly by the Guangxi Development Reform and the Tourism Bureau of Guangxi Autonomous Region).

western Guangxi, southern Guangxi, eastern Guangxi and western Guangdong, with rich ethnic customs and beautiful natural landscape, fully integrating the tourism resources of the three provinces and forming a golden tourism line.

(5) Transforming the advantages of tourism location into economic advantages. Guangxi Province is located in the southwest with a population of more than 200 million, and accessible to the ten ASEAN countries with a population of more than 500 million. Its neighboring provinces and regions like Guangdong, Hong Kong and Macao are its important tourist sources, and Hainan across the sea and its neighbors like Yunnan and Guizhou are all major tourist provinces. Although Guangxi is at the junctions of Southwest Tourism Circle and Southeast Asia Tourism Circle, and has super good location advantage of tourism, its location advantage does not bring high tourism competitiveness. One of the most important reasons is that there is no efficient transport network as external thrust. So, as far as Guangxi's tourism industry is concerned, the first thing to consider is the convenience of tourists entering the region. With its high speed and high efficiency, the high-speed railway may help to improve the strategic and coordinated development of tourism in cities along the route, adjust the tourism location advantages of cities along the route (Liang et al. 2010), and turn them into realistic tourism economic advantages. If we can seize the rare opportunity of "high-speed rail era", Guangxi tourism will achieve more economic benefits.

(6) Promoting the development of tourism-related industries such as floral industry, exhibition industry, and cultural and creative tourism. Because of the high speed of the high-speed rail, more people may visit Guangxi to enjoy flowers and various exhibitions, thus boosting flower industry, exhibition industry, cultural and creative tourism and other industries. In addition, most of Guangxi's high-speed rails adopt the mixed mode of passenger and freight, which enables more fresh flowers to be sold to other places in a shorter time, thus driving the development of the flower industry. Overall, the influence mechanism of Guangxi high-speed rail on tourism industry in economic belt can be shown in Fig. 7.11.

7.3.2 High-Speed Railway and Improving Regional Logistics: Taking Guangxi High-Speed Railway as an Example

Located in the center of the Pan-Beibu Gulf Economic Circle, Guangxi is adjacent to the Pearl River Delta in the east, Vietnam in the southwest, Yunnan-Guizhou-Sichuan in the west and northwest, Central China in the north and Southeast Asia in the south. It is the only province in China that has both land border and sea passage to connect China and the Southeast Asia. Guangxi has a coastline of nearly 1,600 km, which is the only minority autonomous region in China with advantages of coastal, riverside and border areas. Guangxi has three large coastal ports, Fangchenggang, Qinzhou and Beihai, which constitute the most convenient access to the sea in southwest China. In short, Guangxi has a very good location advantage to develop the logistics industry.

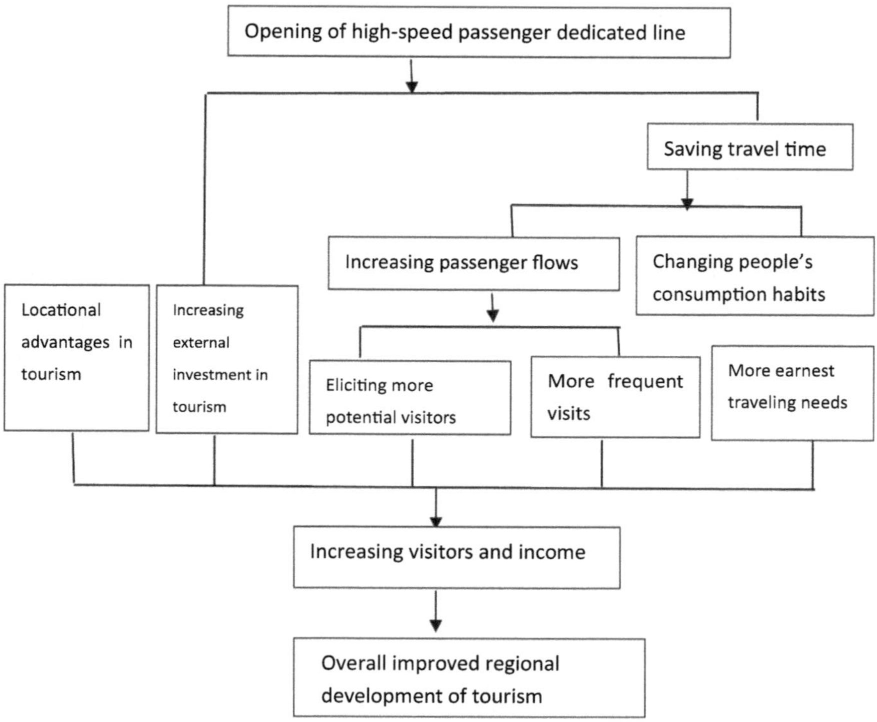

Fig. 7.11 Influence mechanism of Guangxi high-speed railway on tourism industry in the economic belt

According to Guo Fa [2009] No. 42 document, the regional development pattern of "two districts and one belt", namely, Beibu Gulf Economic Zone, Xijiang Economic Belt and Resource-abundant Areas was established. Accordingly, Guangxi actively optimizes the regional distribution of logistics, (1) focusing on building four major logistics areas: Beibu Gulf Economic Zone, Central Guangxi, Northeast Guangxi and West Guangxi, (2) building Nanning into a national logistics hub, and developing six regional logistics hubs, namely Liuzhou, Guilin, Beihai, Fangchenggang, Qinzhou and Chongzuo, and seven professional logistics centers, such as Wuzhou, Guigang, Yulin, Baise, Hezhou, Hechi and Laibin, and (3) establishing a number of logistics distribution centers, thereby forming a modern logistics system suitable for regional economic development.

In recent years, Guangxi Zhuang Autonomous Region has witnessed a rapid development of the logistics industry, with the construction of logistics infrastructure accelerated, the support capacity further improved, and the total amount of social logistics raised. In 2010, the total cargo transportation in Guangxi was 1.208 billion tons, an increase of 22.1% over 2009. The freight turnover was 309.713 billion ton-km, an increase of 23.3% over 2009. The added value of transportation, warehousing and postal services was 45.867 billion yuan, an increase of 16.9% over 2009. The cargo throughput at the port was 186 million tons, up by 25.0%, of which the foreign trade cargo throughput was 73 million tons, up by 22.0%. However, due to the underdeveloped logistics industry, natural environment and other factors, the social logistics cost in Guangxi accounts for more than 20% of GDP, and the logistics activities in industrial production account for 90% of the time, which are much higher than those in the eastern and central regions.[11] In 2008, Guangxi logistics enterprises and related industries have 130 special rail lines, nearly 20,000 sets of freight cars, special vehicles and container handling equipment of major logistics enterprises, and nearly 700 sets of logistics computer information management systems, with a self-owned storage area of about 7.5 million square meters. The scale, structure and service level of logistics enterprises have improved rapidly, and a number of comprehensive logistics and third-party logistics enterprises with industry characteristics have emerged.[12]

Indeed, Guangxi has very obvious location strengths in developing logistics industry. Despite its rapid development over the past few years, there are some things to improve, which are:

(1) Guangxi does not embrace very strong overall economic strength, but a lower level of industrialization and urbanization, few modern industrial enterprises, relatively week high-tech industries, and few large-scale backbone enterprises and central cities. Due to the lack of prosperous industrial economic support, there are some tricky weaknesses in logistics enterprises in terms of scale, management, service and means of enterprises, technical equipment and informatization, professional logistics talents, and knowledge of modern logistics. Take road transport enterprises as an example. There are 161,600 road transport operators in total. The total number of trucks is about 238,800, with a total tonnage of 655,800 tons. On average, each business owner owns only 1.48 trucks, with an average total tonnage of only 4.1 tons.
(2) The comprehensive transport system has not yet been established, with an underdeveloped supporting infrastructure of highway logistics and railway network. For example, there are few large tonnage vehicles, container vehicles, special transport vehicles and other vehicles with advanced performance equipped by transportation enterprises.
(3) Guangxi does not have high technology of management and informatization, and superior road transport enterprises in operation. Third-party logistics enterprises or comprehensive logistics enterprises are in earnest need of learning to make the best use of modern logistics management technologies, barcode technology, material purchasing management (MRP) and enterprise resource management (ERP), to cut down the operational cost of the logistics.
(4) Logistics research is relatively lagging behind, and logistics professionals are in short supply.

[11] Data from *Guangxi Statistical Yearbook 2011*.

[12] Opinions of the People's Government of Guangxi Zhuang Autonomous Region on accelerating the development of logistics industry in Guangxi (No. 74 [2010] of Guangxi Zhengfa).

At present, Guangxi logistics industry has entered an important period of development. According to the national Logistics Industry Adjustment and Revitalization Plan[13] (2009), Nanning has been defined as a regional logistics center in southwest China and a national logistics node city, and an important part of the southwest shipping logistics channel included in the national top ten logistics channel system, which indicates that the Chinese central government has placed Guangxi at a prominent position which plays a key role in the national logistics system. The Development Plan of Guangxi Beibu Gulf Economic Zone approved by China set the goal of building a logistics base of China-ASEAN opening-up and cooperation, made clear the strategic positioning of Guangxi logistics development, which ushered in a rare and significant opportunity for Guangxi logistics development.[14]

However, Guangxi's logistics industry is still in the primary stage of development, with a small total scale, insufficient effective demand for logistics, weak competitiveness and independence of logistics enterprises, weak radiation of logistics services, and inadequate construction of logistics system to meet the requirements of modern logistics services. With traditional basic logistics services as the mainstay, a coordinated mechanism for logistics infrastructure construction and development, and a policy system to support logistics development need to be improved. In brief, there is still a long way to go before it achieves the goal of becoming a regional logistics center. Therefore, it is urgent to establish a sound and reasonable comprehensive transportation system, which first needs to improve the underdeveloped railway logistics supporting infrastructure.

Guangxi high-speed railways will accelerate the circulation of goods among different regions, improve the modern logistics system in its economic belt and reduce the logistics cost. It contributes to the logistics industry in reducing the cost of time and logistics, which constitutes an important support for the modern logistics service system. The construction of high-speed railway has greatly shortened the space and time distance. For instance, the opening of Wuhan-Guangzhou high-speed railway shortened the running time from Wuhan to Guangdong from 10 to 3 h, which greatly shortened the distance between the two provinces with the same initial letters (Hunan province, and Hubei province) and the Pearl River Delta. By integrating into the Pan-Pearl River Delta Economic Zone, Hunan and Hubei are more accessible to advanced management knowledge, science and technology and a large amount of industrial transfer investment. The flow of goods and logistics between the two provinces is speeding up, saving billions of social and time costs every year, and greatly reducing the logistics cost.

The completion and operation of Guangxi high-speed railway will help to release the freight transport capacity of existing lines in its economic belt, and significantly improve its transport capacity. Further, a certain number of passengers will be diverted

[13] For more details, please refer to Notice of the State Council on Printing and Distributing the Adjustment and Revitalization Plan of Logistics Industry (State Council, March 10th, 2009), on the official website: http://www.scio.gov.cn/ztk/dtzt/13/3/Document/524877/524877.htm.

[14] Notice of the People's Government of Guangxi Zhuang Autonomous Region on Printing and Distributing Guangxi Logistics Industry Adjustment and Revitalization Plan (No. 83 [2009] of Guangxi Zhengfa).

from ordinary railways by virtue of its comparative advantages over ordinary railways, thus reducing the passenger load, releasing the freight capacity and improving the freight volume of existing and ordinary railways. In addition, Guangxi high-speed railway undertakes the mixed transportation of passengers and goods, so the newly-built high-speed rails also take the business of carrying goods, which increases the freight volume within the economic belt to some extent. With high-speed rail freight, the overall freight capacity of the logistics industry will be improved. When the freight capacity of existing lines is released, Guangxi logistics enterprises may become more empowered to transport more goods by rail. Therefore, the completion of Guangxi high-speed rail will increase the freight volume in Guangxi high-speed rail economic belt to a certain extent, thus greatly promoting the development of Guangxi logistics industry.

The development of high-speed rail transport in Guangxi is conducive to the development of green logistics in its economic zone. If we compare the energy consumption per unit of "person-kilometer", high-speed rail is 1, while cars are 5, buses are 2 and airplanes are 7. Therefore, high-speed rail has the lowest energy consumption compared to other major freight transport modes. High-speed rail uses electricity for traction, which reduces harmful emissions. Compared with automobiles and airplanes, high-speed rail has the lowest CO_2 emissions and light environmental impact, while not consuming valuable liquid fuels such as petroleum, and also utilizing multiple forms of energy. In addition, high-speed rail is using bridges instead of roads, which can save a lot of land resources. Therefore, high-speed rail transportation is a low-carbon economy and the most energy-efficient and environmentally friendly transport. Today, the choice of next-generation transportation in developed countries is focused on low environmental impact. High-speed rolling stock is in line with this requirement and is obviously better than cars and airplanes. Guangxi's high-speed rail transportation adopts the way of mixed line transportation of passengers and goods, with strong freight capacity, which is exactly in line with the requirements of developing green logistics in China.

The completion and operation of Guangxi high-speed railroads will bring more opportunities and challenges to the logistics industry in its economic belt. On the one hand, with the progressive development of high-speed rails in Guangxi, the cost of people flow and logistics will be greatly reduced, which is conducive to the rapid rise of economic belt along the route and the forming of logistics network. On the other hand, the economic exchanges and cooperation in Guangxi will be accelerated, and so will the economic and technological exchanges and cooperation between Guangxi and other provinces and cities. Some logistics enterprises with abundant capital and rich experience from other provinces will draw to the Guangxi market, so much so that those small logistics enterprises which operate on simple logistics and low informatization in Guangxi may have to meet up with enormous difficulties and challenges.

At present, China's high-speed rail is mainly aimed at passenger transportation, which has little direct impact on the logistics industry. However, because Guangxi's high-speed rail is designed for the mixed transportation of passenger and freight, the completion of Guangxi's high-speed rail will have a direct and profound impact on

Guangxi's logistics industry to a great extent. The opening of Guangxi high-speed rail transportation will reduce the circulation cost and improve the circulation efficiency. Specifically, the opening of the high-speed rail is conducive to the development of the four logistic zones in Guangxi (Beibu Gulf Economic Zone, Central Guangxi, Northeast Guangxi and West Guangxi), to making Nanning into a national logistics node city, to forming six regional logistics node cities including Liuzhou, Guilin, Beihai, Fangchenggang, Qinzhou and Chongzuo, and turning seven cities including Wuzhou, Guigang, Yulin, Baise, Hezhou, Hechi and Laibin into professional logistics centers. Thus, a modern logistics system, which is suitable for regional economic development, will be a professional, socialized and modernized logistics service network system in Guangxi.

High-speed rail may bring important changes to regional location which may draw more investment. In the areas where the high-speed rail passes, the regional conditions will be greatly improved and the investment environment will be optimized. For Nanning, a second-tier city, the high-speed rail will help it fully play its role as the collecting and distributing centre. To be specific, Nanning Station becomes a nodal point where high-speed rail-lines such as Liuzhou-Nanning Line, Nanning-Guangdong Line and Nanning-Pingxiang Line meet, while Nanning East Station will be the locus of the three HSR lines (Yunnan-Guizhou Line, Nanning-Qinzhou Line and Jinchengjiang-Nanning Line). As the high-speed rail greatly facilitates the travel of business people along the route, office properties in the surrounding areas of high-speed railway station are likely to be welcomed by enterprises for the benefit of time and cost. For the third-and fourth-tier cities such as Guilin, Liuzhou, Beihai and Guigang, the coming of new enterprises and factories may also promote the coordinated development of commercial real estate and other housing-related industries owing to the abundant land available for development around high-speed railway station. Of course, inevitably, housing prices in cities along the route will also be raised.

High-speed rail will change the regional house-purchasing preference of residents. In a big city like Beijing, there abound in numerous people who migrant for work on a two-hour trip. Because the housing prices in urban areas of Beijing are too high to afford, many office workers choose to buy their houses in other neighboring cities, for example, Xianghe in Langfang City and Zhuozhou in Baoding City. Accordingly, the housing prices in these cities have been on the rise. Among the Guangxi cities that Nanning-Guangzhou railway passed through, the housing price in Nanning is currently between 5,000 and 10,000 yuan per square meter, about 3,000 yuan per square meter in Guigang and 3,000 yuan per square meter in Wuzhou, while in Guangzhou, many apartments are priced above 10,000 yuan per square meter.[15] Why? After the completion of Nanning-Guangzhou high-speed railway, it takes 3 h from Nanning to Guangzhou, and only 1 h from Wuzhou to Guangzhou, which is included in Guangzhou's one-hour economic circle. Subsequently, some people in Guangzhou may give up high-priced residences in Guangzhou, and then buy their

[15] It is estimated according to the data of SouFun on http://www1.soufun.com/.

houses in Wuzhou or other cities along the Nanning-Guangzhou railway, which will drive the local real estate market.

High-speed rail will change the pattern of urban development. The traffic trunk line has changed the external traffic situation of the city, thus reorganizing the urban development layout. As far as Guangxi is concerned, the present urban distribution in Guangxi is characterized by the locations of rivers and traffic (Zhu et al. 2010), that is, the distribution along rivers and along traffic lines. The completion of the high-speed railway will improve the regional accessibility along the line to a greater extent, and the spatial expansion of towns along the high-speed rail line. Especially, near the entrance of a high-speed railway, the density of cities and towns in the radiation circle with high-speed rail station as the center will increase, driving the development of real estate industry.

High-speed rail will attract local people in Guangxi to invest in real estate in the city circle. In Nanning, Liuzhou and Guilin, which are relatively developed cities in Guangxi, the opening of high-speed rail not only attracts nonlocals to buy their own homes in Guangxi, but also provides a good opportunity for the citizens of these cities to invest and buy their own homes in the metropolitan area. Especially, Nanning, the center of the high-speed rail economic belt, which is located in the Nanning basin, surrounded by almost no mountains, and has very flat open lands in its southbound coastal area, has great potential for urban expansion. After the completion of the high-speed railway, the traffic accessibility between Nanning and other surrounding small cities has been greatly improved. From now on, people in Nanning are more willing to invest and buy their own homes in the surrounding cities, which is conducive to an expanse of the Nanning city circle, promoting exchanges within the city circles and driving the further development of the real estate industry.

In a word, it is predicted that Guangxi high-speed rail will play a more vital role in promoting urban real estate along the line. However, it should be noted that this is not very likely in a short period of time. Especially, when having strong confidence to develop the commercial real estate around the high-speed railway station, we should refer to the geographical location of Guangxi high-speed railway station. Also, the development of commercial real estate should be based on more prosperous economic entities. Generally speaking, the high-speed railway station is located in the suburbs far away from the urban area, with low population density, underdeveloped business and poor infrastructure. Which means, high-speed railway passengers may not stay at the railway station, but rush to the city center. Furthermore, even if the government authorities may plan some high-quality business centers or shopping malls around the high-speed railway station, expecting HSR-driven urban development, there may be a high vacancy rate of commercial real estate due to an insufficient demand. Therefore, the driving effect of Guangxi high-speed railway on the real estate industry in the economic belt should be seen as a long-term one.

At present, the quantitative research on rail transit and land price in theoretical circles mostly focuses on urban planning and transportation. Many scholars have studied the impact of urban rail transit on real estate prices. Since high-speed railway is also a kind of rail transit and a railroad as well, which is mostly used for cross-regional travel, it has the characteristics that the entrances and exits are far away from the urban area, the passenger flow has a timetable, and its radiation extends to a larger area. With these characteristics, the impact of high-speed railway on real estate prices is different from that of urban rail transit.

Three approaches that scholars from China and other countries look at the relationship between rail transit and land price are: comparison and contrast of land price before and after (the completion of a railway), regional comparison, and the land price function. Among them, the function of land price was applied to the evaluation of social and economic benefits of railway trunk line to predict the economic benefits of Beijing-Shanghai high-speed railway (Lin, 2005). Wen et al. (2010) established a simple real estate price model by using the land price function. Furthermore, Duan et al. (2012) examined the influence of Wuhan-Guangzhou high-speed railway on real estate development investment in Changsha by applying this model to research on large-scale intercity high-speed railways and real estate development and investment in cities along the line. By adding the factor of "the distance between a real estate programme and high-speed railway station" as a variable to the real estate price evaluation system, Che (2011) combined this model with the Hedonic model to analyze the influence of Beijing-Shanghai high-speed railway on Zhenjiang's commercial real estate in Zhejiang Province.

According to the model of Che (2011), it is easy to do sampling and calculations. However, by adding a certain variable of high-speed railway to the real estate characteristic price model, Che finds it helpful to calculate the influence of this variable on real estate prices, but difficult to evaluate the influence of high-speed railway on real estate prices systematically. Moreover, his model takes the basic root on multiple regression method and concerns many variables, which may result in multiple collinearities and some good errors. Instead, Chen and Lin (2006) and Duan et al. (2012) adopted the land price function, a method that has been studied and tested for decades in Japan. It is widely examined, and is extensively used in many cities as cases for reference. So, the data is relatively more accessible, more accurate and reliable, which is of great reference significance for this study (Lin, 2005).

The basic idea of land price function method is this: in order to assess the profits of a land resulting from improved transportation facilities, we need to, first, select a number of locations along the rail transit facilities, examine various factors affecting their land price, and assume that the relationship between land price and various influencing factors will not change with the construction of transportation facilities, then figure out the land price function of the areas along the route, and finally, calculate the land price difference of a certain plot before and after the construction of transportation facilities (Ye and Cai 2002). Normally, the land price function is a multivariate function, which reflects the characteristics of a piece of land brought by traffic facilities through the change of independent variables in logarithmic form. According to the typical model of the land price function method, a house price difference function model can be established, as shown in formula (7.5).

$$\Delta p_i = a_1 \ln x_{i1} + a_2 \ln x_{i2} + a_3 \ln_{i3} + \cdots + a_m \ln x_{im} \tag{7.5}$$

Here in this equation, Δp_i denotes the land price difference (real estate price index difference) before and after the railway construction in the main railway route city *i*; X_{ij} is the variable of land price change caused by railway trunk line construction; a_j is the parameter, that is, the influence range caused by the change of each variable; j = 1, 2, 3, …, m. For the land price function, the most important thing is the selection of variables. Choose three variables: the population of cities along the route, the opening time of railway, passenger and freight density of cities using railway lines. These three variables in combined, we now have a formula as shown in (7.6).

$$\Delta p_i = a_1 \ln x_{i1} + a_2 \ln x_{i2} + a_3 \ln_{i3} + u \tag{7.6}$$

where *u* represents a random error; X_{i1} is the population of the city *i* (unit: ten thousand); X_{i2} is the opening time of railway in the city *i*; X_{i3} is the passenger and freight density of the city *i* using railway trunk line; a_j is the parameter of each variable, j = 1, 2, 3.

Following the basic idea of land price function, Wen et al. (2010) make a model to evaluate the impact of rail transit on real estate prices along the route, namely, $Y = a_1 \ln x_1 + a_2 \ln x_2 + C$. Here in this equation, Y is the price difference of commercial housing before and after the opening of a subway, a_1 and a_2 are coefficients, x_1 is the distance from a commercial apartment complex to the subway station, x_2 is the distance from the apartment complex to the main urban district of the city, and *C* is a constant.

Later, Duan et al. (2012) adopted this model to study the impact of Wuhan-Guangzhou high-speed railway on real estate development and investment programmes in Changsha. Firstly, they tried to define the economic central position of Changsha over the years (or the position of a real estate program in the land price function) by applying the economic center of gravity theory and GIS model. Accordingly, the distance from the real estate program to the subway station is represented as the distance from the city economic center to the Changsha station of high-speed railway, while the main urban area in the function of land price is interpreted as the "main urban area" in the virtual economic sense, namely, the average GDP of Wuhan, Changsha and Guangzhou. The distance from the real estate to the main urban area is expressed as the economic distance between Changsha's GDP and the average GDP of the three cities, that is, the gap between them; The investment amount of real estate development in Changsha is regarded as a function of the above two variables. In short, Duan and Zhu they took the high-speed rail as a larger urban rail transit line, and cities along the line as subway stations, and then view the real estate investment as a function of differences in economic strength among cities and the distance between the economic center and high-speed railway station. Their study is innovative, but it is insufficient in theoretical support, and the rationale for comparing high-speed rail to urban rail transit lines.

Admittedly, both the two models have taken into account the distance factor, but as an efficient and comfortable means of transportation, high-speed rails charge higher than ordinary railway. Also, most people who would take high-speed rail have higher requirements for time. Therefore, the two factors of travel time and ticket price are indispensable factors to consider the real estate price or investment of cities along the high-speed railway. Therefore, it is necessary to add variables like time and price into the model, and in the case of Guangxi high-speed railway, the model may be modified and as shown in formula (7.7).

$$\Delta p_i = a_1 \ln x_{i1} + a_2 \ln x_{i2} + a_3 \ln_{i3} + a_4 \ln x_{i4} + a_5 \ln x_{i5} + u \tag{7.7}$$

Here, $\triangle p_i$ is the price difference of urban real estate before and after the completion of a high-speed railway; X_{i1} is the population of city i (unit: ten thousand); X_{i2} is the fare from city i to regional economic center by high-speed rail; X_{i3} is the time needed for a high-speed train from the city to the regional economic center; X_{i4} is the passenger and freight density of cities using high-speed railway; X_{i5} is the opening time of the railway in the city; a_{ij} is the parameter of each variable, j = 1, 2, 3, 4, 5.

According to Guangxi high-speed railway planning, Nanning is a city where multiple high-speed railways confluence, and a regional transportation hub, while Guangzhou is the most important well-developed city along the Nanning-Guangzhou high-speed railway (Nanning-Guangzhou). Judging from our observations, in Guangzhou, the development of real estate in less-developed cities within a certain distance along the high-speed railway may be advancing faster. So, the economic center here could be Nanning or Guangzhou, and different economic centers can be defined in accordance with each line of research. In this model, the data of land price difference, the density and fare of passenger and freight by the high-speed railway will be more accurately accessed only when the relevant high-speed railway is completed and put into operation. Until then can we calculate the specific impact of high-speed railway on real estate prices, and then explain the development benefits of real estate.

References

Che Jichong. Analysis of the impact of high-speed railway on commercial real estate prices and its causes [J]. *China Real Estate Finance*, 2011,11:10-14.

Chen Youxiao, Lin Xiaoyan. Study on Determining Indicator Weight of Investment Estimation Assessment System on Land-Developing Railway [J]. *Journal of Beijing Jiaotong University* (Social Science), 2006,04:8-14.

Duan, Z., Zhu, G., Zhang, S., Ji, X., Kui, X., Zhu, L. Selection of Characteristic Bands in Alteration Information Extraction from Remote Sensing Images for the Heishankouyong Area, Hami, Xinjiang, *Geology and Exploration*, 2012, 48(04): 823–828.

Liang Xuesong, Wang Hejiang, Qiu Hong. The Discussion of the Developing Opportunity for the Conversion of Tourism Spatial Locational Advantages--The Opening Visual Threshold of High-speed Railway Between Wuhan and Guangzhou, Zhengzhou and Xi'an [J]. *Journal of Xi'an University of Finance and Economics*, 2010(03):26-31.

Lin, X. *A Course Book on Economics of Technology*, Economic Management Press, Beijing, 2005.

Wen, C., Chen, Y., Zhang, H. Impact of Urban Rapid Rail Transit on House Prices, Journal of Jiangnan University (Natural Science Edition), 2010, 9(01), 81–85.

Ye Xiafei, Cai Wei. Fundamental study on return method of development benefits for urban rail transit [J]. *Journal of the China Railway Society,* 2002(01):97–103.

Zhang Ming, Liu Xi, Huang Xiang. Strategies of China Tourism Industry in High-speed Railway Era[J]. *China Business& Trade*, 2011 (08):149–150+3.

Zhang Ying, Xue Dongqian. SWOT analysis of tourism integration and linkage development after the opening of Zhengzhou-Xi high-speed railway [J]. *China Economist*, 2010, (04):210+212.

Zhu, C. A brief discussion of the current situation and countermeasures for the development of Guangxi's flower industry, Guangxi Forestry, 2007(2), 47–49.

Zhu Shipeng, Xu Bing, Mao Jiangxing. Fractal Structure of Urban System in Guangxi [J]. *Tropical Geography*, 2010, 30(02):178–182.

References from Chinese Sources

Wang Huimin. Shanghai experience of creative industry and its enlightenment to Tianjin [A]. Tianjin Committee of the Chinese Kuomintang Revolutionary Committee. Materials of seminar on the theme of "solving difficult problems, promoting transformation and upgrading" by Chinese Revolutionary Committee for Tianjin [C]. Tianjin Committee of Chinese Kuomintang Revolutionary Committee, 2010:8.

Chapter 8
High-Speed Railway and Regional Land Development

Land is the foundation of human life, which is a limited and irreplaceable precious natural resource. Transportation infrastructure, such as railway and highway lines, bridges, tunnels and culverts, passenger and freight stations, ports, airports and transportation service areas, all need to consume a lot of land resources. The development of transportation infrastructure should be carried out on the premise of saving land as much as possible and improving land utilization rate. As a new traffic facility in the region, high-speed rails help to meet the traffic demand with relatively economical land resources, which is of great significance for protecting land resources and promoting sustainable development. In addition, the entry of high-speed rail produces a more positive and gradual change on the regional land utilization efficiency, creating higher land use value.

8.1 The Relationship Between High-Speed Railway and Regional Land Development

8.1.1 The Corridor Effect of Rail Transit and Land Development Along the Route

High-speed railway belongs to the rail transit in the major categories. The impact of high-speed rail on regional land development is mainly embodied in the urban areas around the stations. Therefore, the mechanism of the impact of urban rail transit on urban land development is also applicable to high-speed rail.

Corridors are generally of two kinds: artificial corridors and natural corridors. The corridor effect theory refers to the electromagnetic field theory, and its essence lies in the existence of the benefit gradient field around the corridors within a certain range. Through the corridor effect, it can be well understood how the land use is affected by

X. Lin, *High-Speed Railways and New Structure of Socio-economic Development in China*, Research Series on the Chinese Dream and China's Development Path,
https://doi.org/10.1007/978-981-19-6387-2_8

traffic accessibility. The corridor effect consists of two parts, the circulation effect and the field effect. The main manifestations of the circulation effect are: ① Improving the speed of traffic flow along the corridors; ② Increasing accessibility; ③ Strengthening the interconnection of urban interior spaces in traffic corridors. The main performance indicator of the field effect is that the existence of transport corridor must rely on a certain physical geographical scope, within which social and economic activities in some urban spaces are more likely to occur. In other words, these effect fields are the areas with a high incidence of various socio-economic activities. The influence of rail transit on urban land use is dominated by the field effect of corridor.

The corridor effect of urban rail transit gradually attenuates from the center to the outside, and declining with the extension of distance. Its function can be simplified as in (8.1):

$$E = f(T) = e^{-aT},\ T \geq 0,\ a \geq 0 \tag{8.1}$$

where, E is the field strength of the effect field; T is the time needed to get to the rallying point, which is measured by hours; a is the function coefficient. For a residential space, this kind of field strength refers to both land rent and traffic accessibility. Let E be the land rent and T be the travel time in peak hours. Another concrete expression of the corridor effect can be:

$$D = f(e) = a\ln[(a \pm \sqrt{a^2 - e^2})/e] \mp \sqrt{a^2 - e^2} \tag{8.2}$$

In the formula, e is the gradient field effect; a is a constant, representing the maximum corridor effect; D is distance. The function is shown in Fig. 8.1. In Fig. 8.1, with the increasing distance of the rail transit line from d_1 to d_3, the corridor effect decreases from e_1 to e_3, which shows that the intensity of the corridor effect varies with the grading of corridor.

As can be seen from Fig. 8.1, although the overall accessibility of land along the route has been improved after the opening of rail transit, with the increase of the distance to the central area of the city, the accessibility of land will show a decreasing trend, which will reduce the benefit potential of urban land, thus causing the urban land price to decrease from the center to the outside. At the axis of rail transit, its accessibility reaches the peak, but the accessibility on both sides of the axis decreases. Because of the essential differences in the response degree of different types of land use, the spatial attraction intensity of rail transit to different types of land use along its route in different effect fields is different, namely, the spatial differentiation effect.

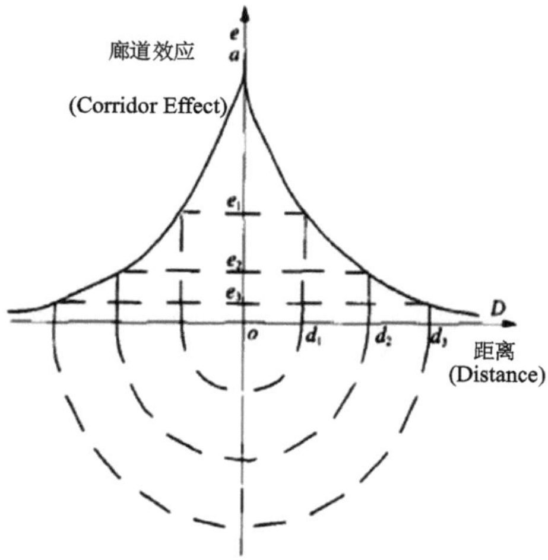

Fig. 8.1 Distance attenuation curve of corridor effect

8.1.1.1 The Spatial Attraction Effect Affects the Urban Spatial Outlook by Increasing the Intensity of Urban Land Use

The construction of rail transit has improved the accessibility along the route. It promotes the high agglomeration of land development to both sides of the track, and increases the intensity of land development in the corridor. All in all, the corridor effect increases the speed of land use and development in the region along the rail transit, thereby increasing the total amount of urban land use, and accelerating the redevelopment of the whole corridor area. Corridor effect changes the spatial layout of the city, creating the spatial attraction effect.

It is a matter of great concern to people that how to increase the total amount of urban land use, for which is obviously a scarce factor of production and living. Expanding urban lands might be the easiest way to solve the problem of limited land resource. But blindly pursuing the urban sprawl will lead to the so called "pancake-frying" development model, which is focused on the expansion of one single center, which is unsustainable. Therefore, a more scientific model is to increase the total amount of urban land use by improving the intensity of land use, that is, to change the original horizontal expansion into a vertical development.

(1) **The rail transit increases the total urban land use by increasing the intensity of land use per unit area** (I_i)

City travelers inclined to choose rail transit when weighing the total travel time because of the characteristics of high accessibility. As the number of inhabitants of the site continues to increase, the enormous commercial needs of the high-density population will drive urban investment and development. Figure 8.2 shows that the improvement of all kinds of supporting facilities in this area will attract office and entertainment owners to settle in so that the aggregation effect of land use intensity in rail transit stations will be improved. For a particular rail transit station, the land use intensity within the influence range of the station is not balanced, but presents a certain level of gradient, which is shown in Fig. 8.3. The corresponding land use intensity decreases with the increase of the distance from the center of the station whether it is based on a horizontal or vertical direction of a city. Normally, the strongest impact of the intensity of land use occurs within 200 m from the rail transit station, while a medium-intensity area within 300–400 m, and a medium–low intensity area within 400–500 m, which will be very weak after over 500 m.

Figures 8.2 and 8.3 are qualitative descriptions. From a quantitative point of view, the plot ratio is one of the important indicators to quantitatively reflect the intensity of urban land use. As is shown in Table 8.1, according to some cities with relatively mature rail transit development at home and abroad, the plot ratio of various types of

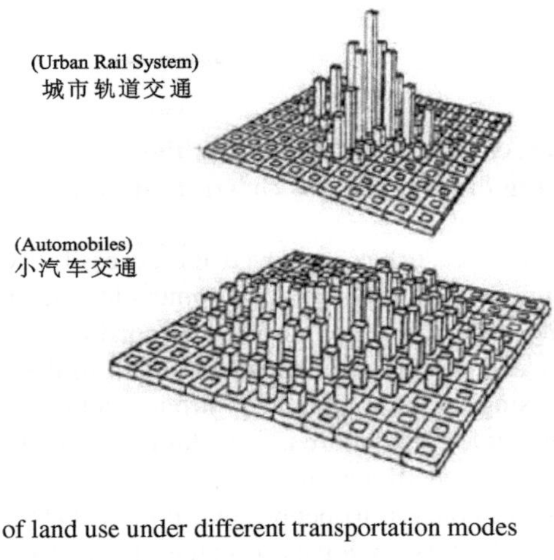

Fig. 8.2 Comparison of land use under different transportation modes

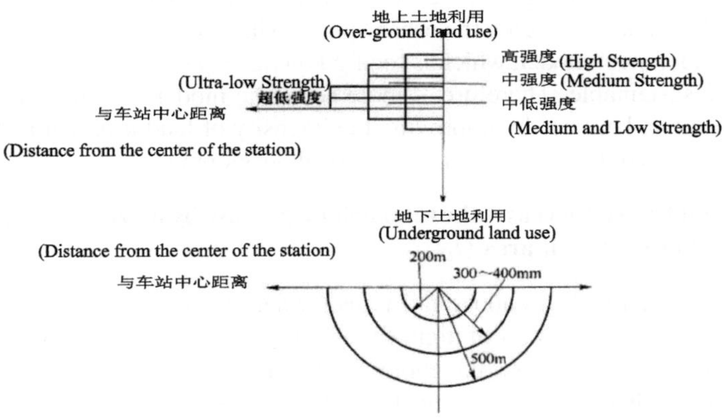

Fig. 8.3 The distribution of land use by rail transit stations

Table 8.1 Comparison of plot ratio of commercial office and residential land use along rail tracks and in general areas

	The plot ratio of commercial office occupancy		The plot ratio of residential occupancy	
	Along the track	General area	Along the track	Along the track
Hangzhou	6	3–4	4	2
Shenzhen	6	3–4	4	2–3
Hong Kong	8–15	5–8	5–9	3–4

land along the rail transit is 30–100% higher than that of the same type of land in other general areas of the city. The intensity of land use is closely related to the economic value of the land, which means the increase in land use intensity will increase the land rent price. Similarly, higher land prices would encourage more intensive land inputs.

(2) **Rail transit improves the total amount of urban land use by expanding the scope of influence (S_i) of urban transportation system**

If a piece of commercial land is located in a single-center city and the plot is located in the periphery of the city center, assuming that the land type of the plot will remain unchanged in the future, and in the absence of rail transit lines, the land price will gradually increase with the continuous improvement of development intensity, and the road traffic load on the plot will also increase, thus forming a dynamic balance between the load of the road traffic network of the plot and the development intensity, as shown in Fig. 8.4. See Fig. 8.5 for the situation after an urban rail transit line is introduced through the plot.

As shown in Fig. 8.5, the introduction of urban rail transit lines will have a series of impacts on this plot and its surroundings. The accessibility of the plot has been improved in the direction of newly introduced urban rail transit lines, and it is assumed that other modes of transportation and urban rail transit can be well connected, so the

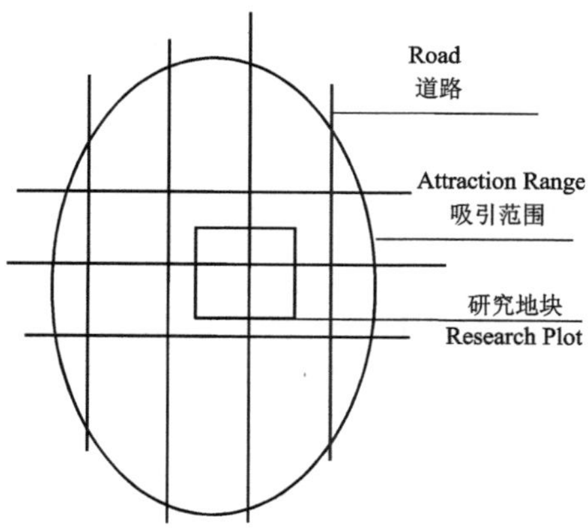

Fig. 8.4 The attraction range of the plot under road traffic conditions

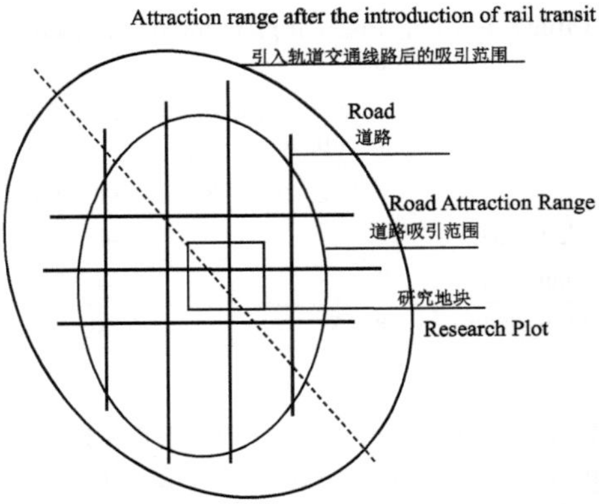

Fig. 8.5 The attraction range of the plot after the introduction of rail transit

attraction range in other directions will also be expanded to some extent. The previous equilibrium state between road traffic and land use will be broken, which will result in an increase in the attraction range of the plot and the scope of construction land. This trend won't stop until a new equilibrium is reached.

As a kind of factors of production, the most important difference between urban land and other factors of production lies in the location of land. Therefore, the discussion of total land use should not only consider its increase in quantity, but also how to make it distributed in proper locations. The change form of total land use distribution and layout in location is a specific urban spatial form. The microcosmic influence of rail transit on urban spatial layout is usually reflected in the change of spatial outlook around stations. At each stop, a compact circular land use layout pattern is formed, which is expanded outward in a concentric circle along the radial direction of the stop. If this effect is extended from the station to the whole rail transit line, it is easy to see that within the influence range of a certain rail transit line, its land use intensity will be significantly higher than that of other sections. When this effect is further extended to the whole network of urban rail transit, the overall spatial layout of the city can be schematized by improving the intensity of land use along the whole network.

The application of TOD (transit-oriented development) in Copenhagen is a typical case of a mechanism. Copenhagen, the capital of Denmark, is well developed through the construction of urban rail transit. As early as 1947, Copenhagen put forward a famous "finger plan" (known as The Five Finger Plan) based on the structural characteristics of the city. According to the plan, the main body of Copenhagen is to develop along several narrow and radial corridors. In between the "fingers", green "wedges" composed of forest and open green spaces are intended to separate the corridors. In the following decades, the Copenhagen government strengthens its implementation of the finger-shape planning through the development of urban rail transit. In the Copenhagen rail network, developed rail networks spread outward from the central city and integrate the development of land and rail traffic along these corridors. With the increase in the density of city construction and the intensity of land development along rail transit, green "wedges" in between the corridors are less developed. With the characteristics of the rail transit system, the regularly uneven development of urban space is formed, thus a "finger-shaped" urban land development model works better than a "pie-spreading" model in Copenhagen.

At present, China's cities are densely populated, and the cohesion of cities is extremely strong, so the expansion of cities is usually difficult to get rid of the strong attraction of urban centers, and an organic urban structure system is also difficult to form under this circumstance. We can take advantage of the characteristic that urban spatial form is significantly influenced by rail transit, to realize the development and transformation of urban spatial form from single-center agglomeration to multi-core by building rail transit network which is suitable for urban spatial development, and get rid of the disorderly spread of "pan-cakes" at present, so that urban spatial layout can be orderly and directionally expanded along the main development axis of the city.

8.1.1.2 The Influence of Differentiation Effect on Types of Land Use

In general, modern cities are normally divided into three major land use types: land for public facilities, for industrial purposes, and for residential purposes. While there are three types of urban rail transit sites according to their functional characteristics: commercial stations, traffic hub stations and residential stations. The three types of rail transit stations have significant differences in the layout, structure and intensity of land use, as shown in Table 8.2.

Land use types around rail transit stations are distributed in concentric circles. The reason lies in two factors: traffic accessibility, which makes living space cluster and land rent that makes residential space spread. In the long run, these two forces will reach a balance. That is:

$$P = f(T) = re^{-aT} \tag{8.3}$$

Here, P stands for the land price, r is the location level of living space, T is the travel time at the rush hour in the center of the residential area. Assuming that there is no rail transit and that the city is concentric, the communities on the edge of the city may be represented as:

$$P_0 = e^{-aT_0} \tag{8.4}$$

where P_0 represents the land price without rail transit affecting the urban fringe. At this time, the location level of residential space is the lowest, T_0 denotes the travel time of automobiles during the rush hour. When the rail transit starts to operate, we assume that the travel time of the rail transit is T_1, and the transfer time between a certain residence and the rail transit station is T_2, then, according to the balance between accessibility and land price, we have:

Table 8.2 Characteristics of three major types of rail transit sites

Types of rail transit stations	Design essentials of land use layout	Land use structure within the 500 m influence circle
Urban commercial	Strengthening the comprehensive utilization of rail transit stations, surrounding integrated facilities and underground spaces; realizing the organic combination of commercial development and public transportation transfer; increasing land revenue by determining reasonable land development intensity indices	Sorting proportion of land use in descending order: commercial lands, square greenings, and residential areas
Traffic hub	Connecting rail transit stations with other transportation facilities	Sorting proportion of land use in descending order: traffic plots (including squares and greenings), commercial lands, public facilities
Residential area	The design of the pedestrian landscape around the station, through a reasonable designed pedestrian system, organically combines the rail transit station with the ground bus to expand the radiation range of the station	Sorting proportion of land use in descending order: residential areas, greenings, urban commercial, and public facilities

$$P_1 = re^{-aT_1}.e^{-anT_2} \tag{8.5}$$

In equilibrium, the boundary of the influence range of rail transit stations may be $P_0 = P_1$, that is, the land price remains unchanged before and after the opening of rail transit. Substituting the above two formulas, we can get the following formula:

$$T_1 + nT_2 - \ln r = T_0 \tag{8.6}$$

where, n is the correction coefficient, which is a modification of the walking accessibility attenuation, and its value is taken as the ratio of the speed of rail transit to the speed of transfer traffic. k is defined as the ratio of the average travel speed of rail transit to urban roads during peak hours. The transfer time T_2 can be obtained by this formula, which determines the size of the affected area of the rail transfer. Typically, the value of r is small and negligible, but the actual meaning of the parameter is important because it represents the different levels of the residential area. Therefore, the impact area of rail transit is jointly determined by k and r. The running time of the automobiles in the city center is close to that of the rail transit, which makes $(k - 1)T_0(kn)^{-1}$ close to zero. However, the value of r may be bigger, which means that the impact of rail traffic on commercial space is greater than its impact on living space. As a result, commercial space will squeeze people's living spaces and make them move outward. On the marginal areas of the city, a larger value of $(k-1)T_0(kn)^{-1}$ usually leads to the formation of a relatively large residential area.

In conclusion, the characteristics of land use re-location along the rail transit line are: firstly, commercial and financial land, such as large supermarkets, stores, and office land, should be distributed in the area closest to the axis of rail transit. Secondly, residential space should be placed in the middle of the impact area of rail transit. Thirdly, industrial land should be set in the most marginal affected area of rail transit. Some scholars took Nanjing Rail Transit Line 1 as an example, and made a quantitative statistical analysis of the land use types in different effect fields along the line. It was found that in 1997, the public facilities land, residential land and industrial land in each circle showed a gradual decreasing trend from the center to the periphery. Within the studied strip range, the highest proportion is residential land, followed by industrial land, and the lowest proportion is public facilities land. However, in 2001, the land for public facilities and residential land in all circles increased obviously, while the land for industrial use decreased to varying degrees. It shows that these three types of urban land will gradually differentiate under the action of rail transit and finally show a regular circle distribution. Therefore, it is necessary to scientifically and quantitatively control the proportion of various types of land to optimize the structure of the land within the influence range of its station by rail transit. According to the relevant regulations of the central government on the proportion structure standard of planned land use, combined with the law of rail transit corridor effect change, we can figure out the reasonable proportion of land use of different types of sites in different circles.

8.1.2 Impact of High-Speed Railway on Regional Land Development

The construction of high-speed railway provides people with rapid access to transportation, thus separating the residential land and industrial land geographically, which has a very significant impact on the axial development of land, leading to the highly concentrated land use nature of commercial and public facilities in the affected areas along the high-speed railway, which will be redistributed according to the laws of traffic economy and optimize the allocation of land resources. That is to say, the construction of high-speed railway has far-reaching influence on the nature of land use, the intensity of land development and the spatial form of urban land, which may drive the high-density development of land along the line.

8.1.2.1 Influence of High-Speed Railway on the Nature of Land Use

The construction of high-speed railway will have a great impact on land use. As a result, the types of land use along the high-speed railway will also change according to market rules, which will not only strengthen the financial, trade, service and other functions of the city center, but also provide strong transportation support for the formation of new cities, and significantly change the nature of land use.

In the area around a station, with the opening of high-speed railway, the nature of land use around the station will change greatly. Due to the increase of land price brought by high-speed railway, some high-profit industries such as commerce, service industry, office buildings, etc. are often arranged in the reasonable walking area, and other industries will gradually change to these industries. In view of the sensitivity of various industries to location changes, the reasonable pedestrian zone off the high-speed rail has the greatest influence on commercial use, followed by office, residence and industry. Theoretically speaking, the business, office and so on within 500 m from the traffic station benefits greatest.

8.1.2.2 Influence of High-Speed Railway on Land Use Intensity

The urban traffic forms are closely related to the land development pattern. The dense urban structure promotes the development of public transportation, while the scattered residential layout promotes the popularity of cars. With the gradual increase of high-speed railway lines and the formation and improvement of urban underground space, high-speed railway will become an organic part of urban land development and old city reconstruction. Many foreign studies show that the intensity of land development is closely related to the transport modes. After investigating 32 major cities in the world, Australian scholars found that when the land use density is less than 40 people/ha, there is a great possibility of relying on car transportation, while when the land use density reaches 60–100 people/ha, the chances of using public transportation will greatly increase. High-speed railway is characterized by rapidity, punctuality and large capacity. If the development density is too high, it is bound to bring a lot of traffic flow, while the road capacity is limited, causing traffic congestion, which reduces the attractiveness of this area and is not conducive to the development of this area. However, high-speed railway brings more people than traffic flow, which will not cause congestion or regional pollution. Therefore, high-speed railway can greatly improve the accessibility of reasonable traffic areas, thus promoting the high-density use of urban land.

8.1.2.3 Impacts of High-Speed Railway on Land Values

The reasons for the influence of high-speed railway construction on the land value along the line can be summarized in three aspects: First, urban traffic can greatly improve the traffic conditions along the land, shorten the space distance between a farther land and the central area of the city, save the travel time and expense of the traveler, and improve the accessibility and traffic convenience of the land along the line. Second, the development of high-speed railway provides the opportunity for government at different levels to adjust their land use planning, and modify their purpose of land development. City planning may be guided by the transport design, which would further change the nature of the land use, and improve land revenue by promoting the intensive use of land. Third, high-speed railway can improve the relative location conditions and strengthen the mutual connection of lands along the line, strengthen the mutual connection of land around the railway and the complementary effect of land. Thus, the lands along the line become an organic whole that supports each other, and become a unity of valuable asset which may uplift the land price.

8.2 High-Speed Railway and Regional Land Price

8.2.1 Influence of High-Speed Railway Construction on Land Price

Land price is affected by many factors, including geographical position, natural condition, population, administrative districts, urban nature, urban development process, social economic conditions, land system, housing system, land use planning, social and national economic development planning, etc., which are the basis of determining the land price. The essence of the relationship between transportation and land use is the complementarity of transportation cost and land value. The closer the land is to the city center, the lower the transportation cost, which attract more people to live, so the higher the land price is. The reasons for the influence of high-speed railway on the land price along the line can be summed up in three aspects:

First, the construction of high-speed railway improves the traffic conditions of the land near the stations, enhances the traffic convenience of the land along the railway, shortens the time-distance between the land along the line and the city, as well as the central area of the adjacent cities, saves the traffic time, and provides high accessibility for the land around the railway.

Second, high-speed railway has changed the land use pattern along the line. The new construction of the high-speed railway and its stations will inevitably make the relevant departments adjust and plan the land along the line and change the nature of land use. Therefore, traffic guides planning and planning changes land use attributes, which promotes intensive land use and improves land use efficiency.

Third, high-speed railway promotes the corresponding adjustment of urban landscape and land use pattern, improves the relative location conditions, the mutual connection and the complementarities of land along the route, which makes the land along the line become an organic whole supporting each other, and the land of adjacent plots value-added, thus increasing the land price.

People's traditional understanding of the relationship between land use and transportation facilities needs to be improved. They overemphasize that transportation facilities are the derivative products of land use, and transportation infrastructure is built to meet the transportation needs generated by land development, but pay less attention to the washback effect of transportation infrastructure on land development. Stimulated by socio-economic and technological development, high-speed railway and land use is a fluctuating process of mutual feedback and continuous interaction. Looking at the construction of high-speed railway and the development process of cities along the line, we can find that they are going through a constant cycle and spiraling process from transient balance to broken balance to new temporary balance. In this process, accessibility and transaction cost play a link role. The imbalance is long-term, and the balance is always temporary. As the backbone of regional transportation, high-speed railway also has a close feedback function with land use. Figure 8.6 describes the process of mutual feedback between the existing transportation facilities and land use after the introduction of high-speed railway (especially the high-speed railway connecting major cities in this study) under the condition that the existing transportation facilities have been unable to bear the huge traffic demand caused by land development.

As can be seen from Fig. 8.6, as an accessibility index to quantitatively study the impact of high-speed railway trunk line on land price, there are three main influencing factors of its change: construction time, urban population, passenger and freight

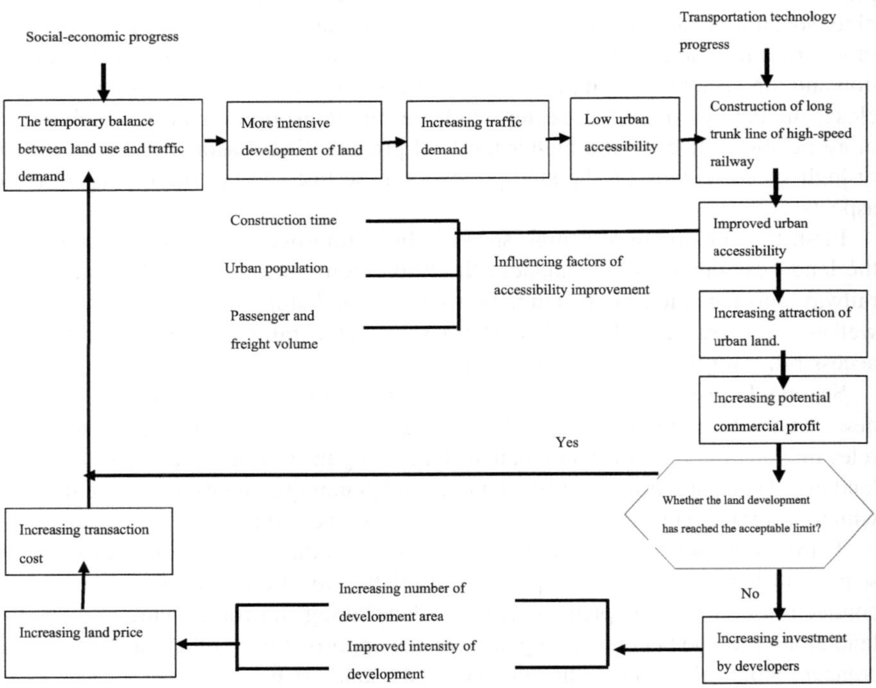

Fig. 8.6 The detailed process of land use between high-speed railway and cities along the line

volume, which is also an independent variable index for constructing an evaluation model of the impact of high-speed railway on land price along the line.

8.2.2 Evaluation of the Impact of High-Speed Railway Construction on Land Price Along a Railway Line

In 1974, Sherwin Rosen (1974) first proposed a functional model based on the hedonic methods to evaluate the impact of suburban train route construction on land prices, and attributed the increase of land price to people's willingness to pay for transportation convenience and urban comfort. After that, some scholars revised and applied this method in practice. Sheppard (1999), Gibbons and Machin (2005) gave realistic observations. Gibbon and Machin argue that ticket prices, rail service levels and distance from stations are the interaction factors to assess the volatility of land prices. In the following spatial regression model,

$$\ln p_{it} = \alpha + d_{it}\beta + x_{it}\gamma + g_t + f_t + \varepsilon_{it} \quad (8.7)$$

p_{it} is the price of a property in postcode unit i in period t, which can be measured by four variables: the distance from the place of residence to the nearest station (d_{it}), the general time effects (g_t), a vector of other property and location characteristics (x_{it}), and the place-specific unobserved components that are fixed over time (f_i). The research on hedonism method has been deep and mature, but it is often used in the evaluation of urban planning, and the comparison and selection of route schemes. Therefore, it is difficult to determine the data in the actual calculation by applying this model (Chen, Lin et al., 2005).

According to Chen et al. (2005), Venables and Gasiorek (1999), and Brocker (1998) established a land use/traffic impact model based on the general spatial equilibrium theory, Simmonds (2001) gave a summative discussion of this approach, while Crampton (2003) analyzed the overall impacts of urban rail transportation on neighborhood areas, and pointed out that the land price under such influence is periodical. According to the sources of different stages of this impact, the income level of different development regions is various. As one of the representative theories to analyze the social benefits of infrastructure construction, the asset value method is based on the experience of the capitalization hypothesis. This hypothesis assumes that the benefits of social infrastructure will ultimately be reflected in the rise of land or property prices. When it is applied to transportation infrastructure, it can be assumed that the development benefits brought by the construction of transportation facilities will eventually be fully reflected in the rise of land or real estate prices in areas along the new routes. The line of thought is simple and clear. It starts with the analysis of the price changes of land or real estate that accompany the transportation infrastructure construction to estimate the benefits of transportation infrastructure development according to the price changes. The asset value method is not only used to calculate the total benefit of transportation infrastructure, but also the benefit of different entities and regions along the route, so it has high practical value. Therefore, in recent years, this method has been widely used by international researchers, and a large number of practical models have been developed, and the accuracy of its quantitative analysis has been empirically analyzed.

The methods of land price measurement include:

(1) Market comparison method

When calculating the price of a land to be evaluated, according to the principle of substitution, the land to be evaluated is compared with similar land transaction examples that have been traded in the recent period. According to the latter's known price, referring to the differences of the land's transaction situation, date, region and individual factors, the method of evaluating the time and land price of the land to be evaluated is revised. On the one hand, this method requires developed market conditions; on the other hand, it is affected by market distortion, which causes deviation of land shadow price. The market comparison method can be only applied when the land market is relatively developed, which can correctly reflect the land price and has sufficient substituted cases of land transaction.

(2) Benchmark land price revision

The benchmark land price revision method is one of the important applied valuation methods in China's land valuation. It is an appraisal method that uses the evaluation results of urban benchmark land price and benchmark land price correction coefficient, compares the regional conditions and individual conditions of the parcel to be appraised with the average conditions of the area where it is located according to the substitution principle, and corrects the benchmark land price by selecting the corresponding correction coefficient according to the correction coefficient table, so as to calculate the price of the parcel to be appraised on the appraisal base day. Here, benchmark land price refers to the average price of land at all levels or regions assessed on the basis of parcel valuation, which includes benchmark land price of urban land and that of agricultural land. The benchmark land price revision method is applicable to the land price evaluation of cities that have published benchmark land prices, especially the evaluation of land assets in the behaviors of land transfer reserve price, land mortgage price, tax land price and merger of state-owned enterprises, and the land for industrial and public facilities in different cities and counties with benchmark land prices.

(3) Cost approximation method

The cost approximation method is a method to estimate the land price based on the sum of various costs of land development, plus certain profits, interest, taxes to be paid and land value-added benefits. The cost approximation method is generally applicable to newly developed land valuation, industrial land valuation, and land that cannot be valued by other methods due to the narrow real estate market short of sufficient cases. Also, the method applies to land for schools, parks, public buildings, public facilities and other special land valuation with no profit and little transactions. However, the cost approximate method is not applicable to the appraisal of developed land, commercial land and residential land in built-up areas.

(4) Revenue reduction method

In fact, the income reduction method is a method that regards land income as an investment aimed at obtaining profits, calculates virtual profits based on the average profit rate, and reduces future profits to the sum of present values. The income reduction method is suitable for calculating the price of profitable land where the income generation is relatively stable and continuous, and at its best use. Because the market comparison method and benchmark land price revision method are highly dependent on the development of land market. Therefore, when measuring opportunity cost under the condition of less-developed market, it is a simple and feasible method to use market information to determine the average market income of the best use of the land and take the average income as opportunity cost.

Some points need to be clarified before analyzing the changing law of high-speed railway and land price:

(1) Changes of urban land prices in different types of cities are various

Different types of cities and regions depend on high-speed railway differently because of their different development modes. The responses of big cities and small villages to long trunk lines are also different. so it is necessary to classify the regions along the high-speed railways before analysis.

First, we divide the areas along the route into cities and villages.

As far as cities are concerned, they can be divided into 7 categories according to their functions within the economic zone, as follows:

Classification of urban functions:

1. Regional Comprehensive Centers : provincial political, cultural and economic centers
2. Sub – regional Comprehensive Centers : regional political, cultural and economic centers
3. Comprehensive Industrial Centers : diversified industries, sub - regional economic centers
4. Cities of Mining industry : mining and processing industries as dominant industries
5. Transportation Hubs : highways, railways, port hubs
6. Tourist Cities : dominant by tourism
7. Port Cities : opening up to the outside

When the trunk line passes through these seven types of cities, it will definitely have different impacts on urban land prices. At the same time, if: (1) the development of the city is highly dependent on the high-speed railway, and (2) the high-speed railway contributes a lot to the urban economy, which can be compared to the land price in this book, we can say that the construction of the high-speed railway trunk

Table 8.3 Evaluation table for various types of cities

City types	Dependence of urban development on high-speed rails	The contribution of high-speed railways to urban economy (taking land price as an example)
1. Regional comprehensive centers	Medium	Weak
2. Sub-regional comprehensive centers	Less Strong	Strong
3. Comprehensive Industrial centers	Less Strong	Strong
4. Cities of mining industries	Medium	Weak
5. Transportation hubs	Strong	Medium
6. Tourist cities	Strong	Medium
7. Port cities	Strong	Weak

line will have a strong correlation with the land price of cities along the line, which is also the focus of our analysis. The evaluation tables of various types of cities are shown in Table 8.3.

Among them, regional comprehensive centers, such as Beijing, Shanghai, Guangzhou and other cities in China, their economic development does not depend on their location, and convenient high-speed railway does not play a prominent role in economic development. Therefore, as far as regional comprehensive centers are concerned, the construction of high-speed railway trunk lines will not bring much impact on urban land prices. Besides, these cities are all hub cities, for which the transportation volume of a high-speed railway trunk line does not account for a large proportion of the transportation volume of the whole transportation system. Therefore, for regional comprehensive central cities, the influence of high-speed railway trunk lines on urban land price may not be taken into consideration.

For 4–7 types of cities, although all four cities have great dependence on high-speed railway in the process of development, the impact of high-speed railway on urban land price is not very significant. Therefore, in these four types of cities, the influence of long trunk lines on urban land price may be not taken into account.

Among these seven types of cities, the sub-regional comprehensive centers and comprehensive industrial centers simultaneously meet the two conditions that urban development depends strongly on the high-speed railway and the high-speed railway contributes greatly to the urban economy. Therefore, among the cities studied, we take the sub-regional comprehensive centers and comprehensive industrial center cities through which the high-speed railway trunk line passes as the research objects.

As far as rural areas are concerned, agricultural land will not have obvious land price changes due to the improvement of high-speed railway conditions (railway stops are concentrated in cities), so it is not within the scope of this study. A comparative study of agricultural land prices will be made in the regression analysis.

(2) The measurement standard of urban land price

This study suggests that the land price index system should be used as the measurement standard of urban land price. Abroad, the research on urban land price index system has a long history. As early as 1930s, developed countries launched land price index. For example, Japan made "National Street Price Index" public in 1936, and China's Taiwan Province and Hongkong Special Administration released real estate price index in 1993 and 1996, respectively, aiming at releasing the price level to the society, promoting the sound development of the real estate market and achieving the purpose of standardized management.

The research on land price index in mainland China is relatively late. In November 1994, China Real Estate Index System, jointly undertaken by China Real Estate Association, Information Center of Development Research Center of the State Council and China Real Estate Development Group, passed the ministerial appraisal. At present, the national land price index has been published regularly. Hebei and other provinces have also begun to set up their own land price index system.

However, due to the late start of the land price index system, it is difficult to obtain the data of previous years, and some provinces do not have their own land price index system. Therefore, in the study of the model, the consumer price index of urban residents in various regions published by the National Bureau of Statistics is used to determine the changing trend of house prices, which has been publicized since 1995.

(3) The attraction range of high-speed railway to passing areas

Attraction scope refers to the functional areas of high-speed railway lines, stations, ports and cities. The attraction range of high-speed railway lines can be divided into direct attraction, joint attraction and indirect attraction. The difference between them lies in that the transportation links within the scope of direct attraction are not related to other associated high-speed railway lines in terms of transportation or loading and unloading, while for the transportation links within the scope of joint attraction, only part of the transportation distance and one of the loading and unloading links are completed through other transportation lines, and for the transportation links within the scope of indirect attraction, only part of the transportation distance passes through this line, while the material loading and unloading has nothing to do with this line.

High-speed railway can be said to be a linear influencing factor of urban land price. From the point of view of regional economics, the underlying rule of linear influencing factors is that the range of influence is limited, and decreases with the increase of distance. This point is clearly reflected in the high-speed rail trunk lines. In foreign countries, the land within two kilometers of the station in the area along the route is usually regarded as the influence area of land price fluctuation.

However, high-speed railway has its particularity. In a city, no matter passengers or other customers who use high-speed railway, they usually don't change their choice according to the distance between their residence and the train station. Usually, the railway is the first or only choice for most Chinese people to travel long distances. Therefore, in the study of the influence range of land price of railway trunk lines, we assume that the influence of railway trunk lines on the land price of route cities covers an area beyond the urban area. In a city, there is no obvious fluctuation of land price as far as the distance to the railway station is concerned. Teng Li's empirical investigation of Nanning-Kunming Railway also illustrates this point.

As mentioned above, high-speed railway is a linear influencing factor. In this sense, the linear factor model could be applied to measure the influence of land price and house price in neighboring cities. Although there are a variety of commonly used linear models, after the test of many models, we hold that the influence of high-speed railway trunk line on the land price of surrounding cities may be modelled by applying the logarithmic increasing model of land price function. The modeling process is as follows:

Assuming that the influence of other factors on urban land price is not considered, and only the influence of high-speed railway trunk line is considered, the influence of high-speed railway trunk line on the neighboring urban land price would topple down exponentially. The basic idea of land price function is: select several locations along the trunk line of a high-speed railway, analyze various factors affecting land price, and assume that the relationship between land price and various influencing factors will not change with the construction of high-speed railway facilities, so as to estimate the land price function of the area along the route. Finally, by calculating the land price difference of a plot before and after the construction of high-speed railway, the development benefits of the plot due to high-speed railway can be computed. The land price function is usually a multivariate function, which reflects the character of land before and after the construction of high-speed railway through the value of independent variables, and its form is generally logarithmic as in:

$$P_i = \sum_j a_j \ln x_{ij} \tag{8.8}$$

where P_i is the land price, a_j is the parameter, and the independent variable x_{ij} is the reflection of different land characteristics. An important premise of the method for calculating the development benefit by establishing the land price function is that assuming that the land price function relationship does not change with the construction of high-speed railway. Under this premise, the land price function can be used for both post-calculation and pre-prediction.

Referring to the typical model of foreign land price function method, we can have the following house price difference function:

$$\Delta p_i = a_1 \ln x_{i1} + a_2 \ln x_{i2} + a_3 \ln x_{i3} + \cdots + a_m \ln x_{im} \tag{8.9}$$

In the Eq. 8.9, Δp_i is the land price difference (gap in residential consumption index) before and after the block in the station circle of high-speed railway trunk line is affected by the long trunk line. x_{ij} is a variable that affects the change of land prices caused by the construction of high-speed railway trunk line, where $j = 1, 2, \ldots, m$; a_j is the parameter, which refers to the quantified magnitude of the effect of each variable, where $j = 1, 2, \ldots, m$.

Three variables are considered in this section: urban population along the routes, construction time, and passenger and freight density of urban high-speed railway trunk line.

Urban population along the route. With the increase of urban population, the influence of high-speed railway trunk line on land price is more significant, but the strength of its influence is decreasing, which meets the requirement of logarithmic increase. This study adopts the data of the National Bureau of Statistics, and takes 100,000 as the unit. As a result of the family planning policy, the growth rate of China's population is stable at around 4/1000 all the year round, with little change. To simplify the study, we assume that the population will remain unchanged in the study year.

The opening time. Generally speaking, when the high-speed railway is completed and put into operation, the land price shows a trend of logarithmic increase, which is mainly due to the steady rise of the city's position in its economic zone under the effect of urban agglomeration, showing a trend that the land price increases with time but decreases in degree.

Passenger and freight volume of urban high-speed railway trunk line. Passenger and freight volume is an index to measure a city's dependence on the transportation capacity of high-speed railway trunk lines. With the increase of passenger and freight volume, the impact on urban land price will definitely show a monotonic increase in a slow turning down.

As such, the regression function of the influence of high-speed railway trunk line on the land price of cities along the line is as follows:

$$\begin{cases} \Delta p = \alpha_0 + \alpha_1 \ln x_1 + \alpha_2 \ln x_2 + \alpha_3 \ln x_3 + \varepsilon \\ \varepsilon \sim N\left(0, \sigma^2\right) \end{cases}$$

where, Δp is the difference in land prices that before and after the influence of high-speed railway on the plots within the high-speed railway station circle (difference in residential consumption index); x_1 is the urban population along the line, taking 100,000 as the unit; x_2 is the opening time (unit: year); x_3 refers to the passenger and freight volume of the city using the high-speed railway trunk line, taking thousands of tons as the unit; a_j is a parameter, which refers to the quantitative amplitude of the influence of each variable, where $j = 1, 2, 3$.

8.3 Land Development Right and Land Development Along High-Speed Rail Lines

8.3.1 *Some Theoretical Underpinnings of Land Development Right*

As Marx (1875) said, "Land is the source of all production and all existence." Land is the most important resource for people to carry out all production and life activities. However, with the rapid development of economy and society and the acceleration of urbanization, the contradiction between people's increasing demand for land and scarcity of land resources has become more and more prominent. How to use land efficiently has become the top priority of all countries in the world.

The land development right system with the highest correlation with land use and its transfer method originated in Britain, which is based on the full and effective use of land and aims at achieving full harmony between environment and economic development. In the late 1930s and early 1940s, in order to evacuate the industrial population and redevelop the crowded urban areas, the British Parliament set up three members to conduct special research. The research pointed out that land use planning should be made at the national level, and suggested that landowners should take public welfare as the restriction condition when using land, thus creating the legal thought of land development right and its transfer. Later, in the British Urban and Rural Planning Act of 1947, it was clearly stipulated that all future development rights of private land should be transferred to the state. Transferred Development Rights were founded on the basis of Zoning system in the United States. The difference is that the land development rights in the United States belong to landowners (including private owners) and are quantitative transferable property rights. Subsequently, countries around the world, including France, Germany, Italy, Canada, Singapore, South Korea and Taiwan (China), have successively established the land development right system.

While tracing the origin of land development right, we should realize that as an important part in land property right, the relationship between land development right and land ownership is both different and interrelated. To some extent, land development right is a kind of property right separated from land ownership, a sub-right of land ownership right bundle, and a property right that can be separated from and independent of land ownership, and also the owner's right to change his useful land to the existing use for greater development opportunities.[1] Originally, because the main purpose of the creation of land development right is to protect agricultural land, it is also called "agricultural land development right", that is, the

[1] The origin of land ownership can be traced back to the relevant content of property law in law. Early Roman law and later German Civil Code and French Civil Code all defined and explained the concept and system of property rights comprehensively. Since the origin of land development right is mainly described here, the related origins of property law and property right will not be described in detail.

right to turn agricultural land into construction land. After the implementation of land development rights, the property rights (ownership) of all other lands are limited to the rights that have been legally acquired at present, that is, the scope of farmland ownership, and the use value of the policies that have been worked out. Since then, the right to change the category of agricultural land use is designated as the right to development.

With regard to the connotation of land development right, Chinese and other foreign researchers have roughly analyzed it from the following perspectives:

Scholars of environmental economics believe that land can be circulated in the market as a unique resource. It is different from the government's behavior of regulating individual land development capacity in the market in that the circulation system of land development rights is a part of tradable permit system. This kind of market circulation with institutional guarantee allows resources to find market players that can exert their maximum utility, thus promoting the high efficiency of land development through the market price mechanism. The circulation of land development rights under the tradable permit mechanism is also helpful to remove the external uncertainties implied in property rights, help to overcome the disadvantages of traditional zoning implementation in land planning, and enable landowners to get partial compensation. From the perspective of environmental resources economics, land development right is a kind of right resource that may better realize the allocation of resources and exert its maximum utility, achieving the two goals of "efficiency" and "fairness" in the process of land development and use.

In the field of land management law, land development right is studied from the perspective of government rights. Scholars in this field believe that the establishment of land development right is a means for the government to improve land use control, which means that when the government collects private land rights and interests, the land development right system may provide a new way of compensation for land ownership in addition to compensation for land expropriation, so as to reduce financial pressure. The right of land development in the field of land management law was originally applied to tackle issues concerning land use of historic buildings and natural resources. The government would set up some institutions that may purchase the development rights held by the owners of historic protected buildings, or allow the owners of immovable property rights to transfer the development density that is not allowed in the place where the protected facilities are located to other areas designated by the government for use. The land development in this sense has embodied some administrative color. That is, only under the premise that the government controls the rule of land development and use can the land development right and its circulation system exist. The land development right as such is an inevitable outcome of the governmental intervention to improve the control of land use and development.

Scholars from the field of property law believe that land development right is the product of socialization of land ownership, and is a way for land owners to enjoy their land use interests. In the traditional concept of property rights, the right to land development is the implied meaning of land ownership, which comes from the land itself, and having land ownership means enjoying land development rights. However, in modern society, the right to land development has been regarded as a right independent of land ownership, which is formed in society and given to the owner. Then the owner is entitled to enjoy the land development potential within the legal norms but outside the special land management measures, and to relieve the influence of strict land control measures on their property rights by transferring the right to land development. But it should be noted that in the field of property law, the discussion of land development right also involves the related issues of land development benefit distribution. That is to say, the transfer mechanism of land development right may provide various development and investment ways for land owners to realize their dominant interests of land, and the land development right itself may also exist in this mechanism as a kind of income (Table 8.4).

Looking at the legislation on land development rights and the research on land development rights by scholars in various countries, we can see that: generally speaking, land development rights can be summarized as the rights to change an area of land into different properties, such as changing residential area facing the street into commercial blocks and agricultural fields into construction land, etc. It is a property right that can be separated from land ownership and disposed separately, which may be controlled by land owners who possess the land ownership or those who enjoy only the land development rights but not the land ownership. It may serve

Table 8.4 An overview of land development right system in Britain, France and the United States

	Land development right system in U.K.	System of legal upper limit density in France	Transferable development right system in the United States
Relevant laws and systems	In 1947, Britain promulgated the Urban and Rural Planning Act, which legally restricted the abuse of private land rights, controlled land development activities and established land development rights. All development activities must be approved by the government	In 1975, France promulgated the Law of Reforming Land Policy, and in 1976, it revised the Urban Planning Code. The new land policy is characterized by the planned expansion of government reserves and the establishment of a regular preemptive system	In 1942, the U.S. Department of Commerce of the federal government promulgated the Authorization Act of Land Use Zoning Control Standards. Under the guidance of land zoning control, since 1974, some States in the eastern United States have designed new rules to transfer development rights among different owners in the region, that is, the transferable development rights system, following the practice of British land development rights

(continued)

Table 8.4 (continued)

	Land development right system in U.K.	System of legal upper limit density in France	Transferable development right system in the United States
Main content	The Urban and Rural Planning Law stipulates that the future development rights of all private land shall be transferred to the state and monopolized by the state, and the land development rights shall be nationalized. Anyone who wants to change the original land use character and carry out development and construction must purchase the development right from the government in advance. The purpose of establishing the land development right in Britain is to establish a mechanism to effectively control the land development and "return the price to the public"	The legal ceiling density limit in France, similar to the land development right, stipulates that the building right, one of the land ownership rights, has a low-level upper limit ratio, and the building right exceeding the limit belongs to the local government. If a building developer wants to build beyond the upper limit, he must buy the building right that exceeds the standard from the government. The direct purpose of this provision is to break the unfairness of land development rights among landowners caused by planning control, stabilize local government revenue and control land price	There exists a close relationship between land development right and land zoning control. On the one hand, the requirements and conditions of land development and utilization of land zoning control are the legal planning basis for setting land development right; On the other hand, land development right, especially transferable development right system, is an effective and flexible institutional means to realize land zoning control

(Contents summarized above are adapted from Liu (2007, 2008))

not only as a means for the government to improve the value of land use, but also as a resource of rights flowing in the market to mediate the trend of land demand in the market, so as to realize the optimal allocation of land resources and the value goals of "efficiency" and "fairness", and solve the practical problems of distribution of development interests and ownership of rights caused by the change of the nature of land use.

As for the structure of land development right, scholars from different fields disagree with each other on the origin of right. Some scholars argue that land development right should be a sub-right in the bundle of land ownership rights, while others believe that land development right is an exclusive property right independent of land ownership. In addition, the right to land development is the right to develop the land for practical use, and enjoying the right to development is very important for the power subjects to participate in social and economic development and share the land increment brought about by development. Therefore, land development right is a right with characteristics of human rights.

8.3.2 Connotation of Land Development Right Along High-Speed Railways

Before we delt into the connotation and structure of land development right along high-speed railways, it is necessary to clarify the relationship between land development right along high-speed rail routes and the proprietorship of construction and operation of high-speed railway. To some extent, the land development right along the high-speed railways should be a sub-right under the proprietorship of high-speed railway construction and operation under certain conditions. For example, the railway law stipulates that land proprietors have the right of land development, that is, the right for land development. Therefore, for lands along the highspeed rail, when their land development right goes into the market as a tradable land right circulation, it should be noted that it is a right derived from and independent of the land ownership along the high-speed railway (Fig. 8.7).

The land development right along the high-speed railway has the function of regulating the management of land development benefits and interests along the high-speed railway, so it needs to be put into the legal protection by law. Therefore, it is critical that the power of land development right along the high-speed railway be examined from the following aspects:

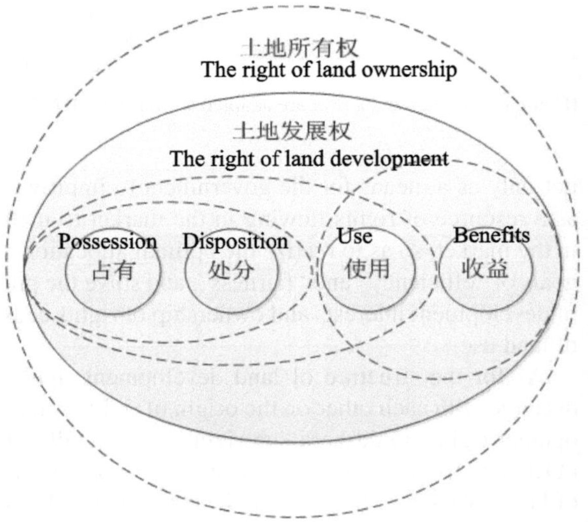

Fig. 8.7 The connotation and structure of land development right for land along high-speed rail routes

Possession: For any kind of real right, an exclusive one-to-one correspondence relationship should be established between the power entity and the object, which is the ownership (possession) of the power entity. Because the land development right along the high-speed railway belongs to the intangible property right, its ownership is mainly in the legal sense, so the legal status of the exclusive power subject can be indicated through the property right registration. Of course, if the power entity of the land development right enjoys both the right of ownership and the right of land use, it may also get the land directly under its control, forming the de facto possession.

Disposition: The disposition power of land development right along the high-speed railway is a necessary system design to realize the function of land development right along the railway, especially its revenue power. Therefore, the disposal (transaction) of land development rights along the high-speed railway includes:

(1) The power subject transfers the land development rights of specific plots to the land developers along the high-speed railway;
(2) The land development rights along the high-speed railway can be transferred between the subjects of development rights within the scope of railway laws;
(3) The subjects of land development rights along the high-speed railway are allowed to have financing cooperation for land development along the high-speed railway by mortgage development rights according to law;
(4) The right holders along the high-speed railway may take back the land development rights along the line.

Among them, the transfer of development rights along the high-speed railway is the most important content of the right to development, and the economic adjustment of land development rights to the development and use of land along the railway is embodied in the institutional design of the transfer of development rights.

Use: The power entity of land development right along the high-speed railway can exercise the development right exclusively, and enjoy the railway legal protection. All kinds of property facilities built along the high-speed rail routes by exercising the legitimate right of land development have legitimate sources of property rights, whereby they can be registered and traded.

Benefits: The land development right along the high-speed railway is a property right with economic value. Owning and exercising the right of land development along the route can bring benefits to the subject of rights. First of all, the subject of rights obtains benefits by transferring land development rights with compensation; Secondly, the right subjects with transferable development rights can obtain benefits by transferring land development rights; Thirdly, land developers can realize the economic value of land development right through land development, and obtain the benefits of land development right through property management.

Table 8.5 Development scale of high-speed railway station areas in five zones of Taiwan

	Area of commercial land use (ha)				Area of the largest commercial building	
	Transport facilities	Ancillaries	Public square	Sum	Square meter	*Ping*[a,b]
Taoyuan Station	0.17	8.55	2.00	10.72	342,000	103,455
Hsinchu Station	0.3	3.13	1.60	5.03	115,500	34,939
Taichung Station	6.14	11.30	–	17.44	542,400	164,076
Chiayi Station	0	3.14	2.30	5.44	87,920	26,596
Tainan Station	1.04	4.02	2.80	7.86	112,560	34,049
Total	7.65	30.14	8.70	46.49	1,200,380	363,115

Source Official website of Taiwan high-speed rail (https://en.thsrc.com.tw/)
[a] *Ping*, an area unit derived from the ruler penetration method of Japanese traditional measurement system, is mainly used to calculate the area of houses and building land. Mainly used in Japan, Taiwan Province in China and Korean Peninsula
[b] One *Ping* is equal to one thirtieth of one acre, or 3.3057 square meters

8.3.3 A Case Study of Land Development in Taoyuan High-Speed Railway Station, Taiwan Province

8.3.3.1 Land Use Policies in Taiwan Province

Enterprises are not allowed to expropriate land in Taiwan Province. However, with the rise of BOT mode in infrastructure, a large number of private capitals enter into infrastructure investment. The new "Regulations on Land Expropriation" in Taiwan Province has revised and stated that "other businesses may expropriate land according to law", which means that private legal persons and natural persons can also apply for land take-over. The "Legislative Yuan" of Taiwan Province passed Act for Promotion of Private Participation in Infrastructure Projects, which grant priority land development rights to operators involved in Taiwan's high-speed railway BOT project. According to the decree, all BOT cases involving land expropriation by sections may apply to the authority in charge, who will prioritize the land within the scope of the zone or section to be allocated (Table 8.5).

8.3.3.2 The Integrated Development of Land in Taoyuan Station Area

Take Taoyuan Station as an example. The land in the station area is owned by the government, but Taiwan High-speed Railway Corporation (THSRC) has acquired the right to develop on the ground. The government has approved THSRC a 35-year period to use it for high-speed railway affiliated businesses such as hotels, conferences, industrial and commercial exhibition centers, restaurants, leisure and entertainment, department stores, financial services, etc. in addition, it is granted a 50-year of the above-ground use right, during which the land can be used and developed by the THSRC itself, jointly with or by other companies.

There are 10.72 hectares of land for other development in Taoyuan Station area, including 8.55 hectares of land for affiliated businesses mainly used for commercial development, with a coverage rate of 60%. The allowed land use projects include hotel facilities, conference and industrial and commercial exhibition centers, catering industry, leisure and entertainment industry, department stores, retails, financial services, general services (special services are prohibited), transportation services, tourism services and offices.

The THSRC acquires the land next to the station by the way of section/zone expropriation in the defined area, and is required to make planes for the public works and affiliated businesses of the station area together with the construction of the high-speed railway. After the land near the high-speed railway station was converted from agricultural land with low value to the construction and commercial land, its value soared up quickly.

References

Brocker, J. Operational spatial computable general equilibrium modeling, Annals of Regional Science 32 (1998) 367–387

Crampton, G. R. (2003). Economic development impacts of urban rail transport. Jyvaskyla, Finland: European Regional Science Conference.

Gibbons, Stephen and Machin, Stephen. (2005). Valuing rail access using transport innovations. *Journal of Urban Economics* 57(1): 148–169.

Marx (1875). https://www.marxists.org/archive/marx/works/download/Marx_Critque_of_the_Gotha_Programme.pdf

Sheppard, S. (1999). Hedonic analysis of housing markets (Chapter 41). In: P. Cheshire, E. Mills (Eds.), *Handbook of Urban and Regional Economics*. Elsevier Science.

Simmonds, D. (2001) The Objectives and Design of a New Land-use Modelling Package: DELTA. In: Clarke G., Madden M. (eds) *Regional Science in Business. Advances in Spatial Science*. Springer, Berlin, Heidelberg. https://doi.org/10.1007/978-3-662-04625-8_9

S. Rosen, Hedonic prices and implicit markets: product differentiation in pure competition, *Journal of Political Economy* 82 (1974) 34–55.

Venables, A. & Gasiorek, M. (1999). The welfare implications of transport improvements in the presence of market failure. In: Report to Standing Advisory Committee on Trunk Road Assessment. DETR, London, pp. 5–48.

References from Chinese Sources

Chen Youxiao, Lin Xiaoyan, Liu Yunhui. Impact of Rail Transport Construction on Land Value, Valuation Model and Empirical Study——a Case Study on Beijing Light Rail [J]. *Journal of Beijing Jiaotong University* (Social Science), 2005 (9): 7–13.

Liu, Guozhen. On the British Land Development Rights System and Its Implications for China, *Law Review*, 2008(04): 141–146.

Liu, Guozhen. On the Land Development Rights System in the United States and its Implications for China, Law Review, 2007(03): 140–146.

Chapter 9
High Speed Railways and Regional Service Quality

9.1 Service Quality Theory

9.1.1 Customer Satisfaction Degree

To study customer satisfaction, it is the most crucial to understand it and give it a proper measurement. Definitions on this concept vary, and no consensus has been reached by scholars all over the world. However, related research papers, be they published in China or the rest of the globe, show that customer satisfaction can be generally defined as a feeling that has two dimensions: The expectations and actual perceptions of a product or a service. By definition, customer satisfaction comes from the subject response that arises from the disparity between these two factors. The smaller the gap is, the more satisfied customers are with their purchases. Here we draw on the view of Philip Kotler, a prestigious scholar in the field of marketing. According to his definition, customer satisfaction is 'a person's feeling of pleasure or disappointment resulting from comparing a product's perceived performance or outcome against his/her expectations'.

There is also no strict definition of Customer Satisfaction Degree (CSD), as American scholars Peterson and Wilson (1992) said: "The biggest feature of customer satisfaction research is that the term lacks a definition." This book, adopting the conclusions of studies in this field, considers CSD as a quantitative statistical indicator of customer satisfaction that can be used to describe the difference between expectations and perceptions of the product, thus serving as an index of CSD. CSD is high when customers' perceptions go beyond their expectations, otherwise it is low.

Customer satisfaction has the following features:

(1) Subjectivity. Customer satisfaction is based on customers' experience of using the product or service. The thing perceived is objective, whereas the perception

X. Lin, *High-Speed Railways and New Structure of Socio-economic Development in China*, Research Series on the Chinese Dream and China's Development Path,
https://doi.org/10.1007/978-981-19-6387-2_9

is subjective—involving factors ranging from customers' knowledge, experience, income, lifestyles and values, to media publicity.

(2) Hierarchy. The hierarchy of needs proposed by the famous American psychologist Maslow shows that there are five levels of human needs, namely physiological needs, safety needs, love and belonging needs, esteem needs, and self-actualization needs. People at each of needs value products or services differently. It can be said that customers' social classes or places of residence may play a part in CSD for a certain product or service, and CSD may also be influenced by individual circumstances.

(3) Relativity. Customers tend to be less familiar with the economic indexes (cost, for example) and technical indexes of a product. They are likely to compare what they purchase with similar products, or relate this purchase experience to their previous one. Satisfaction or dissatisfaction derived from the experience is thus relative.

(4) Dynamics. Due to the variability of customer needs, customer satisfaction changes with the passing of time, technological progress, and the improvement of the whole business and marketing environment. At the same time, the advantages of enterprises will also be completely different. With the development of the society, economy and culture, the needs and expectations of customers will correspondingly increase, and customer satisfaction will change accordingly. Those who are satisfied with a product may dislike it in the future.

There is a motto in the quality world: "You can't manage what you don't measure". Many scholars have done a lot of research on methods of evaluating customer satisfaction, and put forward a variety of models. Some representative ones, such as the PE model, the KANO model, the quarter graph model and the customer satisfaction index, have proven quite reliable and effective in many situations. These approaches are compared in Table 9.1.

Customer Satisfaction Index (CSI), a comprehensive measure of customer satisfaction, is based on customers' evaluation on companies, industries, and even countries in terms of how well these entities respond to customer needs. The main difference between CSI and the productivity index is that the latter focuses on quantity, while the former mainly measures the quality from the perspective of customers. In European and American countries, the application of CSI has been a great success, making it more appealing to an increasing number of countries and regions. The established customer satisfaction indexes at the national or regional level includes the Swedish Customer Satisfaction Barometer (SCSB), the American Customer Satisfaction Index (ACSI), the German Customer Satisfaction Index (DK), Korean Customer Satisfaction Index (KCSI), Swiss Customer Satisfaction Barometer (SCSB), and European Customer Satisfaction Index (ECSI), etc. In addition, with the development of China's market economy, banking, telecommunications, real estate and other sectors have gradually transitioned from monopolies to competitive industries. In order to outcompete their business rivals, companies not only need to react quickly to a series of indicators, including scale, price, and profit, but also are required to manage an important resource: Customers. Meanwhile, as the regulator of the market economy, the government is also the quality supervisor of various products and services. Since customer satisfaction directly contributes to people's quality of life, the theory and practice of customer satisfaction research were developed in

Table 9.1 Approaches to measuring customer satisfaction

Model	Methods	Features
Linear structural equation model	Model fitting, covariance estimation and sampling methods	Has a multivariate normal distribution Applicable to experimental research A minimum of 100–150 samples are required Researchers should have a working knowledge of related theories before they can apply this model
Partial least squares regression	Iterative calculation based on principal component analysis (PCA) and multivariable regression	Applicable to experimental and explanatory research Applicable when the data pattern does not meet normal distributions At least 10 times more complex than the variables it describes Researchers do not need to have a working knowledge of theories related to this model
KANO model	Identifying customers' needs and analyzing them individually	Not applicable to measuring customer satisfaction Often used as an assistive model in the preliminary evaluation stage
Analytic hierarchy process	Quantifying customers' subjective judgments	Effective in dealing with complicated problems Applicable to evaluating the importance of each customer satisfaction index
Quality function deployment	In the form of matrices, illustrating the connection between customers' needs and design specifications of the product; doing quantitative analyses to find out specifications that can satisfy customers' needs most	Market research is crucial The findings of the market research should be accurate Customers' needs are recordable and consistent

China in the 1990s (Liang, 2007). In 2001, China National Institute of Standardization and Tsinghua University first established the China Customer Satisfaction Index (CCSI) model, and conducted two nationwide experiments on the evaluation of customer satisfaction. The CCSI model is formed on the basis of comparative analysis of SCSB, ACSI and ECSI. Previous CSD studies in China, be they at the national, industry or enterprise level, were mostly carried out on the basis of ACSI model. In this book, before several customer satisfaction index models were built, their frameworks and variable selection had been revised according to China's reality.

9.1.1.1 Service Control Theories

The term "service quality" was coined a long time ago. In the 1970s, it was already noted that service and product were different, so were service quality and product quality. However, the exact definition of service quality has been absent so far. Studies on service quality show that scholars at home and abroad have reached consensus on this issue: Service quality is called customer perceived service quality as it can be perceived by customers; customer satisfaction is determined by customer expectations and customer perceptions.

Following are the working definitions of service quality proposed by scholars worldwide: Sasser et al. (1978) believes that service quality involves the outcome as well as the delivery process; Professor Gronroos (1984) from Finland, adopting the basic theories of cognitive psychology, proposes a pioneering model of customers' perceived service quality and clarifies its constituent elements. Most of the subsequent studies on service quality models can be traced to his groundbreaking work. Gronroos (1984) argues that, in psychological terms, service quality is subjective and determined by customers comparing their expectations of service quality with their perceptions of the service's performance (that is, their experience of the service). Gronroos divides service quality into "technical quality" and "functional quality". The former refers to the output of the service process, which is what customers get from the service (what), while the latter is defined as how customers get this kind of service (how). He also points out some factors in the marketing field that affect service quality (Wang, 2006).

PZB (Parasuraman, Zeithaml, and Berry), a service management research group funded by the American Marketing Association, conducted an in-depth study on service quality, and put forward a "difference theory" on how customers evaluate service quality. The study found that customers' perceived service quality dictates the customers' evaluation of service quality, and that customers' perceived service quality is associated with the gap between the customers' experience of the service and the customers' expectations of it. Lewis and Booms (1983) define service quality as "a tool to measure whether a company's service can meet customer expectations." PZB (1988) considers service quality as a concept similar to "attitude", maintaining that service quality comes from customers comparing their perceptions of the various aspects of service quality with their expectations of the service. Gummesson (1988) proposes a model in which service quality is classified as designed quality, production quality, delivery quality and relational quality. Later he (1991) revises the model, dividing service quality into design quality, production quality, process quality and output quality.

ISO 8402 (1994) defines service quality as "the totality of characteristics of an entity that bear on its ability to satisfy stated or implied needs", where characteristics refer to what distinguishes a service from those of other categories or those identical but meeting different standards.

Through empirical research on hospitals, Chinese scholar Wang Chunxiao (1999) concludes that service quality includes environmental quality, technical quality, emotional quality, relational quality and communicational quality. Fan Xiucheng

(1999) puts forward the concept of interactive quality, arguing that technical quality and interactive quality are part of service quality. In this book, Gronroos' notion is adopted, which defines service quality as the comparison between customers' expectations of the service and the service's perceived performance. When service performance exceeds customers' expectations, customers perceive that the service quality is good, otherwise it is bad. Unless specified, the service quality mentioned in this book refers to perceived service quality, since no approach can evaluate services that cannot be perceived by customers and the quality of service depends on the expectations and perceptions of customers.

9.2 Service Quality of High-Speed Railway Trains

This part does not look into safety, which constitutes the very basis of customer satisfaction. A transport service cannot be said to have any degree of customer satisfaction if it is not safe.

9.2.1 Features of the High-Speed Railway Market

9.2.1.1 Characteristics of High-Speed Railway Passengers

With the development of society and economy, there is a growing need for a better quality of life. Customers not only want high-quality tangible goods but also high-quality services, making it necessary to improve the service quality of railway passenger transport.

(1) Short waiting time for passengers

Compared with ordinary railways, high-speed railways have more frequent arrivals and departures, and the number of passengers entering and leaving the station within a unit time has greatly increased. High-speed railways must be efficiently organized so that passengers do not need to wait too long at the station. Stations serving as waiting areas are seen as outdated in the era of high-speed travel: Passengers want to enter and leave the station more quickly, and a simpler procedure is also preferred. Therefore, the service quality of the station and the efficiency and reputation of high-speed railways will be negatively affected if the station layout is not reasonable and the procedure for entering and leaving the station is complex. At present, Railway stations outside China have gradually transformed into urban comprehensive transportation hubs, and waiting for trains is no longer their main function. In developed countries such as Japan, France, and Germany, there is hardly any waiting hall in railway stations. Now railway stations main provide transfer service for passengers between various transport modes.

(2) Rising demand for high-quality travel experience

Passengers who choose high-speed railways generally value their time more and are willing to pay more for services. They want a relatively higher level of comfortableness in their ride, in addition to better refreshments, entertainment and communication services. That requires high-speed trains to be equipped with adequate service facilities. For example, in some countries, catering service on railway trains can already match that on planes. Such services, which have simplified the operations on the train and require fewer attendants, feature better service quality and can generate considerable economic benefits.

(3) Short time for passengers to get on or off the train

Ranging from 3–10 min to 0.10–1 min, the intervals between the arrivals and departures of high-speed trains are relatively short. Therefore, passengers need to get on or off the train more quickly, making clear platform signs very necessary. The absence of such signs will put passengers into great inconvenience, causing delays that impair the efficiency of the whole railway system.

Because of the above-mentioned characteristics of high-speed railway passenger transportation, countries need to make continuous efforts to provide high-speed railway services that cater for the needs of passengers, many of whom want to have a more friendly and comprehensive experience in their travel. Doing so will help increase the competitiveness of high-speed railways.

9.2.1.2 Division of the High-Speed Railway Market

Overall, passenger needs have a close link with the urban and industrial layout of the country and its domestic household spending. Based on the scale and distribution of the entire passenger transport network, passengers can be roughly divided into three categories: The high-income group, the medium-income group, and the low-income group. Passengers in the high-income group, with greater purchasing power, primarily take business trips. At present, such passengers are mainly found in the civil aviation sectors, or often take short-distance road trips. The middle-income group is dominated by corporate employees. As urbanization continues and the occupational structure changes, these people will become the majority of high-speed railway passengers. In addition, with improved quality of life and the resulting shift in people's lifestyles, commuters and tourists, who are also a part of this cohort, will also show a rapid growth in terms of number. The low-income group still takes up a considerable proportion of the population of China, especially in the early stage of the transition of the country's industry and occupational structure. For these passengers, successfully arriving at their destinations, instead of the travel experience, takes priority over other matters of their travel. (Qiang Lixia and Yan Ying, 2007). Specifically, the high-speed railway passenger transport market can be divided by age, income, occupation, and travel expense reimbursement system.

(1) The age group of passengers

Age, one of the attributes of passengers, can reflect the length of travel that they can endure—as well as other factors such as income and travel habits. Studies found that the largest proportion of passengers are under 30, followed by the 30–40 cohort. These two groups of passengers account for more than 80% of the total, a figure which is expected to be higher during peak times (the Spring Festival and summer breaks, for example). Accordingly, it can be inferred that most high-speed railway passengers are aged under 40.

(2) The income standards of passengers

For passengers traveling at their own expense, income can largely influence their travel preferences. A survey shows passengers with a monthly household income between 2,000 and 5, 000 yuan outnumber those whose family earns more than 1,0000 yuan per month or less than 1,000 yuan per month. Transport services of different grades have various degree of appeal to passengers of different income groups. In general, low-income earners prefer lower-grade transport modes, and higher-grade modes are often the choices of those from the high-income group. At the initial stage of operation, high-speed railways (such as Beijing-Tianjin intercity trains) were priced more highly than existing lines, targeting on only passengers who were taking fast railway trains (120 km/h) or express trains (140 km/h) as well as part of airline and road passengers. However, the attraction of high-speed railways will gradually expand as China's per capita income continues to increase and passengers are far more willing to pay than before. (Liao, 2006).

(3) The occupational structure of passengers

Occupation is an important indicator of a passenger's social identity and economic status, thus largely influencing consumer behaviors. According to a survey (Liao, 2006), four types of high-income passengers—namely civil servants, company employees, teachers, and businessmen—constitute nearly 60% of the airline passengers, 38% of road and express train passengers, and only 30% of passengers taking normal passenger trains and fast railway trains. Therefore, it can be inferred that high-speed railways are a favorable option to the high-income cohort, as these passengers prefer high-grade transport service and put an emphasis on speed and comfortableness.

(4) Regulations for travel expense reimbursement

High-speed railway fares are about equal to those of existing sleepers, which means that most companies and governmental bodies can buy high-speed rail tickets for their staff. In addition, except airplanes, high-speed railway trains are unmatched by other modes of transportation in terms of speed and comfortableness. Therefore, high-speed railways will be the first choice of customers on business trips.

9.2.1.3 Basic Characteristics of High-Speed Railway Transportation Customer Satisfaction

(1) Subjectivity and objectivity. Built on customers' experience of using a product or service, customer satisfaction is a subjective feeling, whereas the product or service itself is objective. High-speed railway trains edge out traditional railway trains or other modes of transportation in terms of speed, comfortableness and other aspects of travel services. All these advantages do contribute to customer satisfaction, but the effects are diminished by the remote locations of high-speed railway stations as the time that customers save during their high-speed railway train rides is wasted on the way to the station. This is an existing problem that may concern rational customers.

(2) Complexity and diversity. According to Maslow's hierarchy of needs, human needs can be divided into five levels, and people at different levels of needs have various evaluation standards for products or services. Therefore, a customer's residence and class may affect how he/she rates a certain product or service, and even the individual circumstance is also one of contributor to this. For example, elderly civil servants who ride high-speed railway trains earn the same income as junior company employees do. However, due to their age and occupational disparity, elderly civil servants value such aspects of the service as comfortableness, while young civil servants prefer reduced travel time.

(3) inconsistency and variability. With the development of society, economy and culture, the needs and expectations of passenger increase, giving rise to changes of customer satisfaction. For example, for many civil servants who have resigned to do business, a new career path—with the consequent change in their income and the increase of their age—may cause them to opt for new modes of transportation or raise their expectation for certain parts of the high-speed railway service. Overall, as the high-speed railway passenger service system improves, the level of passenger satisfaction will increase accordingly.

As can be seen from the above, high-speed railway customer satisfaction is closely related to customers' backgrounds. Therefore, by establishing a comprehensive evaluation system for high-speed railway passenger transport service quality—in which data can be collected and sorted to analyze the needs and expectations of different customer groups—targeted improvement plans can be formulated to satisfy most customers.

9.2.2 High-Speed Railway Transportation Service Quality Index

Generally, the demand of railway passenger transportation can be divided into functional demand and derived demand. The passenger satisfaction index is established on the basis of the seven aspects that affect the level of high-speed railway passengers' customer satisfaction, including: ① environmental factors (the environment of the waiting room and cars); ② economical factors (ticket and refreshment prices, etc.); ③ time-related factors (waiting time, the speed of the train, etc.) ④ factors that contribute to comfortableness of the waiting room and cars; ⑤ quick access to ticket purchase, refund and change, quick transfer and easy ways to enter and leave the station; ⑥ the hospitality of service personnel, punctuality, departure frequency, the availability of train information, inquiry service and comprehensive services in case of train delays, and other services of trains as well as service supervision; ⑦ facilities installed on the train, the waiting room's capacity, guide signage, and the capacity and design of the entrance and exit, etc.

Taking into account all dimensions of the high-speed railway passenger transport service system, this book establishes a comprehensive evaluation system for the service quality of high-speed railway passenger transport (see Table 9.2).[1]

9.3 High-Speed Railway Service Quality and Regional Economic Efficiency

The measurement of the service quality of high-speed railways at the regional and societal levels is a long-term concern of the academic community. The measuring indexes proposed by this book are different from those by previous studies. In public investment terms, high-speed railways fall into the category of strategic macro-control imposed by the government, so in planning high-speed railways, the government must take into account the profitability of the transportation sector, and more importantly, the impact of high-speed railways on the overall regional economy. When it comes to technical and economic attributes, high-speed railways increase the connectivity between regions, offering passengers more possible reachable destinations. This can largely promote the clout of the region and facilitate the circulation of resources. Besides, the consequent expansion of the market and the acceleration of

[1] The process of developing the index: In the first stage, a large number of literatures about railway passenger transport service system were studied, and many aspects of the existing problems in China's current passenger transport system were found out. From the perspective of public expectation to improve service quality, each specific observation point of the index system was found and a preliminary index system was formed. In the second stage, according to the high-speed railway passenger service system (Zhao Fei, 2000), many observation points of the high-speed railway passenger service system are selected, and multiple indicators are formulated. The final evaluation method and theoretical support of the index system are also outlined. To study the measurement means and acquisition benchmark of the perceived value and expected value of passengers by groups, mainly according to the data obtained by the index system, it is divided according to age, gender, occupational group, annual income and travel expense reimbursement system, etc., and the data is tested by F test and T test based on SPSS statistical software. In the third stage, on the basis of the index system established in the second stage, the shortcomings of China's current passenger service system are integrated, and observation points, that is, specific indicators, are added.

Table 9.2 The evaluation index system of passenger service quality of high-speed railway

Level 1 indicators	Level 2 indicators	Level 3 indicators	Main customer satisfaction contributors
Passenger services in the station	Ticketing	Automated ticketing	The availability of automated ticketing service
		Ticketing in the station	The spatial locations of ticket windows, time spent on ticket purchasing, hospitality of ticket agents, update of ticket information, and the environment of the ticket office (including its air quality, temperature, humidity, lighting, hygiene, tidiness, capacity)
	Passenger guidance	Equipment for passenger guidance	Whether the high-speed station is equipped with a well-equipped passenger guidance system, multimedia technology and eye-catching signs are used to point out the direction and location of various passenger service facilities at the station, to guide passengers to conveniently use various service facilities at the station, and to conveniently get on and off trains and change trains according to prescribed routes
	The waiting hall and platforms	Check-in and check-out	The ability of the station to ensure efficiency and smooth passenger flow; the width of entrances and exits of the station; the visibility of signs for train cars
		Signs for on board/off board	
		Services during the waiting	1. The environment of the waiting hall, including its air quality, temperature, humidity, lighting, hygiene, tidiness and capacity 2. The availability of refreshments, shopping and hot water, restrooms 3. Clear signage in the station, broadcasting, wireless network, and quick access to the train 4. The provision of platform tickets
	Self-service inquiry	Information updates	Whether the station deploys advanced inquiry systems that provide various types of information, ranging from train departures and arrivals to the city transport and accommodations
		the number of systems available	
	Automated check-in	The availability of e-tickets	Whether there are enough e-tickets and automated check-in systems

(continued)

Table 9.2 (continued)

Level 1 indicators	Level 2 indicators	Level 3 indicators	Main customer satisfaction contributors
		The deployment of automated check-in systems	
	Other services	Responding to customer feedback	Whether customers' inquiries related to other issues are handled properly; whether customers' suggestions and complaints are managed
		Other Reservation services	Whether the station can help customers reserve hotels, restaurants and entertainments
On-board service	The operating of the high-speed trains	Public security on the train	Whether public security is maintained during the ride
		Smoothness of the train	whether the train is stable in operation; whether passengers feel comfortable about the ride
		Punctuality of the train	Whether the train arrives or depart on time
	Services provided by the crew on the train	Friendliness and hospitality of train attendants and hygiene workers	Whether train attendants and hygiene workers are friendly and welcoming
		Emergency management	Whether emergencies (sudden illness and conflicts, for example) are handled properly and timely
		Responses to complaints	Whether complaints about the ride are managed properly
	Food services	Food safety	The organization of food service; food processing and delivery; in-time food delivery for passengers with first-class tickets; the availability of vendors dining cars
		Price	
		Hospitality	
	Telephone and Internet	The number and types of facilities	The availability of high-quality international and domestic calls, fax and wireless network

(continued)

Table 9.2 (continued)

Level 1 indicators	Level 2 indicators	Level 3 indicators	Main customer satisfaction contributors
		Signal strength	
	Train information updates	Information display	The display of information relating to the train (its speed, temperature, arrival and departure, stations it arrives at, and the time it stays at the station, etc.)
		Information accuracy and timeliness	
	Accommodations and restrooms	Whether facilities are well-arranged and people-centered	Whether the seats and beds are clean; whether the cars are quiet; whether environmentally-friendly portable toilets are deployed; whether benches for parents with infants are available
		The sanitary condition	
Supporting transport facilities	The coordination between high-speed railways and city public transport	The networking of city public transport services	The connectivity of city transport modes with the high-speed railway station; the availability of quick, convenient bus services; the development of moving walkways and ramps for easy transfers

Note (1) Description of evaluation indexes: All the indexes, with a total score of 5 for each one, were divided into three categories: 3 first-level indexes, 13 s-level indexes, and 28 third-level indexes; (2) How the indexes were determined: participants were asked to answer a question designed for each index and give their scores. (3) participants included passengers and railway staff; (4) the meaning of scores: 1 point for very poor, 2 points for poor, 3 points for satisfactory, 4 points for good, and 5 points for excellent resource circulation have largely marginalized the effectiveness of repeated construction of facilities. To a certain extent, high-speed railways enable the integration of the previously dispersed industrial layout, which can impact the overall economy of the region (particularly its spatial connectivity with other regions, its economic and demographic structure). Therefore, this book uses social utility as indexes of the quality of high-speed rail services, and divides these indexes into three categories: Spatial connectivity indexes, regional economy indexes and demographic structure indexes, as is shown in Table 9.3.

Spatial connectivity refers to the ability of active nodes in an area to influence each other. This book chooses accessibility as the index of spatial connectivity. Accessibility is defined as the easiness of using a specific transportation system to reach an activity location from a given location, and is the ratio of the transport distance (D) of an active node to the average transport distance of all nodes in the system. The formula is as follows:

Table 9.3 Social-utility-based high-speed railway service quality evaluation system

Types of indexes	indicators selected	Specifications
Spatial connectivity	Accessibility	Accessibility
	Railway capacity (in passengers and freight transported)	Passenger turnover Rotation volume of goods
Regional economy	Per capita GDP, Economic contribution of primary, secondary and tertiary industry; Urbanization rate	Per capita GDP, Economic contribution of primary, secondary and tertiary industries, Urbanization rate
Demographic structure	Employment figures for primary, secondary and tertiary industry; Population density	Employment figures for primary, secondary and tertiary industry; Population density

Table 9.4 Classification of control factors of high-speed railway service quality

Control factors	
Speed	Technical factors
Safety	
Price	Economic factors
Convenience	
Comfortableness	
Tangibles	Managerial factors
Empathy	
Service commitment	
Random factors	
Passengers' past experience	
Corporate image	
word of mouth	

Table 9.5 Items of the questionnaire on high-speed railway service quality control factors

Item code	Service control factors	Item
Q201	Tangibles	Do the station and high-speed railway trains look attractive?
Q202		Do the station service crew and train attendants follow the dressing code?
Q203	Speed	Do the train travel fast?
Q204	Safety	Are high-speed railway stations safe, almost free from criminal activities?
Q205		When in operation, is the train stable?
Q206	Empathy	Are train attendants friendly and welcoming?
Q207	Comfortableness	Is the waiting hall of the train station cozy and comfortable?
Q208		Are the seats on the train comfortable? Are the air quality and temperature on the train satisfying?
Q209	Convenience	Is is easy to buy tickets?
Q210		Are the entrance and exit signs clear and easy to understand?
Q211		Can passenger transfer easily to buses or the subway at the station?
Q212		Is the train timetable friendly to all passengers?
Q213	Prices	Are the tickets affordable?
Q214	Service commitment	In general, is the service consistent with media publicity?
Q215	Random factors	Are high-speed railway trains safer than other transport modes when the weather is bad
Q216	Corporate image	Are high-speed railway trains environmentally friendly?
Q217	Previous travel experience	Were the past journeys enjoyable?
Q218	Word of mouth	Are high-speed railway trains recommended by your colleagues and friends?
Q219	Service value	Do the passenger transport service have good value for money and time?
Q220		Are the ticket price and travel time reasonable?
Q221	Complaints	Do you know where to file a complaint
Q222		Are the complaints handled properly and in time?
Q233	Customer satisfaction	In general, the service is satisfying
Q234	Customer loyalty	Next time will you still choose high-speed railways?
Q235		Will you recommend high-speed railways to your colleagues and families?

Table 9.6 Reliability

Cronbach's alpha	Item number
0.921	23

$$A_{di} = D_i/(\sum D_i/n) \quad (i = 1, \cdots, n) \tag{9.1}$$

In Eq. (9.1), A stands for accessibility and D is the distance between each node. To simplify the calculation, the distance is rewritten using the reciprocal of time. The lower the value of accessibility is, the more accessible the region is. A city with accessibility value below 1 is more accessible than China's average.

The regional economy is reflected in the change in its total gross regional product (GDP), the GDP of its primary, secondary and tertiary industries and their proportions, and the urbanization rate brought about by the construction of high-speed railways. Although high-speed railways cannot directly affect the economy, but they can exert their impacts by activating related industries and changing the regional spatial layout. Therefore, analyzing the relationship between regional economy and high-speed railway is an important way to evaluate the specific utility of high-speed railway. The urbanization rate is the ratio of urban population to the total population of the region:

$$\mathrm{CHL} = \mathrm{CP}/\mathrm{AP} \tag{9.2}$$

In formula (9.2), CHL is the urbanization rate, CP represents the city population, and AP stands for aggregate population, which is the sum of agricultural and non-agricultural populations.

Demographic structure is also transformed by the construction of high-speed railways, which plays a great role in promoting the circulation of regional resources. This paper selects the number of employees in the three industries as an indicator for analysis. In population migration, high-speed railways have comparative advantages over other modes of transportation in terms of factors such as price, time, and comfort. In addition, high-speed railways help reorganize the industry and geographical layout of the region, as well as the employment figures for the local primary, secondary and tertiary industry.

After determining the indicators, the multiple regression prediction method is used to carry out regression analysis and fitting prediction on the indicators selected, and obtain the weight relationship between the economic development and various indicators. This relationship plays a decisive role not only in the classification of cities by size, but also in the division of urban functions.

Let y be the dependent variable, and × 1, × 2, ···, xk be the independent variables. When there is a linear relationship between the independent variable and the dependent variable, the multiple linear regression model is:

$$y = b_0 + b_1x_1 + b_2x_2 + \cdots + b_kx_k + e \tag{9.3}$$

In formula (9.3), b_0 is the constant term and $b_1, b_2 \cdots, b_k$ are the regression coefficients.

According to China's " Methods and Parameters of Economic Evaluation of Railway Construction Projects" (Third Edition) (2006), the economic evaluation methods of China's railway projects can be divided into financial analysis, economic cost benefit analysis, cost effect analysis, regional economy and macroeconomic impact analysis, uncertainty and risk analysis, program economic comparison and selection, etc. Adopting the "yes or no" method, this book studies the impact of the construction of high-speed rail on the economy along high-speed railways. The formula for calculating the social utility of is:

$$U_t = \sum \frac{1}{2}(Q_{ij}^0 - Q_{ij}^1)(A_{ij}^0 - A_{ij}^1)$$

In formula (9.4), 0 on the upper right corner of the variable indicates planned high-speed rail, and 1 indicates that the high-speed rails under construction;

U_t: social utility at time t;
Q_{ij}: The transportation quantity from city i to city j before and after the construction of the high-speed railway.
A_{ij}: the accessibility of city i to city j.

In formula (9.4), the weight of transportation quantity Q is obtained through the multiple linear regression model, which is used to predict the impact of the high-speed railway on the change of Q as well as the change of the economy. After that travel time before and after the operation of high-speed railways is determined by related files, and its value T can be plugged in the formula (9.1) to determine the value of accessibility, which is pivotal to finding the social utility U generated by each section of the high-speed railway. Once the social utility U is determined, it will be easy to identify the area most significantly affected by the service quality of high-speed railways (Chen & Lin, 2014).

9.4 High-Speed Railway Transportation Service Quality Control

9.4.1 Characteristics of High-Speed Railway Transportation Products

In reality, services and products are often inseparable —when consumers buy products, they will definitely purchase some services as well. Thus, in this book, the inalienability of products and services is proposed to help managers take a holistic view towards marketing and quality control while avoiding thinking only from the perspective of products.

The inalienability of products and services is dynamic, making it crucial for managers to have a good understanding of the consumption process of customers, throughout which companies must manage to increase the overall value perceived by customers and enable these people to achieve their consumption goals. This can improve the service quality perceived by customers.

The products of railway transport companies can be classified as passenger transport and freight transport. The core of the former is passengers' movement to their destinations. This type of product is made accessible by passenger trains and facilities such as railway overhead line equipment, all of which can dictate the punctuality, comfortableness, speed, price and other dimensions of high-speed railway services. In addition, advance tickets, pre-boarding service, on-board service, post-arrival service and brand effects can also be seen as part of high-speed railway service.

Seeing high-speed railway passenger transport as a combination of products and services, Fig. 9.1 shows the three layers that constitute its very basis.

Technological layer: It is the inner layer of high-speed railway passenger transport, involving transporting passengers to their destinations. High speed, one the noticeable feature of high-speed railway trains, lies in this layer. Generally, the quality of passenger transport is directly proportional to the speed of the train—the faster the vehicle travels, the better the product quality is. However, it comes at a huge cost to increase the speed of the vehicle as the technological and economic conditions of transport companies remain fixed. For example, in contrast to a total length of 1,318 km, the Beijing-Shanghai high-speed railway cost more than 200 billion yuan

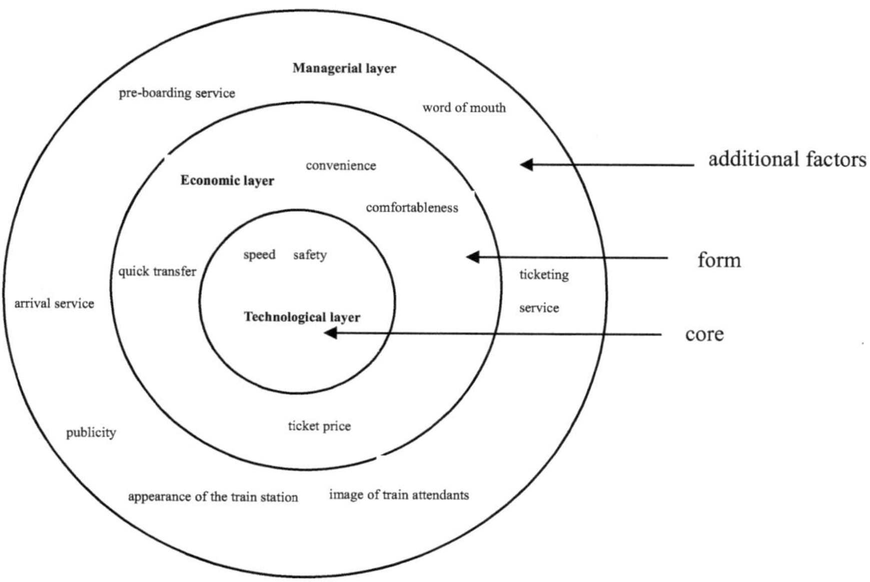

Fig. 9.1 Schematic diagram of passenger service of high-speed railway transport

(150 million yuan per kilometer). Higher speed promotes the perceptions of better service quality, but more money has to be spent on building faster vehicles.

Economic layer: In this layer there are travel time and travel expenditure. Factors in this area include convenience, safety, reliability, and comfortableness of high-speed railway services, all of which affect passengers' perceptions of the high-speed railway service. Travel purpose, cost, information, travel time, time value, comfortableness and even congestion can have a great impact on the decision-making of passengers.

Managerial layer: In this layer are the appearance of the station, the images and hospitality of train attendants, branding, publicity, etc. These also affect passengers' perception of service quality.

As is shown in Fig. 9.1, the biggest difference between a transport product and other products is that the former—a service provided by railway transportation companies to change passengers' spatial location—lacks a physical form. Therefore, each layer in the figure, especially the inner and the peripheral layer, are actually services. Railway passenger transportation companies must carry out marketing through promoting high-quality services—in which they can showcase product quality—so as to increase their competitiveness.

When customers purchase railway passenger transport products, they are willing to pay extra for the high-quality services in the core and middle layers. Thus their perceptions of the railway service very important. The railway passenger transport service department has established a certain image for passenger transport products in terms of quality, service, safety, advertising, grade, brand and price, to achieve the social and economic benefits of related enterprises. Passengers' willingness to pay is determined not only by the transport product itself, which is moving people from one place to another, but also by its service quality which is dictated by customers' perceptions.

9.4.2 The Modelling of Service Quality Gap of High-Speed Railway Transport

The five gaps proposed by the (PZB, 1988) are: The gap between customers' expected services and the managers' understanding of them, the gap between the managers' understanding of the customers' expected service and how such understanding is reflected in service design, the gap between service design and service delivery, the gap between service delivery and the publicity of service, the gap between perceived service and expected service. The five gaps analyzed by the passenger service gap model are shown in Fig. 9.2.

Gap 1: The gap between customers' expected services and the managers' understanding of them

Managers of railway passenger transport companies sometimes do not know what passengers' real needs are. For example, sometimes they think that their customers

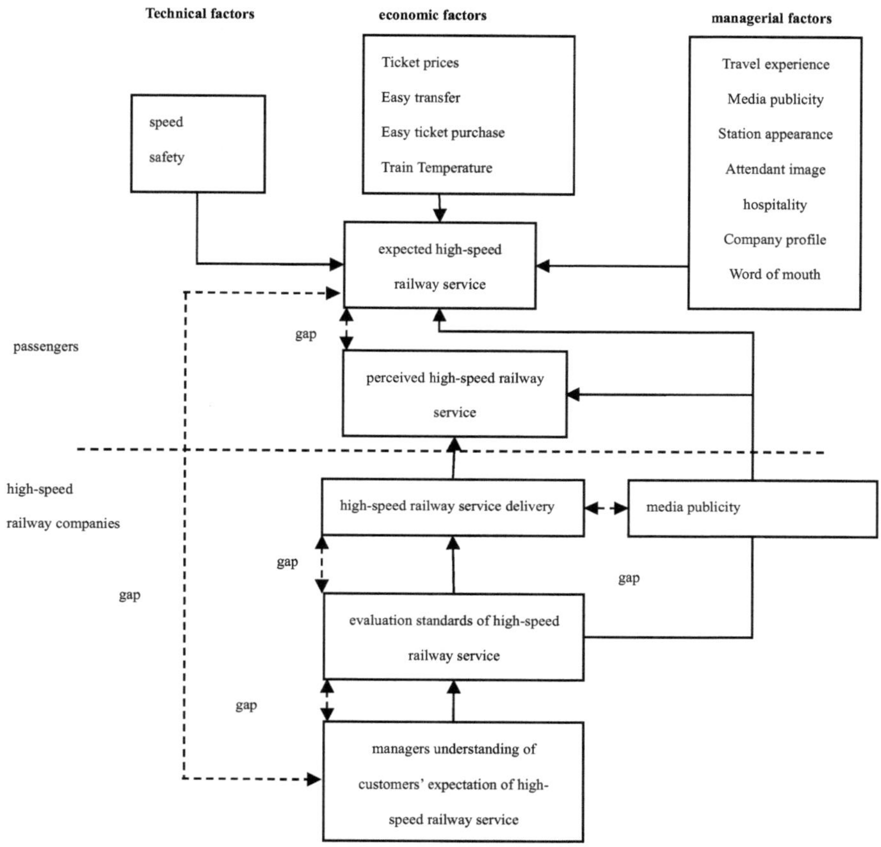

Fig. 9.2 A gap model of service quality of high-speed railway transportation products

prefer fast transport, while some fare-sensitive passengers are reluctant to spend more money for higher speed.

Gap 2: The gap between the managers' understanding of the customers' expected service and the making of passenger railway service standards

Even though the managers can recognize the real needs of passengers, they may be unaware of how to draft the corresponding standards that aim to satisfy such needs. For example, these executives know that the service staff and the crew on the train should friendly and welcoming, but lack a clear definition of these adjectives.

Gap 3: The gap between service design and service delivery

Service staff of the train station and train attendant are the majority of the service crew who need to engage customers. Therefore, the actual service provided by these employees may not meet the standard if they are incompetent or deviate from their job requirements. In addition, specific standards of service quality—for example, punctuality and safety—sometimes conflict with each other, which can also affect the provision of standardized services.

Gap 4: The gap between service delivery and the publicity of service

Passengers' expectations are often influenced by publicity. Certain passenger stations boast about their tidiness and hygiene, disappointing passengers who have great expectations of the station environment.

Gap 5: The gap between perceived service quality and expected service quality. Passengers' judgments on the quality of railway passenger transport services are affected by many seemingly peripheral factors (the inquiry service offered by a patient station conductor, for example), all of which can influence their perceptions of the quality of railway passenger transport services provided.

Among the above gaps, gap 5 comes from the perceptions of passengers, while gaps 1 to 4 are the internal deficiencies of the service sectors of railway passenger transport enterprises. Therefore, gap 1, 2, 3, and 4 can give rise to gap 5.

9.4.3 Plans for Service Quality Control

9.4.3.1 Measurable Dimensions of Service Quality

Based on the analysis of the gap model and the SERVQUAL model in the previous two parts, the measurable dimensions of passenger transport service quality are summarized in Fig. 9.3.

The standard SERVQUAL model has five dimensions, namely reliability, assurance, tangibles, empathy, and responsiveness.

Tangibles here mainly refer to the station environment and the image and attitude of the train attendants. An improvement in this dimension can lead to good perceptions of service quality.

Empathy means engaging the customers attentively to satisfy their needs so that caring and distinguishing passenger transport service can be delivered.

Assurance requires transport service providers to send passengers to their destinations in accordance with the train schedule, avoiding delays or cancellations without sufficient reasons. Besides, the railway passenger transport enterprises must transport passengers to their destinations in accordance with the provisions on passenger tickets, so that passengers will not take the wrong train or get off the train at a wrong station.

Reliability can be defined as the courtesy and competence of passenger service personnel. A company with reliability can win the trust of passengers for the quality of its passenger service. Poor attitudes or untidy clothing of the service staff will make passengers unhappy. In addition, without proper training, these employees may disappoint passengers and therefore negatively affect customers' perceptions of the service quality.

In the revised model, reliability and assurance are considered as safety, the overlap of the two dimensions. Emotional or physical injuries to passengers should be avoided

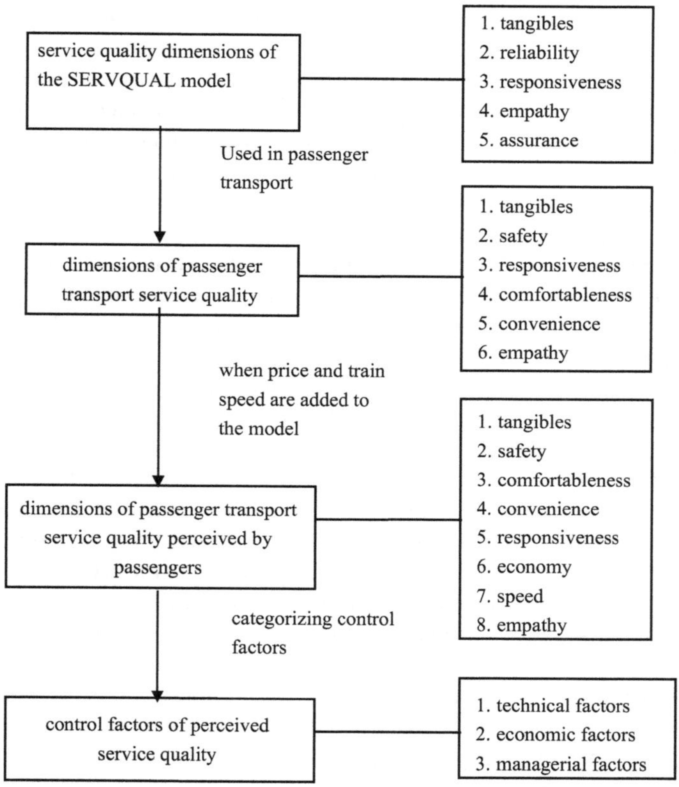

Fig. 9.3 Dimensions of high-speed railway service quality

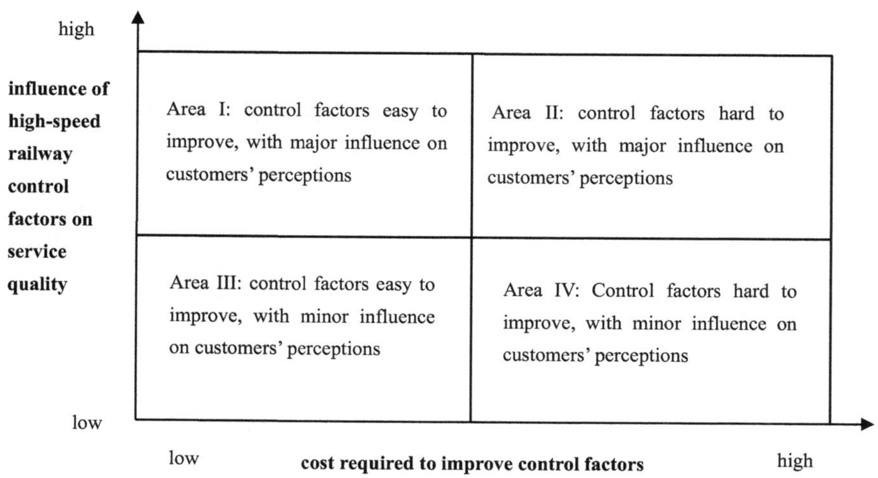

Fig. 9.4 Categorization of high-speed railway service quality control factors

during the ride, unless there are unpreventable natural disasters or other emergencies beyond their control.

Comfortableness, the requirement of high-quality services as well as the extension of safety, is also added to the model. It is unlikely that passengers will perceive the service quality as high if they have to risk their health standing in an overloaded, noisy, vibrating train in summer. Passenger transport companies should make every effort to improve the environment and facilities of the train and reduce passengers' fatigue so as to improve their travel experience.

Convenience, which is ensured by simple check-in and check-out procedures and the transport network, is also crucial to measuring the service quality of high-speed railway transport as this dimension indicates the time and money that passengers spend on their travel.

Economy, which refers to the prices of passenger tickets, reflects both the affordability of the service for the majority of the public and the profitability of related enterprises. On the one hand, companies must price their products in accordance with the law of value so as to cover their operation cost. On the other hand, they must bear in mind that public transport is also part of social welfare—an undertaking that requires them to cut down the cost while refraining from pursuing excessive profits. When formulating ticket prices, companies must take into account the economic capacity of passengers. Improving the service quality of high-speed railway transport can effectively increase passengers' acceptance of relatively higher prices. Better passenger transport service quality can be achieved by raising the value of high-speed railway transport, improving the competence and profile of employees, as well as reducing the amount of time, money and energy that passengers spend on their travel.

Convenience means transporting passengers to their destinations safely and quickly. Apart from comfortableness and safety, time saving is also favored by passengers. In short, being able to meet the various requirements of passengers in a timely manner shows that railway companies are customer-oriented and that they can deliver their service efficiently.

9.4.3.2 Control Factors of High-Speed Railway Transport Service Quality

The control factors of high-speed railway service quality refer to factors that can affect passenger expectations and perceptions, including those in service quality dimensions and those in the gap model. In this book, all these factors are categorized into technical, economic and managerial factors. See Table 8.22.

Technical factors are speed and safety. A rise in the speed of a transport mode will inevitably involve technical transformation; Safety, although influenced by management as well, is also integrated with technical features.

Economic factors are price, convenience, comfortableness.

Comfortableness is defined as the degree to which the travel cost influences passengers' expectations and perceptions of service quality. Price refers to the ticket prices high-speed railway trains. For high-speed rail companies, lowering ticket prices to meet the expectations of passengers is very difficult as companies must risk their profitability doing so, and a large number of departments and procedures will also be involved to achieve this goal. These are the two major challenges of improving service quality in terms of ticket price 9.4.

Managerial factors are tangibility, empathy, service commitment, responsiveness, passengers ' past experience, corporate image and word of mouth.

Among these factors, some are controllable, while others are not.

Tangibles, empathy and service commitments are classified as controllable: Tangibles refer to the image of the services and the environment of the station; empathy means the attitude of train attendants; service commitments denotes the consistency between the service promoted by the high-speed railway companies and passengers' perceptions of it.

On the other hand, responsiveness, passengers' past experience, corporate image and word of mouth fall into the category of uncontrollable: Responsiveness indicates how fast high-speed railway companies react to emergencies; passengers' past experience of other transport modes are often used to rate the high-speed railway service; an corporate image of energy-saving and environmental-friendliness is attractive to the public; word-of-mouth—previous passengers' rating of the high-speed railway service—can also influence passengers ' expectations.

9.4.3.3 Selection Model of High-Speed Railway Transportation Service Quality Control

Since many factors can affect the service quality of high-speed railways, it is very difficult to examine all of them with a limited number of customers. Therefore, in this book, these factors are ranked by importance so as to determine those crucial to service quality. After that, a selection model was established to find out areas for improvements so that targeted actions can be taken to deliver better service.

Analyzing the gap model has helped to determine the disparity between the high-speed railway service and the passenger's perceived service quality, which is crucial to identifying the factors that affect the passenger service quality. The classification of these factors was done using the SERVQUAL model and the PZB service quality model—so that key control factors of service quality could be found.

The close examination of these contributors allows us to figure out the key problems in current high-speed railway passenger service as well as the reasons behind these challenges. Therefore, a selection model for improving the quality of high-speed railway passenger service could be built. In this model, customer attraction refers to the ability to attract customers, while competitive advantage means the circumstances that put products and services in a favorable position.

Service quality control factors in Area I need to be improved first. In this area, the factors are attractive to passengers, and the cost of quality control remains at a low level. However, resources should be reallocated at an appropriate time to prevent over-investment.

At a competitive advantage that needs to be secured, service quality control factors in Area II are also highly attractive to passengers. However, companies should take into account the relatively higher cost involved in improving these factors.

Passengers favor service quality control factors in Area III, and it is cost-effective to improve them. Therefore, actions are not urgently needed in this area, but some resources still need to be allocated to consolidate this advantage.

Service quality control factors in Area IV do not appeal to passengers, but to ameliorate them is very easy. To a certain extent, they affect the competitiveness of high-speed rail companies, so a lot of resources need to be put into the area so as to make rapid improvements here.

After the model is established, we can do evaluation on each factor using questionnaires (Fig. 9.4).

Targeted measures to improve service quality can be taken with simplified control standards. Therefore, only two standards of were applied, namely practicality and cost: The former is chosen because some measures to improve service quality appear promising at first but in fact are hardly implementable; Cost is also a crucial aspect of service quality control as companies must ensure their normal operation before they deliver better service. The objective of building this model is to categorize the services and prioritize those with controllability and low cost of improvement.

The above analysis of passengers' expectations helps us to distinguish four types of service quality control factors. High-speed railway companies need to focus on the first type of control factors as these are crucial to their survival. The second type of control factors, despite the challenges in improving them, should be put on emphasis as well. Meanwhile, high-speed railway companies need to strengthen their advantage in the third type, and stay alert over the fourth type, where their weaknesses lie.

The importance of high-speed railway service control factors can be reflected in the data collected from questionnaires, and it can be made clear which factors are the most concerned by passengers. By summarizing service quality control factors into technical, economic, and managerial, this book measures the cost of quality control. Knowing the category that a factor falls into, we can make an estimate of how much it costs to improve it. Generally, the cost needed to improve technical factors is higher than economic and managerial ones.

9.4.4 Demonstration of Quality Control of High-Speed Railway Transportation Service——Taking Beijing-Tianjin Intercity Railway as an Example

9.4.4.1 Questionnaire Design and Data Collection

(1) Questionnaire design

The questionnaire design is shown in Fig. 9.5. First, relevant empirical research was studied to find out the operational methods and scales of the structural variables used in this questionnaire. Under the design principle of multiple items, the book used

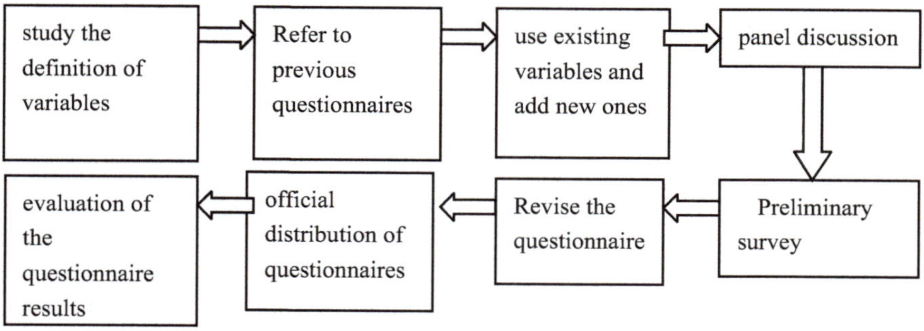

Fig. 9.5 Questionnaire design

variables of previous questionnaires, with some new items consistent with such variables. Then, the questionnaires were translated as part of the questionnaire content, and were discussed and revised later.

The items of the questionnaire are shown in Table 8.23, Table 9.5.

(2) Data collection

The samples studied in this book were passengers of the Beijing-Tianjin intercity high-speed railway trains. With a recovery rate of 94%, a total of 300 questionnaires were distributed. Among 282 collected copies, 17 of them were invalid and the rest of them valid, response rate reaching 88.3%. Gender, age, education, occupation, income, places of residence, purpose of travel, reasons for choosing high-speed railways were surveyed in the questionnaire, and the data are shown in the bar chart below.

Figure 9.6 shows that 71% of passengers have a bachelor's degree or above, and 26.4% of them receive a high school education, indicating that most high-speed rail passengers are well-educated.

Fig. 9.6 Educational background of high-speed railway passengers

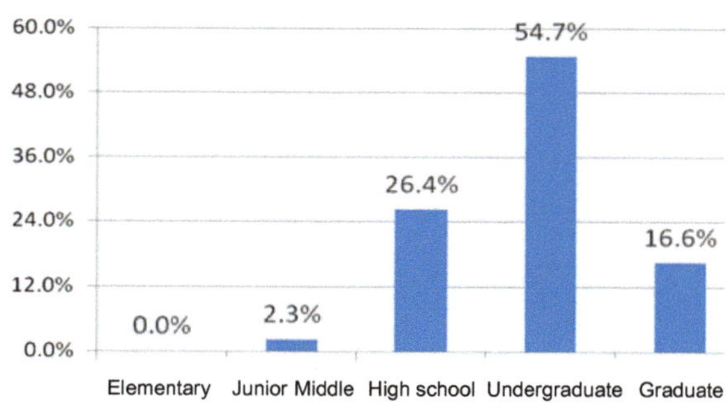

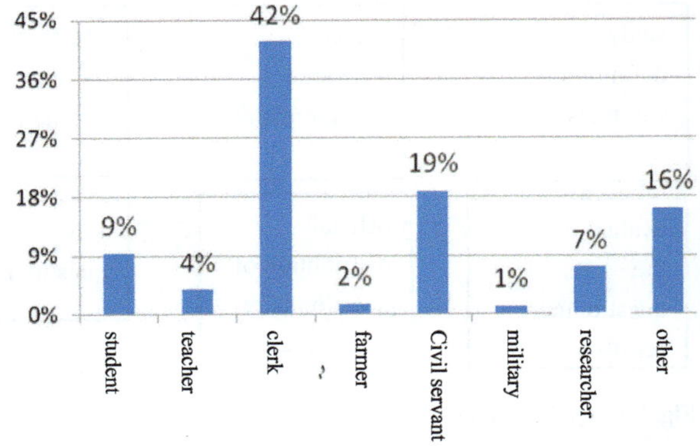

Fig. 9.7 Passenger occupations

Figure 9.7 shows that company employees and civil servants account for 42% and 19% of high-speed railway passengers respectively, indicating that passengers who take high-speed railway trains mainly travel for business reasons.

Figure 9.8 shows that high-speed rail passengers are mainly from the middle-income group (with a monthly income of 1,000 to 5,000 RMB yuan). At the same time, high-income passengers earning more than 5,000 yuan per month also took up a large proportion. The middle- and high-income groups account for 92.5% of all passengers.

Figure 9.9 shows that the majority of passengers (63%) are from Tianjin and Beijing. As a typical intercity railway, the Beijing-Tianjin high-speed railway is a direct driver of the local economy.

Multiple-choice questions are designed to find out the reasons why passengers choose high-speed railways. It is found that speed, comfortableness, safety and convenient transfer are the top four features that appeal to them Fig. 9.10. Beijing South Railway Station and Tianjin Railway Station already have an advantage in transfer,

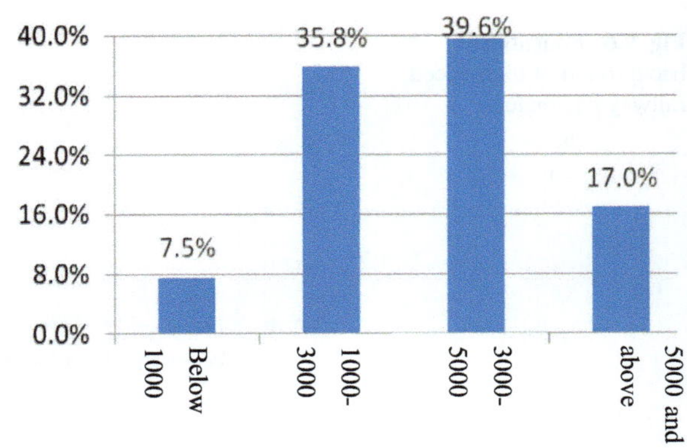

Fig. 9.8 Monthly income of passengers

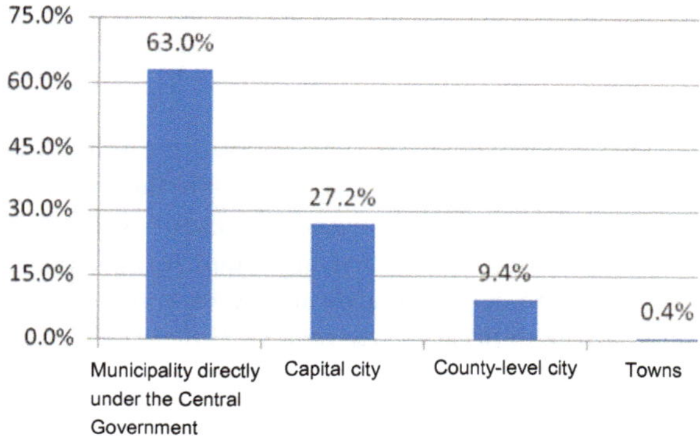

Fig. 9.9 Passengers' places of residence

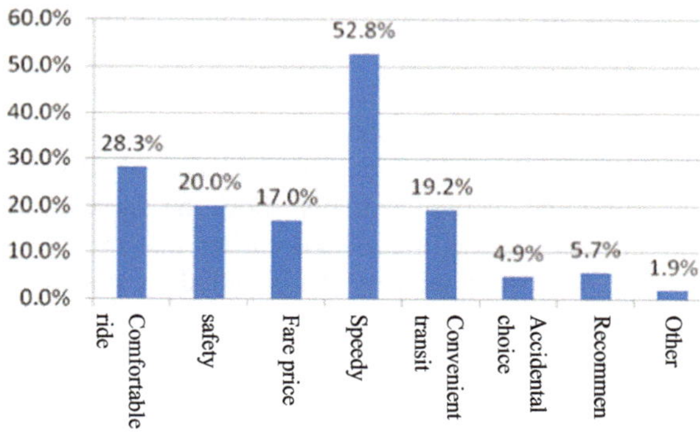

Fig. 9.10 Reasons why passengers choose to travel by HSR

and high-speed EMUs with a speed of 350 km/h are also favored because of their speed, comfortableness and safety.

9.4.4.2 Data Analysis

(1) Reliability

Reliability refers to the degree to which the result of a measurement, calculation, or specification can be depended on to be accurate. The scores obtained in the actual evaluation of the questionnaire results are called the "real scores", denoted as X. These scores are composed of the true scores (denoted as T), the deviation scores (denoted as B) and the measurement error scores (denoted as E). See Formula 9.4

$$X = T + B + E \tag{9.4}$$

Cronbach's Alpha can be determined using Formula 9.5:

$$\alpha = \frac{K}{K-1}\left(1 - \frac{\sum_{i=1}^{K}\sigma_i^2}{\sigma_T^2}\right) \tag{9.5}$$

This coefficient is used to analyze the reliability of the questionnaire results, where K denotes the number of items in the questionnaire, σ_i^2 denotes the variance of the scores of the item i, and σ_T^2 describes the variance of the cumulative scores, which are often considered important in multiple item scales. Internal reliability refers to whether each scale measures a single idea. If the Cronbach's Alpha for internal reliability lies between 0.70 and 0.99, it means that the questionnaire is highly reliable.

The book uses the reliability analysis function of SPSS statistical software to analyze the reliability of the questionnaire results. The book finds that the Cronbach's Alpha of the passengers' responses reaches 0.921, much greater than 0.7. Thus, the questions in the questionnaire are highly reliable (Table 9.6).

(2) Validity

Validity refers to the quality of being logically or factually sound. Statistically, a measurement method is considered to be effective if the survey results of a certain measurement method actually show the features of the variables intended to be investigated. it was mentioned earlier that the scores obtained in the actual evaluation of questionnaire results is called "real scores", which are composed of true scores (denoted as T), deviation scores (denoted as B) and measurement error scores (denoted as E). In fact, the true scores can be further broken down, as is shown in Formula 9.6:

$$T = T_x + T_0 \tag{9.6}$$

Where T_x is the score of the content required to be evaluated as the target in the service quality evaluation, while T_0 is the score obtained by deviating from the target content.

If $X = T = T_X$, or if the actual evaluation score = true score = service quality evaluation target score, then this measurement is considered to be effective. The validity of the evaluation can be defined by validity: The ratio of the variance $\sigma_{T_S}^2$ of the score related to the measurement target to the variance σ_X^2 of the actual score:

$$Validity = \frac{\sigma_{T_x}^2}{\sigma_x^2} \tag{9.7}$$

Or the validity can be defined as in Eq. (9.8)

$$Validity = \sqrt{\frac{\sigma_{T_x}^2}{\sigma_x^2}} \tag{9.8}$$

In order to test the rationality of the questionnaire, it is necessary to find out the influence of each item on the overall questionnaire results. This book uses the factor analysis function of SPSS to analyze the validity of the evaluation index system. Each evaluation index is a variable, and the sum of the squares of the factor loads of each evaluation index is the common degree of the measurement index. The greater the common degree, the greater the mutual dependence of the evaluation index on the common factors, that is to say, the more effective these common factors are to explain the evaluation index. The output results are sorted as shown in Table 9.7.

Generally speaking, when the common degree is greater than 0.4, the common factor can explain the evaluation index well. According to the above output results, the commonness of each evaluation index factor in the questionnaire is greater than 0.4, which shows that the evaluation indexes set in the questionnaire have significant influence on the passenger service quality, and the index system design is reasonable.

Table 9.7 Validity analysis

	Initial value	Output value
Q201	1.000	0.591
Q202	1.000	0.667
Q203	1.000	0.620
Q204	1.000	0.592
Q205	1.000	0.482
Q206	1.000	0.712
Q207	1.000	0.446
Q208	1.000	0.404
Q209	1.000	0.620
Q210	1.000	0.538
Q211	1.000	0.592
Q212	1.000	0.732
Q213	1.000	0.431
Q214	1.000	0.678
Q215	1.000	0.545
Q216	1.000	0.669
Q217	1.000	0.650
Q218	1.000	0.466
Q219	1.000	0.670
Q220	1.000	0.634
Q221	1.000	0.563
Q222	1.000	0.664
Q223	1.000	0.633

9.4.4.3 Application of Beijing-Tianjin Intercity Transportation Service Quality Control Scheme

(1) Result analysis

As can be seen from Table 9.8, high-speed railway fares saw the largest difference between passengers' perceptions and expectations, with.... This shows that passengers would like to see a further decrease in the ticket prices. High-speed railway passenger service is expensive at this stage, because passenger and freight revenues, the major source of income for high-speed railway companies, have to spent on covering the construction and operation cost of high-speed railway systems. As a major national transportation infrastructure, railways play a major role in promoting social and economic development. The state should adopt a more flexible transportation policy and subsidize high-speed railways—so as to realize a further reduction in their fares. This will help to attract more passengers from lower-middle-income groups as they will have access to more affordable high-speed railway transportation services.

The second largest disparity between the actual perceptions and expectations of passengers was found in the easiness of high-speed train ticket purchase (0.17). This means that high-speed railway companies need to further improve their ticket services. At present, air ticketing has widely adopted services such as telephone booking and online booking, which can be applied in the railway ticketing system to make inquiries, ticket reservation, ticket changes and ticket refunds easier.

The third largest gap between passengers' perceptions and expectations was found in the waiting environment. with a difference of 0.16. Today's high-speed railway stations were built as places for passengers to get on and off the train, but in reality, there are relatively few departures, so passengers often have to wait quite a long time. Therefore, in addition to reduced noise, high-speed railway stations should also have better lighting, ventilation and hygiene.

The fourth largest disparity between the perceptions and expectations of passengers is the passenger was found in entrance and exit signs, which is 0.08. As high-speed railway passenger stations are usually integrated transportation hubs in local cities, it is necessary to set up signs at the entrances and exits—so that it will be easier for passengers to enter or leave the station or transfer to other means of transport modes.

In terms of empathy (services provided by train attendants) and service commitment (consistency between high-speed rail service and its media publicity), the perceptions of passengers is lower than their expectations by 0.06 and 0.10 respectively. This shows that passengers want to see the quality of services provided train attendants further improved— so that the actual service may match the media publicity.

(2) Determination of high-speed railway service quality control area

In order to facilitate the implementation of high-speed rail service quality control, the 12 dimensions of high-speed rail service quality are numbered in Table 8.27.

Table 9.8 Service quality control factors analysis

Item	Service quality control factors		Average expectation value	Average perceived value	Perceived average values-Expected average values
Q201	Tangibles	The appearance of the station and the train	2.33	2.58	0.25
Q202		Employee dress code	2.53	2.66	0.13
Q203	Speed		2.83	2.87	0.04
Q204	Safety	Public security	2.78	2.84	0.06
Q205		Train stability when under operation	2.80	2.83	0.03
Q206	Empathy		2.7	2.64	−0.06
Q207	Comfortableness	The waiting hall	2.72	2.56	−0.16
Q208		The train	2.81	2.76	−0.05
Q209	Convenience	Ticket purchase	2.77	2.6	−0.17
Q210		Signs for entries and exits	2.73	2.65	−0.08
Q211		Transfer	2.76	2.77	0.01
Q212		Travel time	2.74	2.77	0.03
Q213	Price		2.60	2.13	−0.47
Q214	Service commitment		2.71	2.61	−0.1
Q215	Random factors		2.61	2.62	0.01
Q216	Corporate image		2.58	2.59	0.01
Q217	Past travel experiences		2.55	2.51	−0.04
Q218	Word of mouth		2.49	2.51	0.02

The statistical results of the questionnaire show that, according to the gap between the actual feelings and expectations of the passengers, the larger the gap, the lower the satisfaction of the passengers in this item, the greater the room for improvement, and the higher the attraction to the passengers. The room for improvement is relatively small, and the attraction to passengers is small. For example, the speed factor, passengers' perception of this item exceeds or equals expectations, the cost of improvement is huge, and the improvement of this item can be delayed Table 9.9.

In terms of high-speed railways fares (controlling factor serial number B1), ticket service (B2), waiting environment (B3), passenger station signs (B2), and station and train attendant services (C2/C5), the perception is lower than expected, indicating that passengers It is hoped that the high-speed rail service can be further improved; after sorting the above five control factors according to the improvement cost, the position of each control factor in the control area is shown in Fig. 9.11.

Table 9.9 High-speed railway service quality control factors

S	Control factors	
A_1	Speed	Technical factors
A_2	Safety	
B_1	Price	Economic factors
B_2	Convenience	
B_3	Comfortableness	
C_1	Tangibles	Managerial factors
C_2	Empathy	
C_3	Service commitments	
C_4	Random factors	
C_5	Previous travel Experiences	
C_6	Corporate image	
C_7	Word of mouth	

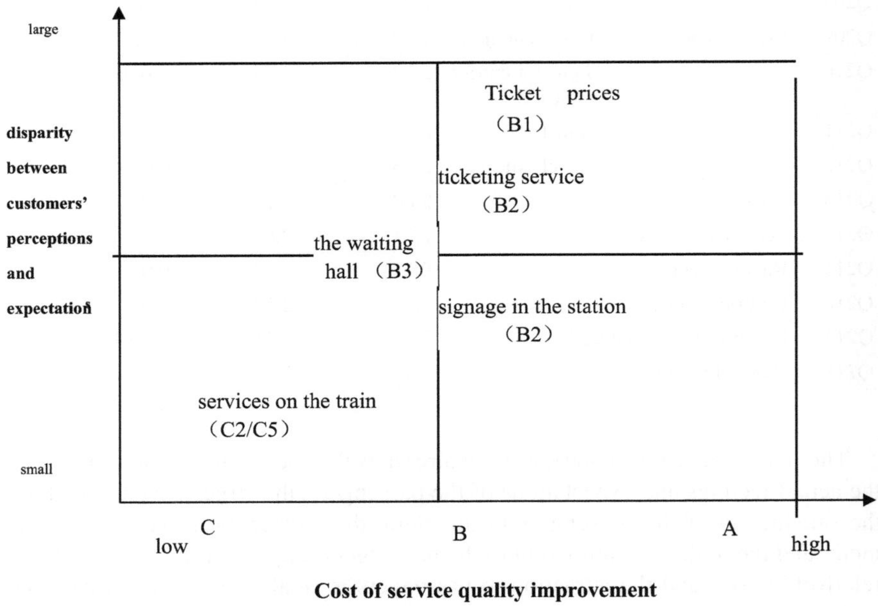

Fig. 9.11 A model of high-speed railway service quality control areas

(3) Ranking high-speed rail service quality control

After comprehensively considering the control cost and service quality gap, according to the control idea of large gap in service quality and low cost of improvement, the order of service improvement can be: waiting environment (B3), ticket service (B2), passenger station identification (B2), station, train Flight attendant service (C2/C5) or fare (B1).

9.5 Summary

On the basis of extensive review of studies on models formed on the customer-satisfaction-based evaluation system of railway passenger service quality, this book proposes a model that categorizes service quality determiners into certain areas, studying the high-speed rail service quality from the perspective of service control. With high-speed railway passenger services at the center of this empirical research, the data is obtained by questionnaire survey for statistical analysis. Followings are the research findings:

(1) By dividing the 12 control factors into technological, economical and managerial, this book builds an exploratory service quality gap model which can be referred to when managers of high-speed railway companies consider investing in improving service quality. This book also finds that it will be easier to improve and manage HSR service quality when the control cost of service quality determiners and their room for improvement are identified.
(2) The data analysis shows that the station environment and services provided by train attendants should be further improved. Entrance and exits signs should be made more visible; ticket fares should be reduced, with more flexibility in ticketing services.
(3) After comprehensively considering the cost of quality control and the gap in service quality, the book concludes that a service needs to be improved first if its passengers' perceptions are far below expectations. Based on that, the book suggests that service quality determiners should be enhanced in the following order: The station environment, ticketing service, station signage, services provided by train attendants or ticket prices.

The high-speed service quality control model and working ideas proposed in this part optimize the high-speed rail service quality improvement steps and provide a useful reference for the high-speed rail service quality management improvement.

References

Chen, Xiaojun and Lin, Xiaoyan. Emergency Evacuation Management of Transportation Infrastructure: Based on Analysis of Prospect Theory and Temporal Perspective, *Research on Economics and Management*, 2014 (08): 71–80.

Fan Xiucheng. Management of service quality: interactive process and interactive quality [J]. *Nankai Business Review*, 1999,01:8–12+23.

Gronroos, Christian. (1982) *Strategic Management and Marketing in the Service Sector*, Helsingfors: Swedish School of Economics and Business Administration.

Gronroos, C. (1984) A Service Quality Model and Its Marketing Implications. *European Journal of Marketing*, 18, 36–44.
Lewis, R. and Booms, B. (1983) *The Marketing Aspects of Service Perspectives on Services Marketing.* American Marketing Association, Chicago.
Liang Yan. A review of the customer satisfaction studies [J]. *Journal of Beijing Technology and Business University* (Social Science), 2007 (02):75-80.
Lin Xiaoyan, Regional Quality and Social Effect of High-speed Railway-Research on the Timing of High-speed Railway Construction, working paper, 2014.
Lin Xiaoyan, Technical Economics [M], Tsinghua University Press, Beijing Jiaotong University Press, 2014.
Liao, Zhen. 2006, Study of the theory and application about market research in railway passenger traffic, Master's thesis, Beijing Jiaotong University.
Parasuraman, A., Zeithaml, V.A. and Berry, L.L. (1988) SERVQUAL: A Multiple-Item Scale for Measuring Consumer Perceptions of Service Quality. *Journal of Retailing*, 64, 12–40.
Peterson, Robert A. and William R. Wilson. (1992). Measuring Consumer Satisfaction: Fact and Artifact. *Journal of the Academy Science* 20 (Winter).
Qiang Lixia, Yan Ying. Research on the passenger flow organization and marketing management strategy of express railway passenger transport network [J]. *China Railway*, 2007,05:60-62.
Sasser, W. Earl, Ir., R. Paul Olsen, and D. Daryl Wyckoff. (1978), *Management of Service Operations: Text and Cases*, Boston: Allyn & Bacon.
Wang Haixiang. Research on Quality Evaluation System of Railway Passenger Service [D]. Central South University, 2006.
Zhao Fei. Passenger Service System of High-speed Railway-Talking about the Seventh High-speed Railway (VII) [J]. *Railway Knowledge*, 2000(06):15-17.

Part IV
Strategic Options for More Emphasis on the Significant Role of High-Speed Railway in Regional Development

Chapter 10
The Development Models and Policies of High-Speed Railway

The emergence of high-speed railways has had a competitive impact on transportation methods of aviation and highways. For short-haul transport, high-speed rail possesses even more advantageous than aviation. The construction of high-speed railways and cost recovery are closely related to the size of the population and the level of economic development, and such projects can promote the development of the national economy, and on the other hand, impact the national territorial planning to a very large extent. In recent years, with the passenger headcount increasing steadily, China's high-speed railways underwent massive expansion, offering people a new, more convenient mode of transport. In terms of construction investment, operation and management. High-speed railways are very different from traditional lines. it is important to study how high-speed railways in China are developed and the policy-making around them, as further research in these areas can not only help fully utilize the advantages and benefits of railway transportation network but also accelerate the development of other related industries.

10.1 The Role of High-Speed Railway Policy

It has been agreed that the railway sector, a network-based, naturally monopolistic industry, requires the government to deliver suitable policies to support its development. However, the selection of railway policies and studies about their roles remain almost absent Feng and Xue (2011) propose that the central government and local governments should participate in making the subsidy policy for high-speed railways. The two researchers argue that subsidies from the central government should be increased, and local government subsidies should be allocated in a more all-round way so as to satisfy all railway-related companies. The government should build a better environment for the education and research of high-speed railways. Financing policies should be accelerated, and reforms and innovations should be implemented.

X. Lin, *High-Speed Railways and New Structure of Socio-economic Development in China*, Research Series on the Chinese Dream and China's Development Path,
https://doi.org/10.1007/978-981-19-6387-2_10

Tian and Liang (2010) and other scholars analyze the infrastructure planning in developed countries, and develop a menu-based subsidy method based on the price ceiling model. Used to calculate public finance subsidy, this model is based on the Pareto optimal and therefore meet the needs of the government, business and passengers. Wang (2011) proposed that high-speed railways in cities should be a part of a multi-level transportation system that impacts urban development and regional layout. This study provides the policy framework for decision makers and later researchers. Some researchers have analyzed the economic and social benefits of high-speed railways. Drawing lesson from the advanced achievements of foreign countries, other scholars believe that high-speed railway projects require a huge amount of investment to cover the costs of maintenance and operating, advocating that the central government and local governments should improve the subsidy policies for high-speed railways.

Through empirical research, Li (2012) studied the high-speed railways of Japan, South Korea and France from the perspective of the impact of high-speed railway policies on the economy. He studied cities in these countries with high-speed railways, analyzing the effects of high-speed railways on urban and regional economic growth of these countries. Later, he concluded some high-speed railway policies beneficial to China' regional economic planning. He argues that as high-speed railways are rather expensive, national finance will be negatively affected if appropriate planning is absent. He suggests that the construction of high-speed railway network should benefit cities along the line, accelerate the growth of their real estate markets and cultural and educational industries. In addition to these objectives, He also argues that high-speed railway projects should make medium- and small-sized cities more attractive to enterprises. Wu and Lin (2001) analyze how Taiwan's high-speed railways attract private capital. They believe the project legislation, establishment of full-time institutions responsible for project planning and governance, implementation of sub-tender system and joint tender contracting project management system can be applied to high-speed railway projects of the mainland. Based on the success and history of Japan's high-speed railways, Takatsu (2007) summarizes the strategic goals and actual policy effects of Japan's high-speed railways, and blueprints the future high-speed railway policies. He proposes that the construction of high-speed railways should be emphasized, as it promotes tourism and economic agglomeration. He also puts forward the one-day Travel Circle, finding it favorable to improve the speed, convenience and other aspects of high-speed railways. He thinks the development of magnetically levitated trains and the provision of diversified railway services should be put on the agenda, so that high-speed railways can better serve the economy.

From the perspective of the impact of high-speed rail policy on regional planning, Ye and Tang (2010) analyze the planning of European high-speed railway integration. They look into the policies involved, and elaborated on how these policies promoted the revitalization of railway transportation and environmentally friendly transportation, accelerated the process of European integration and optimized urban spatial structure and coordinated the development of national space. They suggest that China should explore and coordinate new mechanisms of high-speed railway planning and integration planning, and that China should attach importance to high-speed railway integration planning as a political process and focus on the following issues: regional co-ordination of high-speed railways, integration planning, urban circle structure optimization and environmental protection.

Jiang (2009) makes an in-depth study of Japan's territorial planning, explaining the reasons and course of the seven stages in its evolution. Among them, high-speed railways are key to achieving the strategic objectives of "two-comprehensives", which are to reallocate over-intensive urban resources and halt rural population outflow. As for the "three-comprehensives" which are to transform resource-intensive industries into labor-intensive and high-tech ones, high-speed railways have also played its role. All this has reflected how the high-speed railways can influence macro-territorial planning policies. Hou (2008) analyzed the effect of the world's major high-speed railways and railway stations on regional development, and summarized how the related regions were developed and how transport hubs in these regions were built under the influence of high-speed railways. and proposed that China should increase the connectivity between urban and transport facilities and high-speed railway hubs. integrate transport system. He argued that the construction of a high-speed railway comprehensive transport hub must take into account regional development to promote regional comprehensive development. Wang (2011) studies the impact of high-speed railway on city layout as well as impact of transportation facilities on the city from three spatial levels: the periphery of the station, the district that the station lies in and the city, put forward an analytical framework of the impact of high-speed railway development on urban spatial structure, and made a preliminary empirical analysis with Beijing-Shanghai high-speed railway as a case, After doing these, he concluded the principles of making policies for the planning of the station area: 1. The scale and functions of the planning and development of the site area need to be evaluated by considering multiple factors, including the relationship between the site and the built-up areas, the distance between the site and the urban center, the linkage of high-speed railway linking with other means of transport, etc. 2. In the planning and design around the site, registering land of multiple uses should be encouraged to enhance the popularity and vitality of the area. 3. It is necessary to clarify the characteristics suitable for the development stage of the city itself and locate the high-speed railway station area accordingly.

10.2 Objectives of Policy for Developing High Speed Railways

10.2.1 Objectives of Policy for Economic Development

It is undoubted that high-speed railway construction investment can directly drive economic growth in the short term. A large number of studies have a positive view on how high-speed railways will promote regional economic development in the medium and long term: in the process of high-speed rail operation, people can improve their economic efficiency by saving time, and improved transportation capacity can also be translated into greater value. High-speed railway can also promote regional economic integration and industrial structure upgrading. For policies for high-speed railways, one of the important objectives of high-speed railway development is to promote and serve regional and national economic development. Economic benefits generated by high-speed railways are not limited to ticket revenues. They also include the development of industries and cities nearby.

10.2.1.1 Goals of Direct Economic Benefit

The direct economic benefits of high-speed rail refer to the investment return of high-speed railways and the time value of high-speed railways. Taking Beijing-Shanghai high-speed railway as an example, the construction cost per kilometer of high-speed railway reaches 130 million RMB. As a high investment, the high-speed rail directly create the economic benefits. The total length of the high-speed railway from Beijing to Shanghai accounts for 2.8% of the total length of the national railways, but it accounts for 14% of the national railway passenger headcounts and 10% of the cargo transportation. The passenger number of high-speed railway from Beijing to Shanghai has reached more than 30 million every year, and the revenue generated exceeds 20 billion RMB. Secondly, the speed of high-speed railway is higher than that of ordinary railways. The transportation price of high-speed railway is lower than that of airplanes. The high-speed railway has improved people's travel efficiency and produced great time value. The high-speed railway from Beijing to Tianjin is 113.54 km, and the travel time between the two places is within 30 min; the high-speed railway from Beijing to Shanghai is 1,318 km, and the travel time is less than 5 h. The scale effect of the railway industry, and the growth of the number of passengers has greatly reduced transportation costs, and further improved direct economic benefits.

10.2.1.2 The Goal of Economic Benefits of Cities Along the High-Speed Railway

One of the economic goals of the high-speed railway development policies is to promote urbanization through high-speed railways. By 2020, China's urbanization rate will increase from 46.6% to 60%. China will have 200 million people flocking to cities, which make a strong transportation system necessary. The construction of high-speed railways provides transportation support for the urbanization. As the population increases, the rise in the number of enterprises will produce more fiscal revenue. Such a growth will also promote employment and encourage consumption, which can stimulate production and improve the employment figure. The flow of these economic factors will greatly facilitate the optimal combination of urban resources and markets.

10.2.1.3 Increase the Economic Efficiency Target of Related Industries

The construction of high-speed railways will promote the development of electronics, construction, machinery, steel and other industries. On average, a Chinese high-speed railway train has 100,000 parts high-speed parts, and there are more than 260 independent subsystems related to high-speed railways. More than 100 core-level enterprises are involved in the production of high-speed rail parts, and more than 500 companies are found in the tight layer. More than 20 provinces and municipalities, participate in high-speed railway projects, forming a huge technology and manufacturing chain. The flourishing of high-speed railway enterprises has also effectively promoted the development of the high-speed rail industry.

In addition, the policy objectives of high-speed rail development include high-speed railway services in the tourism, cultural and real estate industries. First of all, as far as tourism is concerned, high-speed rail has diversified travel modes and made travel time flexible. Compared with airplanes, high-speed railway stations are mostly close to cities. Passengers do not need to spend a lot of time going to the airport, and does not need to wait at the train station for a long time, which makes the journey more convenient.

High-speed railways are safe, convenient and, at most of the time, unaffected by bad weather. In terms of travels within 1000 km, they are highly favored, which means there will be more tourists as well as tourism revenue. For example, Since August 1, 2008, the Beijing-Tianjin inter-city railway had transported about 4.5 million passengers in a hundred days, which increased the retail sales of consumer goods in Tianjin by 0.31%. During the Chinese National Holiday of 2008, the Beijing-Tianjin Intercity Railway transported 70,000 passengers daily. The increase in the number of high-speed rail passengers has pulled up the number of visitors who chose Tianjin's boutique tour routes by 3–5 times. Secondly, in terms of the cultural industry, high-speed railways increase the accessibility between cities, allowing people to appreciate different cities. In addition, high-speed railways increase the number of employees in the cultural industry. Cultural enterprises flourish along high-speed railways and competed with each other, which has enabled the entire industry to achieve rapid development.

Besides, high-speed railways can increase the clout of central cities, enhanced the capacity of the city transport network and urban functions. With high-speed railways, people who originally lived in the city may choose to live in the suburbs, and the increase in secondary business centers and white-collar buying houses would greatly promote the real estate industry in the central city. For the satellite cities, high-speed railways can make commercial centers more accessible, office areas more concentrated, ands the sales volume of small and medium-sized apartments along the railway line will also increase. All this can enhance the city functions and improve people's living environment. Land use around the high-speed rail site will also change, which will increase the number of business office properties, make land more valuable. As a result, more facilities will be built around the area, helping to pull up business office rents and the number of recreational facilities.

10.3 The Policy Objectives of Serving the National Land Planning

The high-speed rail construction has produced huge economic benefits. In addition, it has greatly changed the land planning and use between cities., The local impact of high-speed railway construction is the improvement of urban functions, and the large-scale impact is that it promotes national land planning.

10.3.1 High-Speed Rail Hub and Metropolitan Development Goals

China's urban form is in transition, accompanied by industrial re-aggregation, suburbanization, and urban extension. With an increase in average travel distance, travel frequency and cross-movement, regional and intercontinental travel has become one part of people's daily lives in the world. High-speed railway hubs now appeal to more and ore people. The high-speed rail station played a key role in promoting the development of the metropolis. The high-speed railway line from Beijing to Shanghai has a total of 21 stations, including 7 first-class stations. The construction of high-speed railways means huge business opportunities for cities along the route.

Transportation hubs are highly accessible, which will concentrate on the advantages of development for the surrounding areas. The construction of the transportation hub can lead to the building of more residential areas, commercial areas and other facilities, which can translate into increase traffic volume, and increased urban accessibility, increased the intensity denser traffic and higher rents. The construction of the transportation hub will optimize the functional structure of the city. For example, Japan's Shinkansen high-speed rail uses a transportation hub to establish a business circle, attracting tourists and businesses successfully and centralizing urban functions The construction of transportation hubs will also help the dispersion and fixation of the population. On the other hand, high-speed railways have greatly reduced the time of travel, and the number of passengers commuting in urban areas and non-urban areas through high-speed railways will increase greatly, and the population will gradually move to the outside of urban areas. The rapid information flow in urban areas can influence the industrial layout, making them centers of decision-making for organizational departments and large companies. As high-speed railway trains are fast and convenient, the organizational department which was able originally located in urban areas may be able to move out, which will help to reduce cost and expand the market. For the city, the construction of the high-speed railway hub can effectively impact the periphery of the city and upgrade the internal functions of the city.

10.3.2 The Development Goal of the High-Speed Railway Hub for Regional Planning

Transport construction and regional development are closely related. Better transport shortens travel time and reduces travel costs. The improvement of the transportation system will increase the accessibility of its influenced area, turning upside down the advantages and disadvantages in location. Furthermore, it can impact the distribution of the population and industrial area, as well as the spatial layout of the region and the way the land there is used. In addition, the changes of population, regional development patterns of industrial and land use can contribute to the growth of transportation demand, giving rise to the changes of traffic distribution and construction in return. High-speed railways help upgrade the tertiary industry, a sector with high turnover and high mobility of production factors. The tertiary industry can expand the channels for product sales. This will benefit the area outside of the city and form an industrial belt driven by high-speed railways, increasing the income and bringing job opportunities for residents in the region. These advantages can improve the competitiveness of manufacturers and help develop regional economies along the route.

The policy objective of high-speed railway construction and operation is to serve economic development and land planning. As an infrastructure construction investment, high-speed railways can not only produce huge direct economic benefits, but also facilitate production and people's lives. Meanwhile they can generate benefits to cities and related industries along them. High-speed railways promote regional development in urban areas, population movement, and the balance of business gathering.

10.3.3 High-Speed Railway Development Model and Policy Serving Macroeconomics: A Case Study of Taiwan, China

One of the goals of Taiwan's high-speed railway construction is to expand the demand of market, stimulate the economy, and come back from the economic downturn. The construction of high-speed railways activates the development of the western corridor economic belt. since their operation, Taiwan's high-speed railways have become an important way of transportation for the people in western Taiwan as it is fast and comfortable. After the completion of the high-speed railways, the number of passengers who chose to take the high-speed rail increased by 10% within five years, and the railway lines attracted more people to use public transport for commuting between cities, creating around 100,000 jobs.[1]

[1] Dr. S.K. Jason Chang. Roundtable on the Economics of Investment in High Speed Rail, 2013 [C]. High Speed Rail in Taiwan, 2013.

In terms of planning, the starting point and the terminal of the Taiwan High-speed Railways are the largest cities in Taiwan Province and the high-tech industrial zone respectively. The north and south parts of the island are highly heterogeneous, and the construction of high-speed railways will further connect the two regions, promoting their overall economic growth.

In project financing, through legislation, the rights and responsibilities of the government and investment enterprise are clarified. The government provides a series of preferential policies for private investment in financing, taxation and land rents and land development. The government adopts the BOT financing model, which mobilizes of private investment construction and management.

10.4 China's High-Speed Railway Development Model Selection and Policy

10.4.1 Dual Attribute Positioning for High-Speed Railway Development

10.4.1.1 Industry Attribute

High-speed railways are part of national infrastructure and means of transportation. Their roles are vital to the national economy and people's lives. From an industrial perspective, high-speed EMU train operations and services are the core of high-speed railways, including high-speed rail technology research and development, passenger line construction, rolling stock equipment manufacturing, modern communications and signal technology, product design and services, and other strategic industries. The reasons why the high-speed railway is recognized as a national strategic industry is because:

The development of high-speed railways relies on and represents the most advanced and modern railway transportation technology in the world today;

High-speed railways, as an advanced mode of transportation, have played a huge role in promoting China's social progress. They accelerate urbanization and promote regional economic development. In reality, China is in the post-transportation period of the new industrialization era. The transportation industry is an important infrastructure that determines the level of national economic development. China has a vast territory and a large population. The uneven distribution of resources and economic development make railway transportation the backbone of comprehensive transportation. Therefore, the development of high-speed railways is an inevitable choice, as it greatly reduces the cost of production and transaction costs of inter-regional economic factors. The development of high-speed rail creates more job opportunities and generates investment, effectively promoting urbanization and industrialization. At the same time, advanced high-speed railway network can help optimize the regional industrial layout, as well as adjust and upgrade industrial structure. It plays an active role in coordinating the balanced development of regional economy. Thirdly, with the construction and operation of high-speed railway, a huge high-tech industrial cluster is gathered. The new generation of high-speed EMUs has more than 100,000 components and 260 independent subsystems, involving machinery, materials, electronics, electrical, metallurgical, chemical and other industries. It integrates the related enterprise resources of upstream and downstream. It takes more than ten domestic key enterprises of locomotive and rolling stock equipment manufacturing as its core. It constitutes the high-speed train technology and equipment manufacturing industry chain, which involves hundreds of enterprises. The fourth is that the high-speed rail industry has high environmental benefits and is a green industry. In summary, the development of high-speed railway is not only in line with China's strategic industrial technology, high comprehensive efficiency, large market potential, and large driving coefficient, but also reflects the transformation of industrial structure and green economy in the world today. And the development characteristics and development trends of energy-saving economy.

10.4.1.2 Enterprise Attributes

From the perspective of market competition, high-speed railways, as suppliers of high-speed transportation services, must follow the laws of the market economy. However, what does the high-speed railway transport enterprise which has been involved in the fierce competition of market economy rely on to compete in the market? Without exception, it relies on its own sales to the transportation market service products, using its own transportation products to participate in the market competition. Therefore, as a branded product, high-speed railway has become an important indicator to measure the economic strength and core competitiveness of Chinese railway enterprises. It is also an important symbol of the powerful new modern image of China Railway. The above points, we can get evidence from the situation, since the emergence of the high-speed railway, the railway market share reversed the declining situation which last for many years.

10.4.2 Circular Economy Development Model Under the Orientation of Industrial Attribute

10.4.2.1 Basic Connotation

Academia and industry generally believe that transportation industry is the key area of resource utilization and environmental protection, and the development of circular economy in transportation industry is also an important part of the development of economic and social circular economy. As Lenin said, transportation is the main foundation of our entire economy, and the improvement of circular economy capacity of transportation industry is also an important aspect of the development of circular economy in economy and society.

The circular economy development of high-speed railway is an effective allocation of transportation production factors within the permit of carrying environmental resources in order to meet the needs of economic and social development for transportation. It is a cyclic process to realize the balance of supply and demand and production of high-speed railway. From this definition, it embodies the inheritance of the essential concept of circular economy development from the development of high-speed railway transportation circular economy, that is, the development of high-speed railway circular economy needs to achieve a balance between supply and demand, constantly adjust the supply capacity and service level with the change of economic and social demand to meet the needs of the current development of transport services; at the same time, it also needs to achieve a balance of production, but not excessive development. Beyond the carrying capacity of resources and environment, there is room for future generations to develop and meet the needs of future generations. Specifically, there are the following points.

Firstly, the development of high-speed railway circular economy must be able to meet people's growing demand for travel services in terms of capacity and service quality. That is, it must develop, maintain a moderate growth in the total amount of infrastructure and the scale of transport vehicles and equipment, at the same time, it needs to constantly optimize the development mode and realize the optimal allocation of transport elements.

Secondly, the cycle process of realizing double balanced development of high-speed railway must be a dynamic cycle process, which is a cycle process of continuous optimization, demand optimization or sub-optimal solution. This circular process is also a process of constantly adapting to the needs of economic and social development. Therefore, the development of high-speed railway circular economy must consider its leading, basic and service functions, pay attention to its relevance with the stage of economic and social development, and moderately advance the development.

Thirdly, high-speed railway will inevitably consume a large amount of energy resources in the process of achieving production balance, which will have a negative impact on the environment. The development of circular economy is to balance scientifically and reasonably the relationship between high-speed railway and the carrying capacity of resources and environment, to develop moderately and green, rather than to pursue growth blindly.

Fourthly, the development of high-speed railway circular economy is related to intergenerational equity. It is a long-term development, not a short-term behavior. Therefore, in planning and layout, in the supply of service capacity and level, we must consider meeting the needs of long-term development, and we should avoid short-sightedness in behavior.

Fifthly, the sustainable development of high-speed railway must consider the coordinated relationship with other modes of transport, and should continue to develop under the perfect comprehensive transport system, so as to promote the comprehensive utilization of traffic resources, meet the carrying requirements of resources and environment and meet the needs of transportation.

Sixthly, the circular economy development mode of high-speed railway is not only a production mode, but also a consumption mode. Therefore, the choice of circular economy development mode of high-speed railway should not only pay attention to the allocation of production factors, but also emphasize the management of travel demand, change people's travel behavior, use economic and administrative science and technology to guide travel consumption, and advocate green travel and low-carbon travel.

In summary, with the support of technological progress and capital supply, high-speed railway should provide high-speed railway infrastructure and high-speed railway transportation services for the society to meet the requirements of economic and social development for human and logistics movement and achieve the "balance of supply and demand" between economic and social systems. Meanwhile, high-speed railway should consume various resources and produce a pair of rings in production and operation. Under the influence of negative externalities, how to achieve the "production balance" between high-speed railway and environment, resource system is an important basis for the realization of circular economy development of high-speed railway.

10.4.2.2 Guarantee Mechanism

According to the basic connotation of China's high-speed railway development under the circular economy, we need to discuss the problems existing in industrial development and formulate relevant measures to promote the circular development of China's high-speed railway industry.

Firstly, improve the efficiency of resource utilization. Each project of the high-speed railway is invested in tens of billions or even hundreds of billions of dollars, most of which is used in the early stage of infrastructure construction. The resource input standard and the upper and lower floating limits are set for each construction situation, which can effectively improve resource utilization, control resource waste, and have a great effect on resource conservation and utilization. At the same time, we should actively utilize the achievements of scientific and technological innovation, apply existing applied technologies to improve work efficiency and resource utilization, and save human resources investment. Scientific and technological progress has also made it possible to utilize new materials and energy resources. The utilization of new materials and energy resources should be strengthened to eliminate materials and energy with low economic benefits and great impact on the environment. Moreover, it is also necessary to pay attention to the unified planning, layout and construction of high-speed railway network, so as to avoid duplication of construction and waste of resources.

Secondly, strengthen profitability. First of all, we should broaden the existing financing channels of high-speed railway, and formulate policies to promote local and social capital and other available funds, so as to maximize the enthusiasm of local governments, social investors and other investors. And then, reasonable fares should be set to ensure the revenue of high-speed rail. At the same time, it needs to be accepted by the public, attract enough passengers and ensure the volume of high-speed rail.

Thirdly, determine the reasonable development speed. Firstly, in view of the problems caused by the rapid development of high-speed rail, it is necessary to formulate a reasonable development plan for high-speed rail and promote the construction of high-speed rail in an orderly manner. Secondly, change the management concept and innovate the management system. The government supervision should be reasonable and non-interference, and the management within the industry should be scientific and efficient. The representatives of state-owned assets investors can only be fully utilized. We will improve the laws, regulations and policy system of the high-speed rail industry and determine a scientific and rational project organization so that the management of the high-speed rail industry can follow the rules. At the same time, efforts should be made to break the traditional management mode in transportation organization, route and site layout, institutional system and financial management.

Fourthly, accelerate technological innovation. First of all, we need to recognize the importance and urgency of railway independent innovation ability, and establish a good science and technology policy environment, so as to ensure the sound and rapid development of high-speed railway industry in the direction of science and technology development. Secondly, we should accelerate the achievement of technology transformation. We should innovate the transformation mechanism of scientific and technological achievements of high-speed railway, establish a new mechanism with market orientation, wide participation of enterprises, business schools and scientific research institutions, and joint production, learning, research and utilization. At the same time, we should strengthen the construction of other mechanisms for the transformation of scientific and technological achievements, such as investment and financing, technology transfer and diffusion.

Fifthly, strengthen the training of talents. Talents are an important force to promote the development of high-speed railway productivity. Therefore, we need to improve the incentive mechanism for talents. Based on the basic principles of material and spiritual incentive, incentive and restraint, and timely incentive on demand, we should adopt various incentive means to ensure the rationality and effectiveness of incentive mechanism to establish a team of talents to meet the needs of economic development.

10.4.3 Brand Construction Development Model Under the Orientation of Enterprise Attribute

10.4.3.1 Basic Connotation

There are several opinions on the brand building of high-speed railway. One is to equate train number with brand, which train number is what brand; the other is from the substantive point of view, the railway transport brand is divided into high-quality trains, high-quality stations, high-quality lines, such as "Harmony Number". These views are not unreasonable, but they are worth discussing. In the former view, the number of trains is regarded as a brand, which can be temporarily substituted in the case of high-speed railway transport products that have not yet been widely named. However, we should understand that the number of cars can not be equated with the brand, because this approach is very intuitive, but it only grasps the displacement function of transport products. The latter view, such as the "Harmony" train entity as a brand, also only pays attention to the product itself or the name of the product. The disadvantage of the above two views lies in the neglect of the development of brand concept, which has evolved from a narrow overflow product concept to a modern overall product concept.

With the development of productive forces, especially in the era of science and technology, people's living standards have been greatly improved, and the level of demand for transport products has been raised, that is, from the past as long as we can go, and then we need to go well to meet the needs of people's development and enjoyment. In this way, passengers and transport products are required to meet the material needs of displacement and spiritual needs; they are required not only to transport products economically, but also to transport products with cultural tastes. Therefore, the high-speed railway should not only be safe, economical, fast and punctual, but also be comfortable and provide family services, so as to make the passengers' boring and fatigued riding process become the enjoyment process of being edified in train culture, being respected in spirit, enjoying relaxation, enjoying pleasure and beauty in psychology. That is to say, high-speed railway transportation products are required not only to provide displacement as a core product, but also to obtain more added value in the process of displacement. This has become the main aspect of passenger evaluation and purchase of transport products. Because there is a strong substitution between different modes of transport and their products when providing the core products of displacement, the passenger decides which mode of transport to choose and which products to buy ultimately depends on his evaluation, trust and satisfaction of the added value of the transport products. Therefore, the brand of high-speed railway refers to the brand of the whole product of high-speed railway transportation, which reflects the overall interests of high-speed railway transport enterprises to buyers, including meeting the basic interests of passengers' safety, punctuality and speed, as well as the high-level needs of enjoying relaxation and enjoying cultural edification in the process of displacement. The basic demand of the displacement process is the foundation of the high-speed railway brand, and the core of the high-speed railway brand is that it can provide more added value to passengers than homogeneous products, such as aesthetic value, emotional value, knowledge value, social value and other spiritual value added to passengers.

10.4.3.2 Construction and Maintenance Measures

At present, China's high-speed railway has taken the pace of enterprise brand building, and achieved good results. However, the construction and operation of high-speed railway brand in China is still in the exploratory stage, so it is necessary to formulate corresponding safeguard measures to provide road signs for the construction of high-speed railway brand in China.

Firstly, the brand positioning of high-speed railway. Brand positioning of high-speed railway in China refers to the commercial decision-making of high-speed railway brand on the basis of market positioning and product positioning. It is the process and result of establishing a brand image related to the target market. Generally speaking, high-speed railway brand positioning can take the following measures: firstly, we should analyze the advantages and weaknesses of the transport market environment and competitors, and find the general position of competitive brands in the minds of passengers; secondly, we should find out the difference, which is to find out the core concepts and values that distinguish high-speed railway from competitors; thirdly, we need to find the difference support points, and we must prove them. Show the passengers; finally, focus on the promotional brand advantages and characteristics to the passengers. It can be said that only such an accurate brand positioning can promote high-speed railway to find its core brand value and play a huge brand utility.

Secondly, high-speed railway brand shaping and dissemination. In the process of high-speed railway brand shaping and dissemination, first of all, we need to refine the brand concept. Because the brand concept can reflect the difference between high-speed railway and other modes of transport, and show the different value to passengers. Secondly, build brand value. Brand value, as a comprehensive projection of the overall strength of passenger brand provided by high-speed railway, reflects the sum of brand assets of high-speed railway. Therefore, the key content of high-speed railway brand value will be the core content of high-speed railway brand construction. Thirdly, develop brand relationship. The development of high-speed railway brand relationship is to enhance the service awareness and skills of high-speed railway and enhance the brand experience of passengers. Thirdly, design brand image. High-speed railway brand should have a clear and rich brand image besides intangible assets. Otherwise, passengers will not be able to have a deep impression on high-speed railway. Finally, expand the channels of communication. In order to stand out from the fierce competition in the transport market, high-speed railway enterprises need to use various effective forms to publicize and display the image and strength of high-speed railway in addition to being skilled in service quality, technological innovation and management.

Thirdly, the maintenance and adjustment of high-speed railway brand. In the process of brand maintenance of high-speed railway, service quality is the core of brand. At the same time, due to the quasi-public nature of high-speed railway service, it should assume the corresponding social responsibility, and take the initiative to assume the responsibility for the environment and stakeholders, play the role of high-speed enterprises in the social environment, so that it gradually becomes an important part of influencing the brand of high-speed railway. In addition, with the change of market structure, high-speed railway also needs to adjust its brand strategically according to the actual situation, which can be reflected as brand extension. The extension of this brand can help high-speed railway enterprises maintain market share and form economies of scale. Of course, the extension of this brand is also a double-edged sword, which may damage the original brand image and weaken the relationship with the core competitiveness of high-speed railway enterprises.

With the continuous improvement of the transport market, building a strong brand is the development direction of China's high-speed railway enterprises. Brand building will become an indispensable part of high-speed railway enterprise management. Whether in the international market or in the domestic market, if high-speed enterprises want to base themselves on development, they rely on their brand competitiveness. As a necessary guarantee for China's high-speed railway enterprises to become market-oriented and internationalized, the development of its procedural and standardized will make China's high-speed railway enterprises occupy a place in the future international and domestic market competition.

References

Feng Hua, Xue Peng. Analysis of comprehensive benefits and supporting policies of China's high-speed railway [J]. *Social Sciences in Guangdong*, 2011,03:12-19.

Hou Mingming. Research on the Construction of HST Transportation Hub and Urban Development [D]. Tongji University, 2008.

Jiang, Ya. Historical Evolution of Land Planning in Japan and Implications, *Land Resources Information*, 2009(12):2–6.

Li Jiandong. Impact of high-speed railway on regional economy and industry along the line [J]. *Co-Operative Economy & Science*, 2012, 09:8–10.

Tian Zhenqing, Liang Hengyi. On the Asset Management Mode of Beijing Urban Rail Transit [J]. *China Transportation Review*, 2010, 07:27–30.

Toshiji Takatsu, The History and Future of High-speed Railway in Japan, *Japan Railway& Transport Review*,2007(8): 20.

Wang Jixian. Urban and regional impacts of high-speed railways: A preamble [J]. *Urban Planning International*, 2011, 26(06):1–5

Wang Lan. Research framework of the impact of high-speed railway on urban space [J]. *Planners*, 2011(07):13-19+8.

Wu, L. & Lin, X. Learning from the financing of Taiwan high-speed railway and its construction system [J]. *Railway Economics Research*, 2001(05):45-47

Ye, B. & Tang, J. Decoding the Integration Planning of the High-speed Train in Europe: a Perspective of Public Policy [J]. *Urban Planning International*, 2010, 25(02):97-100

Chapter 11
High-Speed Railway and National Macro-strategy

At the opening of the Second Session of the Twelfth National People's Congress, Premier Li Keqiang pointed out in his government work report that it was necessary to create a new level of opening up to the outside world and encourage high-tech equipment such as high-speed rail and nuclear power to go abroad. During his visit to other countries, Premier Li Keqiang also vigorously promoted China's high-speed rail. Therefore, the development of high-speed railway has become an important part of the country's implementation of several macro strategies. This chapter studies the relationship between high-speed railways and the country's macro-strategy and its role in four aspects: new urbanization, resource asset management, energy conservation and environmental protection, and technology transfer.

11.1 High Speed Railway and New Urbanization Strategy

11.1.1 High-Speed Rail and Urban Planning and Development Along the Route

The construction of high-speed rail will bring changes to the layout and planning of cities along the route. Judging from the development process at home and abroad, each major breakthrough in transportation and the adjustment of transportation hubs have profoundly affected the development and spatial evolution of the city. With the gradual implementation of the "four vertical and four horizontal" basic road network plans for China's railways, China will step into a new era of transportation marked by high-speed railways. Transportation technology determines the strength and breadth of the interaction between cities and is one of the important factors in changing economic activities. Therefore, the rational analysis of the high-speed railroad's important impact on regional development along the railway line, especially in the

X. Lin, *High-Speed Railways and New Structure of Socio-economic Development in China*, Research Series on the Chinese Dream and China's Development Path,
https://doi.org/10.1007/978-981-19-6387-2_11

area where the station is located, is crucial for the timely planning of future urban development strategies and the formulation of effective countermeasures.

Take Beijing–Shanghai high-speed railway as an example. The operation effect of the Beijing–Shanghai high-speed railway since its opening on June 30, 2011 fully demonstrates that the Beijing–Shanghai high-speed rail has begun to play its gateway role in the eastern part of China and has strongly promoted urban planning and urban development along the route, with urban architecture, functional support and urban commerce greatly improved. The overall level of urban economy such as housing, tourism, and employment has been greatly improved, and has been cascaded to the surrounding districts and counties, forming the "one-hour quality of life circle", "two-hour high-quality service circle", and "eastern central city". The 3 h industrial cluster circle's pan-city pattern has successfully integrated and expanded the development space and radiation intensity of the regional economy such as the Yangtze River Delta, the Beijing–Tianjin–Tangshan economic circle, the Bohai Economic Circle, the Shandong Peninsula, and the Huaihai Economic Circle.

From the perspective of resources, the eastern part of China has concentrated population, industrial base, very rich human and natural resources, and a very intensive investment situation. The Beijing-Shanghai high-speed railway will first create a north-south "co-located" public transport corridor in the most economically developed eastern region of China, and then radiate and drive industrial advantages from east to west, integrating the advantages of talent, infrastructure, natural resources and investment channels. This will enable China's eastern regions to seize new opportunities in the new round of international division of labour, and lay the groundwork for the smooth international take-up and transfer of China's new industries to the west after industrial restructuring, and for intelligent integrated transport trunk lines and urban transport hubs with advanced resource advantages.

The changes in urban planning along the Beijing-Shanghai High Speed Railway (HSR) line show that all cities, large, medium and small, where the stations are located, have re-planned their urban functions due to the construction of the HSR. Specifically, the urban planning of local governments has seized the opportunity of siting new HSR stations to move them out of the overcrowded old urban areas, which are full of "urban diseases", and to launch new cities, new planning, new business centres and new residential clusters. The new urban plan aims to create a modern and ideal city with "high starting point, quality construction and management"—such as a convention centre, a university town, an administrative building, a business and financial district, and a commercial street full of chain shops and long-established brands. In addition, their plan place emphasis on sustainable development programs such as ecological wetlands and waterfront landscapes. Taking the "high-speed rail station as a point, high-speed rail line as the line, and new city planning as the surface", emphasis is placed on replacing the existing planning tradition of the "old city stalls" with the integrated urban planning concept of the independent functional areas and seizing the "high-speed rail concept". This new opportunity for urban planning focuses on humanity, gathering popularity, shaping the humanities, and bringing a new look to the city's appearance along the high-speed rail line. Therefore, this is an opportunity for urban planning in China, a leap in China's economic growth,

Traffic Positioning	Development Goals	Urban Positioning	Core Functions	Industrial structure	Support conditions: traffic, infrastructure
Existing Lines	Old City Terrace Pie Expansion	surviving living developing	Industrial, commercial and agricultural facilities	Recycling industry, comprehensive services, integrated living	Single-channel flow of people, logistics
	Undertake industrial transfer				
	Urban-rural integration				
High-speed Railway	Low-carbon city	Eco-urbanization	Industrial chain logistics chain intensive agriculture	Circular economy, service economy, ecological living	Rapid transit system
	Industrial agglomeration				
	New pattern of regional economy				
High-Speed Rail-network	Smart City	Smart urbanization	Web-based business exhibition industry integration	Network Economy Global Positioning Sustainable Ecology	Intelligent Integrated Transportation System
	Featured commerce cultural creativity special tourism smart living modern logistics				

Fig. 11.11 Effects of the Beijing–Shanghai high speed rail on the urban planning along the line

a ladder for the renewal of the Chinese people's living standards, and a new reservoir for the urbanization of Chinese society. It truly verifies the evolutionary truth from the uprightness of ape to humanity and the traditional Chinese saying—"To get rich, build roads". There is no doubt that the high-speed rail will bring more beauty and well-being to the Chinese people. From the perspective of the new planning, urban management has realized the concept of modernization due to the high-speed rail effect, as shown in Fig. 11.1.

Such influence is accompanied by the evolution of the Beijing–Shanghai railway technology. As we can see from the existing cable line, the Beijing–Shanghai high-speed rail line and the future high-speed rail network, such technology already has had a fundamental and transformative impact on the surrounding cities. With the operation of the HSR, the evolution of the city will move from a place for residents to an ecological city and then to a sustainable smart city.

11.1.2 *City Effect of High-Speed Railway*

Railway transport has contributed greatly to the development of industrialization in China in the past. However, the transport capacity of China's railways is now far from being able to meet the accelerating industrialization process, and has instead become a bottleneck for it. The construction and development of high-speed railways will divert a large part of China's passenger traffic, greatly increase the capacity of the existing railways, enable the rapid transportation of energy, raw materials, and manufactured goods in the east and the west, and greatly promote the industrialization of China. Further, the increased industrialization links between east and west and central China help to promote the industrialization process in the central and western parts of the country and narrow the industrialization gap in China.

Urbanization is a process in which the population continues to accumulate in cities and towns, and it is an inevitable historical stage in the process of industrialization in all countries of the world. Since the reform and opening up, the process of urbanization in China has developed rapidly. The level of urbanization has increased from 10.6% in 1949 to 51.27% in 2011. It is estimated that by 2020, the urbanization rate in China will increase from 46.6 to 60%, which means that more than 200 million people need to move to cities. According to statistics, the national traffic volume will reach 50 billion passengers by 2015. This will undoubtedly require a strong transportation system as a support, but China's existing railway system is far from meeting current and future needs. High-speed railways have a great advantage as their one-way transport capacity is four to five times that of ordinary railways, and therefore provide strong capacity support for the massive migration brought about by urbanization.

The high-speed railway is in line with the current new trend of urbanization in China, promoting the process of urbanization and promoting the coordination of transportation and land use. At present, the level of urbanization in China is evolving to a higher level. The spatial pattern of urbanization continues to diversify, and urban-intensive areas such as urban agglomerations and metropolitan areas continue to emerge and develop. Seven cities along the Beijing–Shanghai high-speed railway line cover the Yangtze River. Delta, Beijing–Tianjin–Hebei, Shandong Peninsula and other towns gathered. The high-speed railway not only strengthens the links between these towns, but also influences the regional and transport development between and within towns and cities. High-speed rail has realised the urbanization of the rural areas around the railways in one step, with a number of modern new towns springing up from scratch in each site area.

For a long time, there have been huge differences in the economic structure, production layout and resource allocation between urban and rural areas in China. Taking education as an example, according to statistics, the proportion of the urban population in China with high school, secondary school, college, undergraduate and postgraduate education is 3.5 times, 16.5 times, 55.5 times, 281.55 times and 323 times higher than that of the rural population respectively. This is mainly due to the unbalanced economic and social development in China, which in turn becomes barrier to education. In regions where economic development is relatively backward, the investment in education is seriously insufficient. Secondly, the institutional gap caused by the dualistic economic structure between urban and rural areas has directly contributed to the subordinate and secondary status of rural education, which lacks effective institutional protection. The gap between urban and rural education is widened by the different educational assumptions of urban and rural residents. For a long time, rural areas in China have been in small-scale production and relatively closed. A considerable number of farmers in rural areas do not see the long-term benefits of education and investment. In contrast, the sense of education investment of urban residents in China is relatively strong. In particular, with the high rate of return seen from higher education, the education and investment enthusiasm of urban residents has been greatly improved. The opening of high-speed railways has not only brought the distance between cities closer, but also between urban and rural areas, more communication between residents and connection, making all residents more aware of the importance of education. With increased investment in rural and local education, there will be more coordinated, comprehensive and sustainable educational progress as a whole so that both rural and urban residents could receive better quality educational services.

High-speed railways play an obvious role in co-ordinating urban and rural development and balancing public transport services. Take Beijing–Shanghai high-speed railway as an example. At present, even in areas with relatively developed economic levels such as the seven provinces and cities along the Beijing-Shanghai high-speed railway, there is still a large gap between urban and rural transport infrastructure so much as that some rural areas are not enjoying good public transport services. One of the government's priorities during the 12th Five-Year Plan period is to increase public expenditure, and investment in transport infrastructure in rural areas so as to improve the accessibility, travel conditions and public services in rural and remote areas. After the opening of the Beijing–Shanghai high-speed rail line, all the stations along the route have formed a central circle of different sizes. The high-speed rail not only serves the site city, but also provides convenient conditions for the nearby rural residents to travel within the circle. It also drives the development of public transportation in these areas.

Regional economic coordination and integration are the major trends in international economic development. High-speed railways have greatly reduced the distance between places and accelerated the regional economic integration in China. The Beijing–Shanghai high-speed railway has shortened the train journey between Beijing and Shanghai to about 5 h, with 90 pairs of trains running daily at the initial stage of operation, with a mixed mode of running at 300 km/h and 250 km/h. Among them, 76 pairs of G trains and 14 pairs of D trains will depart every 16 minutes on average, linking up 24 stations in 7 provinces and cities along the way, driving the economic and social development of the provinces and cities along the way. The Beijing–Shanghai high-speed rail will promote not only the rapid development of tourism, trade, catering and other tertiary industries along the line, but also the urbanization process, drive the prosperity of the rural economy, and promote the investment environment along the route and the appreciation of real estate properties. While shortening the distance in time and space, it will also create a "co-location effect", giving rise to a "one-hour or two-hour economic circle" along the route, narrowing the development gap between cities, city clusters and economic zones, and promoting the transfer and rational layout of industries between regions. The reduction in distance between cities accelerates the flow of people, capital and goods between cities and the formation of integrated transport infrastructure between cities and regions, leading to the integration of industrial layouts, market, urban and rural construction, resource use and environmental protection, and ultimately the formation of city circles. The Beijing–Shanghai high-speed railway has improved the industrial division of labour and the concentration of factors in the cities along the route between Beijing and Shanghai, and the rapid flow of capital, talent, technology and materials between the two places, reshaping the urban development between Beijing and Shanghai. It will enable the two mega-cities to become a "one-day economic and business circle of Beijing and Shanghai", and create a co-location effect. It can be expected that in the next 5 to 10 years, the Beijing–Shanghai high-speed railway economics will come into being.

11.1.3 High-Speed Railway and the Urbanization Along the Line

Urbanization is a complex socio-economic phenomenon. Traditionally, urbanization refers to the process of rural-urban migration that occurs as the industrial economy concentrates in cities. In a more recent sense, urbanization highlights the diffusing process of lifestyles and urban civilization while emphasising population transfer, occupational shift and industrial agglomeration. The postmodern definition of urbanization places particular emphasis on the deeper connotations of changing lifestyles and the permeation of urban civilization, and even questions the need for the concentration of population, territory and factors of production in the traditional definition. For post-modernists, the place of residence should not be an indicator of urbanization due to the advancement of communications and transport in contemporary society, which has greatly reduced the distance between people.

With rapid economic development, China is changing from a production-based to a consumption-based economy, with the central task of cities shifting to consume or serve consumption. As we enter the twenty-first century, and with the expansion of the economy, the economic activities of many cities have expanded beyond their administrative boundary. Economic links between cities have become closer and there is a trend towards coordinated development. The integration process has accelerated especially in dense urban areas such as the Yangtze River Delta, the Pearl River Delta and Beijing–Tianjin–Hebei. Meanwhile, cities in the central and eastern regions are also clustering and are becoming city circles and urban agglomerations, such as the Yangtze River Delta City Cluster, the Pearl River Delta City Cluster, the Bohai Rim City Cluster, the Changsha-Zhuzhou-Xiangtan City Cluster, the Central Plains City Cluster and the Wuhan City Circle. The scope of urban clustering generally covers an area with a radius of 100–250 km.

According to data released by the National Bureau of Statistics, the urbanization rate of cities along the Beijing–Shanghai high-speed railway is shown in Table 11.1.

Accordingly, the cities along the route can be ranked according to their urbanization rates, as shown in Table 11.2. We can see that the Beijing–Shanghai high-speed railway creates a differentiated development space and provides opportunities for cities along the route to develop.

Along with the technological evolution of the Beijing–Shanghai railway, the existing line, the Beijing–Shanghai high-speed railway line and the future high-speed railway network will have a transformative impact on the surrounding cities. For the

Table 11.1 GDP and urbanization rate of cities along the Beijing–Shanghai high-speed railway

Index	GDP (billion yuan)		Urbanization rate (%)
Cities along the rout	Year		
	2008	2011	2011
Beijing City	10,488.00	16,000.40	86.23
Langfang City	1061.50	1612.00	47.80
Tianjin City	6354.38	11,190.99	79.55
Cangzhou City	1716.16	2600.00	–
Dezhou City	1400.91	1950.71	30.23
Jinan City	3017.42	4406.30	63.21
Tai'an City	1513.30	2304.30	35.11
QufuCity, Jining District	200.53	265.76	31.57
Zaozhuang, Tengzhou City	477.13	728.13	34.77
Zaozhuang City	1089.12	1561.68	36.16
Xuzhou	2007.36	3551.65	55.40
Suzhou City	511.10	802.40	33.10
Bengbu	486.39	780.24	46.60
DingyuanCounty, Ganzhou City	53.40	103.30	
Shengzhou City	520.11	850.49	43.40
Nanjing City	3775.00	6145.52	79.73
Zhenjiang	1491.83	2311.45	63.02
Danyang City, Zhenjiang City	425.45	724.90	53.30
Changzhou City	2266.32	3580.99	65.20
Wuxi City	4460.62	6880.15	72.32
Suzhou City	7078.09	10,716.99	65.20
Kunshan City, Suzhou City	1500.26	2432.25	75.10
Shanghai City	14,069.86	19,195.69	89.00

first time in the developed eastern regions, the Beijing–Shanghai high-speed railway will play a logistical support function in the industrialization process, bringing new opportunities and new models for urbanization in the east and the mid-west. It has a mutually beneficial and polarising effect on urban space—moving out of old urban areas and developing new ones—and it plays a clear role in co-ordinating urban and rural development and improving public transport services. The evolution of the city will evolve from a simple residential area to an ecological city along with the operation of the high-speed rail, and then into a sustainable, smart city.

Table 11.2 Urbanization rates and business cards of cities along the Beijing–Shanghai high-speed railway

Evaluation index	City name	Urbanization rate (%)	City card
High	Shanghai	89.00	The richest Crowns
	Beijing	86.23	
	Nanjing	79.73	
	Tianjin	79.55	
	Kunshan	75.10	
	Wuxi	72.32	
Secondary high	Changzhou	65.20	Rising stars
	Suzhou	65.20	
	Jinan	63.21	
	Zhenjiang	63.02	
Upper middle	Xuzhou	55.40	The soaring eagles
	Danyang	53.30	
	Cangzhou	53.00	
Middle	Langfang	47.80	Walking stars
	Bengbu	46.60	
	Chuzhou	43.40	
Low	Zaozhuang	36.16	Ground start
	Taian	35.11	
	Tengzhou	34.77	
	Suzhou	33.10	
	Qufu	31.57	
	Dezhou	30.23	

11.2 High-Speed Railway and Resource Management Strategy

The first 20 years of the twenty-first century are a period of important strategic opportunities for China's economic and social development, but also a period in which the contradictions between resource and environmental constraints are intensifying. Therefore, it is necessary to fully understand the challenges facing China in land use and management: the situation of agricultural land, especially arable land protection, the contradiction between supply and demand for construction land, and the right use of land. Of course, it is also necessary to see the favourable conditions for solving land use problems in the current and future period. From the perspective of land use, the use of land for construction in China is generally extensive and there is a large space for economical and intensive use. Economic globalization, scientific and technological revolution and upgrading of industrial structure are conducive to China's transformation of economic development and building a resource-saving society. The in-depth implementation of the Scientific Outlook of Development and the great importance attached to land management by the Party Central Committee and the State Council are conducive to strengthening macro-control of land and the harmonious relationship between people and land. Land use planning in China must follow the following basic principles: (1) Strict protection of arable land. In accordance with the requirement of stabilizing and improving the basic status of agriculture, arable land, especially basic farmland, will be strictly protected. During the "12th Five-Year Plan" period, the amount of arable land retained will be maintained at 120 million hectares; (2) Economical and intensive use of land. In accordance with the requirements of building a resource-saving society, based on the protection and promotion of scientific development, and reasonable control of the scale of construction, the transition should go from extensive and inefficient to intensive and efficient; (3) Coordinate the use of land for various industries. In accordance with the requirements of implementing the overall national regional development strategy, and a new structuring of land development, it is necessary to optimise the allocation of different types of land for different industries, guide the reasonable flow of population, industries and production factors, ushering the urban-rural integration and coordinated regional development; (4) Strengthen ecological protection of land. In accordance with the requirements of building an environment-friendly society and based on the construction of a good living environment, it is very necessary to arrange land for living, ecology and production in an integrated manner, giving priority to the protection of natural ecological space.

As can be seen from Table 11.3, the proportion of land for transportation in China will be on the rise in the next 10 years, which means that intensive use of land is a must for future transportation development strategies. With land resources becoming increasingly tight, especially in the eastern region, where existing high-speed railways already occupy a considerable proportion of land, it is crucial to make effective use of land resources and increase the transport capacity on the lines and nodes. The rational layout of transport route corridor strips and stations should be accelerated, with priority given to options that can maximise land conservation and protect arable land. Large transport hubs and integrated transport corridors, such as large deep-water terminals, large stations and corridors, will become important models for transport efficiency and intensive resource use.

Table 11.3 Traffic conditions in China

Years	Land for planning and construction (10,000 hectares)	Land for transportation (10,000 hectares)	Proportion of traffic land to construction land (%)	Proportion of land used for transportation to national land (%)
2005	3192.90	230.90	7.23	0.24
2008	3305.80	249.60	7.55	0.26
2010 (expected)	3374.00	264.80	7.85	0.28
2020 (expected)	3724.00	357.60	9.60	0.37

For high-speed railways, the economic characteristics of high-speed rail are more favorable than others in the rational use of land resources. Due to its highly closed nature, the land occupied by its tracks is basically a direct impact which can be measured. The land occupied by the station has a ripple effect, affecting the land market in the local area, especially in the surrounding areas. The significance of high-speed rail is to improve accessibility, which mainly refers to the time savings. Due to the reduction of train running time, people will be able to extend their travel mileage or increase the number of trips without increasing the original travel time, thereby changing the distribution of laborers among the regions, stimulating the local market, especially the real estate market, and affecting land use. In addition to the principle of land saving, the modernization of railways should also emphasize the integration of railway planning and land use planning. The advanced model of integrated railway and land use, such as the TOD model, can both reduce transportation demand and reduce land use. The impact of railways on land use includes the direct occupancy, influence and change of surrounding land by railways, and indirectly the land use and value of the entire region. High-speed rail meets the requirements for the use of land resources in China.

11.2.1 China's Railway Land Use Policy

11.2.1.1 The Scope, Origin and Characteristics of Railway Land

(1) The scope of railway land

Railway land in China refers to the land that the railway sector has obtained the right to use in accordance with the law, including reserved and expropriated land for transport production, auxiliary production, living and other land. It is owned by the government and is an important part of state-owned land, which is utilised and managed by the railway sector and protected by national laws.

Land for transport production refers to land for railway lines, stations and yard facilities, land for water supply, power supply, communications and signalling facilities, land for safety protection and epidemic prevention, environmental protection and water conservation facilities, land for railway ferries, terminals and other required facilities, land for transport production and equipment repair and maintenance facilities, land for special equipment production facilities, land for storage depots and guard and war reserve facilities, etc. In 2009, China's railway transportation used 432,085.9 ha. of land. Auxiliary production land refers to land used for transport and production organization and commanding organs, land for organ propaganda, staff education and training, land for forestry and quarrying, land for public prosecutors, and apartment houses and guest houses. The land for living facilities refers to land for cultural, recreational, educational, medical, and other facilities, and residential land that has not yet been stripped. Other land use refer to land used for railway transport production and services and other supportive land use that are not included in the above site types.

(2) Sources of railway land use

The source of railway land refers to the way in which the right to use railway land is acquired. According to the relevant laws and regulations on land management in China, the acquisition of state-owned land use rights mainly includes the following five: ① Allocation, which refers to the act of being approved by the people's government at or above the county level in accordance with the law and delivering the land to the land user for use after the land user has paid compensation, resettlement and other fees, or delivering the state-owned land use rights to the land user without compensation. For land acquired by administrative allocation, the enterprise has the right of possession, the right to use and a certain right to income, but the disposer of the land belongs to the State, and the enterprise has no right to transfer, lease, mortgage, or change the purpose of the land used. ② Transfer. This refers to the act of the state granting the right to use state-owned land to the land user for a certain number of years, with the land user paying the state for its right to use the land. According to the law, all construction land that does not conform to the Catalogue of Allocated Land use shall be supplied in the form of a grant or other paid use, and the land shall revert to the state at the end of the specified period of use. ③ Leasing. It refers to the leasing of state-owned land by the state to users for use, who sign a land lease contract with the land administrative department of the people's government at or above the county level for a certain period of time and pay rent. State-owned land leasing is a form of paid use of state-owned land and is a supplement to the transfer method and is not applicable to land for business real estate. ④ Capital contribution to shares. It refers to the State's use of state-owned land for a certain number of years as an investment into a joint-stock company or limited liability company, and the corresponding land use rights are transformed into the State's capital or equity capital contributed by the State to the enterprise. According to the requirements of the separation of government and enterprises, the state equity formed by the land use rights as capital contribution to shares should generally be held in accordance with the main body of investment in state assets by the land administration department of the people's government which has the right to approve the commissioning of qualified state equity holding units.

Up to now, most of the railway land in China has been obtained through administrative allocation. The land originally owned by the country is directly allocated to the railway; the land originally owned by the collective is transferred to the railway for use after the state-owned land is acquired, and the railway unit pays the requisition (disbursement) fees according to the standards prescribed by the state. However, in recent years, with the continuous advancement of the strategy for building a harmonious railway, there have been some new changes in the source of railway land in China. On the one hand, most of the new lines (including passenger lines and intercity railways) have been built under the project responsibility system in accordance with the ministry-province agreements on railway development signed between the Ministry of Railways (which has been abolished) and the various provinces, municipalities directly under the Central Government and autonomous regions of the country respectively. The relevant joint venture companies have been or will soon be formed as project legal persons, responsible for the financing of the construction process, construction management and the operation and management of the lines after completion. According to the division of labour between the former Ministry of Railways and the local government, land acquisition and relocation, compensation and resettlement for the land required for the construction of new lines is mainly the responsibility of the local government, with the relevant costs included in the project estimates according to the prescribed standards and as a contribution by the local government. Although these railway sites are still not allocated, the costs of land acquisition, demolition, compensation and resettlement are ultimately borne by the enterprises. On the other hand, corporate restructuring and stock reform are one of the important measures for comprehensively building a modern enterprise system and accelerating the development of the financial and securities market. This is reflected in the railway enterprises in that the number and scale of joint venture railway companies are expanding, and the number of joint stock companies is rising and will soon be listed one after another, which places higher demands on the independence of the enterprises and the financial structure of their assets. Therefore, the railway land acquired by the allocation method should also be reflected in the accounting of corporate assets.

(3) Characteristics of railway land use

Railway land is an indispensable material basis for railway transportation production, which has the general characteristics of land, and at the same time has the characteristics of corporate capital as an integral part of the assets of railway enterprises. Although the railway transportation production land is still used to obtain the right to use the state allocation, this is the preferential policy for the state to provide railway and other infrastructure construction, and the general trend of the value orientation of land use marketization will not change. Specifically, railway land has the following characteristics:

Natural features: Railway land is a place of work that serves the production and construction of railway transportation. The characteristics of its production layout determine the relative independence of its land use and the extensive and continuous characteristics of the land area. As far as the national railway is concerned, the distribution of railway land is characterised by linear features and the railway lines are distributed in a ribbon pattern, spanning provinces and cities through counties and towns, with many points and long lines, covering a wide area of land, both in the countryside and in the cities, with plains, hills and mountains, rivers, forming an interwoven network system nationally.

Structural features: The diversity of uses determines the complex layout of its land use planning. The complex and diverse use of railway land determines the professional characteristics of its layout structure. No matter which bureau has a production and operation area, an auxiliary production area and a living and residential area, it is subdivided into inter-area line land, station land, plant (segment) land, institutional land, cultural, educational and sanitary land, cultural and recreational land, residential land, and other land use. Since the railway transportation lines are continuously arranged in a certain direction and fixed always, the configuration, land occupation and scale of the railway land are limited by its transportation operation flow system, and the land use function is also special, a kind of land use. It can be used by multiple departments, cross-layouts are interconnected, and adjustments can be made at any time with the development and changes in the transport operations organization.

Economic characteristics: Due to the particularity of its sources, railway land has the following characteristics in terms of economy: First, the source is low-priced or unpaid. In recent years, the continuous tension in railway transport capacity has had a major impact on the development of China's national economy. Whether it is from China's own development or foreign experience, it is a necessity for the development of the railway for a long period of time. The prerequisite for the harmonious development of China's railways is to ensure the supply of land for railway construction. The second is intangible assets belonging to railway companies. Railway enterprises only enjoy state-owned land use rights for railway land, and land use rights do not have physical form. Therefore, in the category of assets, railway land should be attributed to the scope of intangible assets, and its value is usually manifested when the land use rights of railway lands are transferred, leased, mortgaged, used for equity, or invested.

Legal characteristics: The railway land is a state-owned land that has obtained the right to use according to law. The State grants the right to the use of land to railway enterprises by way of retention or transfer. The legal certificate for railway land is the land use right certificate issued to the railway enterprise by the state land administration department.

Taxonomic characteristics: Railway land is a dedicated land, which belongs to construction land. It has the characteristics of construction land: non-nurturing and space. Non-nurturing refers to the use of land for carrying and non-productive functions. This means that when there is a contradiction between construction land and agricultural land, it is first necessary to consider giving priority to the use of land with higher functions for farming. Spatiality means that in order to obtain more load-bearing space and project foothold, construction land can not only be used flat, but also be multi-layered, high-level, or underground.

Target characteristics: The safety of railway transportation is the first goal of railway management. The management of railway land must revolve around and adapt to this theme and put safety first. Any company must achieve safe production and ensure product quality, otherwise it is difficult to avoid production accidents and tort damages caused by product quality disputes. Therefore, no safety will detract from economic indicators such as income and profits, and there will be no economic benefits. From this perspective, the objectives of railway land management are consistent in terms of safety and efficiency, that is, to fully realize the economic benefits of land under the premise of ensuring safety.

11.2.1.2 Policies and Status Quo of Railway Land-Use Management

(1) Changes in Railway Land Use Policy

The external influence system of railway land management in China includes the state land property laws and land management policies. Since the founding of the People's Republic of China in 1949, these laws and policies have undergone tremendous changes, which have also changed the management system of railway land in China. See Table 11.4 for details.

(2) Basis for management of railway land

Land management is one of the basic functions of the state. The state adopts legislative bodies to standardize the will's performance and fixes it in legal form. The national administrative organ, that is, the people's government at all levels and the land administration departments, guarantees the implementation of laws and regulations so as to achieve the purpose of the management function. The management of railway land is a management function conferred by national laws and regulations.

Authorised by national laws and regulations. Land is the carrier of railway transport, and the railway authorities can only fulfil the railway transport management functions conferred by the State if they actively carry out railway land management activities and make efforts to improve the management of railway land. Article 2(2) of the Land Management Law of the People's Republic of China stipulates that "The ownership of land owned by the whole people, i.e. owned by the State, shall be exercised by the State Council on behalf of the State." Article 3 of the Railway Law of the People's Republic of China promulgated by the Standing Committee of the National People's Congress provides that "The competent railway authorities under the State Council shall be in charge of the work of the national railways, implement a highly centralised and unified command transport management system for the national railways, and provide guidance, coordination, supervision and assistance to local railways, special railways and special railway lines. State railway transport enterprises exercise the administrative functions conferred by laws and administrative regulations." Article 3 of the Administrative Measures for Railway Land stipulates: "Railway land is owned by the state and is used and managed by the railway department and is protected by the national laws." Article 1 of the Circular on the Protection of Railway Facilities to Ensure the Safety and Smoothness of Railway Transport promulgated by the State Council provides that "all land legally reserved and requisitioned by the railway departments in accordance with the Regulations on Land Reserved for Railway Use promulgated by the State Council of the Central People's Government in 1950 and the Regulations on Land Requisition for State Construction promulgated by the State Council in 1982 are railway land and shall be managed by the railway authorities". "The above laws and regulations are the legal basis for the railway department to exercise its land management functions". Article 5(2) of the Land Administration Law of the People's Republic of China stipulates that "The establishment of land administrative departments of local people's governments at or above the county level and their duties shall be determined by the

Table 11.4 Changes in China's railway land policy

Time	External constraints	Policy summary	Policy features	Policy evaluation
Resource-based stage: From the founding of the People's Republic of China to the late 1980s	Planned economy, single land use, lack of external conditions reflecting land value	Guidelines on property rights, security and job responsibilities	Most of the land management departments are attached to the technical and construction departments, using the technical division of labour as a criterion for setting up the functions; simple assessment methods	Low agency costs, internalized the division of labor revenue, and promoted the economies of scale in railway land management
Transitional stage: From the late 1980s to early 2006	Economic transition, relaxation of external property rights constraints, enhanced plasticity of land resources, asset value began to show	Land management involves a legal dimension, emphasizing the efficiency of land use, and progressively with some specific management norms in terms of property rights, management functions, approval and enforcement	The connotation of land use, disposal and revenue rights is complex, and the functions and assessment system of railway land management are correspondingly complicated, diversified and constrained	The asset-based value of land has been exploited to some extent for efficient management, while reducing agency costs associated with increased resource plasticity, but due to a lack of motivation for comprehensive institutional change, management functions are confused and many sectors are inefficient

(continued)

people's governments of provinces, autonomous regions and municipalities directly under the Central Government in accordance with the relevant provisions of the State Council." Article 7(4) of the Measures for the Administration of Railway Land provides that "It shall be responsible for the guidance, inspection and supervision of the utilization of railway land; and shall be entrusted by the land administration departments of the people's governments at or above the county level to carry out

Table 11.4 (continued)

Time	External constraints	Policy summary	Policy features	Policy evaluation
Production stage: Since 2006	Land appreciation, rapid increase in the cost of railway land acquisition, shortage of railway construction funds	External preferential policies were gradually withdrawn, and restrictive policies on railway lands increased	The reforms in terms of responsibilities, management mechanisms, technical standards, and technical measures have achieved results and formed a relatively clear organizational system	The reform has achieved results, but due to the inertia of the system and the lack of an appraisal system that reflects asset efficiency, there is still much work to be done in further comprehensive reforms

Source Zhang Chun, "Study on the Change and Reform of Railway Land Management System in China", Beijing Jiaotong University

supervision of railway land." Paragraph 6 states, "Responsible for relevant matters entrusted by the national and provincial land administration departments." In accordance with the above-mentioned legal provisions, there are currently a number of railway land (land) administration bureaus or management sub-bureaus set up by the 18 railway bureaus across the country. These railway land (land) administration bureaus or management sub-bureaus are all authorised by the local provincial people's governments or entrusted by the land administration departments under the leadership of the local provincial people's governments to take credit for railway land management.

The need for railway reform and development. Railway land, as a professional land allocated by the state, provides indispensable material conditions to ensure the smooth flow of railway transport and maintain the safety of the national economic arteries. Its role in railway transportation is summarized as follows: ① Railway land provides the material conditions for the development of railway construction. During the period of socialist revolution and construction, especially when China entered a new historical period of reform and opening up, railway land, as the country's professional land, provided a strong guarantee for the development and long-term planning of railway, and provided material conditions that could not be replaced by other resources to ensure the safety and smooth flow of railway transport. ② Railway land is a huge asset that is constantly appreciating. In terms of preserving the value of state-owned assets, the existing fixed assets of the railway are affected by various factors and can be said to be on a declining trend year by year. However, the economic value of railway land, an irreplaceable and non-renewable resource, has been rising year on year as a result of the market economy. This is attracting great attention from the relevant railway land management departments. ③ Railway land provides a strong material guarantee for railway rescue and reconstruction. The land for railway use has a long history. There is land for line, station and field, water supply, factory section and living area. And as reserved land, it provides a strong land resource guarantee for the entire railway transport in organising the construction of drainage systems, daily road repair, afforestation and greening, stabilization of the roadbed and emergency repair of lines. ④ As railway land is accompanied by railway transport in all directions, crossing provinces and cities and connecting towns, it constitutes a large number of points, long lines and wide areas, presenting the characteristics of multiple neighbours, multiple boundaries and continuity. It is due to this special feature of railway land that an invisible bridge has been built for the smooth flow of the national economic artery.

(3) China's railway land management agencies and functions

The management of railway land is based on national laws and regulations, and the internal management of the use of land under the jurisdiction of the railway is an important part of railway transportation management. The railway land management organization is different from the national land resource management department and the land asset management department. It protects the land and resources to protect the rights of state-owned assets and organizes the realization of land functions and maintenance and appreciation of value.

According to Article 6 of the Measures for the Administration of Railway Land Use issued jointly by the State Land Administration Bureau and the Ministry of Railways (revoked): "All railway administration agencies of the Ministry of Railways, Railway Bureau, Railway Sub-bureau and other units of the Ministry of Railways shall be under the unified guidance of the land management department of the people's government at or above the county level, the management of railway land will be carried out." The Land Administration Department of the Railway Bureau is specifically responsible for the management of land within the jurisdiction area, and its business is regulated by the land management department of the local people's government and the land management department of the Ministry of Railways. The dual leadership and supervision of the exercise of railway land management authority. The approach also has clear provisions on railway land management functions. It has six functions: "1. Promote, implement and enforce national laws, regulations and policies concerning land management, and formulate regulations on railway land under the direction of the land management department. 2. In accordance with the unified national regulations, it is responsible for the investigation, declaration, registration, statistics, and planning of railway land use. 3. Undertaking the application for the approval of state-recognized railway construction land acquisition (distribution). 4. Responsible for the use of railway land. Guidance, inspection and supervision; entrusted by the land administration department of the people's government at or above the county level to carry out the supervision of railway land use. 5. Cooperate with the land administration department of the people's government at or above the county level to handle land disputes. And related matters entrusted by the provincial land administration department."

(4) Content of Railway Land Management in China

Railway land management is the internal management of the use of railway land by railway land management agencies in accordance with national laws and regulations or commissioned by the government, in accordance with the characteristics of railway transport organizations, using administrative, economic, legal, and scientific methods of comprehensive management. Railway land management mainly includes cadastral management, planning management, utilization management, and supervision and management.

Cadastral management, which is the basic work for making railway land use legal basis, mainly includes conducting cadastral surveys, reporting and registration, and obtaining authorization for certification; establishing and improving cadastral archives and carrying out statistical work on land use; conducting surveys on the use of conditions, and mapping work.

Planning and management is based on the needs of railway operations and development, and the overall coordination and reasonable use of railway land. The objects include two types: one is the planning of new railway lands to be acquired due to the construction of new lines, the upgrading of existing lines, etc. The second is the useful development and utilization planning. Its work includes: preparation of site-based overall utilization plans and preparation of annual land-use plans in accordance with railway line change plans; timely integration of plans into government land-use master plans, submission of annual land-use plans to government land-administrative departments, and implementation of plans. The situation is supervised and checked. In short, planning and management is a prerequisite for guaranteeing the scientific use of railway land and guaranteeing the legal status of railway land.

The use of management refers to the management of the use, disposal and protection of railway land in accordance with the relevant laws and regulations of the State and the relevant provisions of the railway. Including: establishing relevant rules and regulations and technical standards; guiding the use of land in accordance with approved planning and utilization plans; mastering the dynamics of the use of railway land, taking charge of the investigation and utilization of railway land, and examining and approving; formulating the method for the division and disposal of railway land in railway reform And organize implementation work. It is the core content of railway land management. Supervision and management is an effective means of regulating the use of railway land and protecting the rights of railway land. The management in this area mainly includes the promotion of national laws and regulations relating to land management and the regulations governing the management of railway land; and the supervision and inspection of the use of railway land in accordance with laws and regulations and the government's entrustment, and the prevention and investigation of violations of laws and regulations, and coordination Land ownership disputes, etc.

In short, the four aspects of railway land management are necessary to realize the effective and efficient use of railway land. In recent years, China has made great achievements in the management of railway land. China's railway land management department has gradually established a land management talent team with rich experience in railway land management, formed a relatively complete set of railway land management system, and summarized some good land management experiences.

Especially in the construction of the railway land management system and the organization and construction, a law and regulation system based on the railway law and based on the railway land management method was established, which played an important role in managing the railway land according to law and realizing the legalization of the railway land management. According to these regulations, the organization of railway land management has been improved, railway land management has been incorporated into the national land management system, and it has won the support and assistance of the national land administration department. It is of special significance to strengthen the management of railway land and maintain the order of railway transportation production.

(5) Management of railway land-use charge

In recent years, with the acceleration of the pace of railway construction in China, the area of land used for railways has been increasing. At the same time, with the rapid development of China's economy, due to the construction of special railways and the construction of water supply, power supply, and communications facilities, the situation in which railway neighbors or individuals occupy railway land is getting more and more serious and the problems are becoming more and more prominent. In order to save valuable land resources and ease the contradiction between railway departments and adjacent units and individuals for occupying railway land, some localities have strengthened the management of railway charges.

Occupied railway land use fees refer to the usage fees or management fees collected by the railway department when it legally occupies the railway land for its neighboring units or individuals. According to the legal system of neighboring relations and easements, railway neighboring units or individuals have the right to occupy railway land in accordance with law. The railway department has the right to charge fees according to law, and the use of railroad land charges complies with the legal system of neighboring relations and easements in China. In order to strengthen the management of railroad land use fees and save land resources, various local governments have issued some regulations in succession.

On July 1, 1983, Article 24 of the Interim Measures for the Administration of Railway Land, promulgated by the General Office of the People's Government of Heilongjiang Province, stipulates that the off-road units and individuals (including employees) shall temporarily occupy the railway station area (excluding the land used for the goods yard) and the interval. For land use, the real estate section and the public works section shall be responsible for signing the "Temporary Occupied Railway Land Use Agreement" in quintuplicate with the occupiers and individuals (The two parties shall each hold one copy, each of which shall be filed with the Bureau of Roads, the Land Management Bureau of the Branch, and the Finance Department.), and calculated monthly according to the following rate standards, and charged in two phases each year. 1. All national and collective units and self-employed households with a business income are charged five cents per square meter per month (but the government office and public facilities can be charged according to two standards); 2. Individuals living in the urban station area, monthly per square meter will be collected at the rate of 6%; 3. Individuals living in the rural commune station area will be charged at the rate of 3% per square meter per month.

On July 27, 1992, the Shandong Provincial Bureau of Pricing and Shandong Provincial Department of Finance promulgated the "Circular on the Promulgation of the Notice of the Ministry of Finance of the State Pricing Administration on the Issuance of Centrally Managed Administrative Fee Projects and Standards for the Railway System." In the past, the regulations concerning administrative service charges and standards concerning the railway system were all abolished. In the future, they will be reported in accordance with the Administrative Regulations on Administrative Charges in Shandong Province. In the "Regulations on Railway Fees Collection" attached to the "Notice", regulations are stipulated on the charges for occupying railway land[1]: (1) Units, enterprises, individuals outside the Jinan Railway Bureau system, and collective units within the railway bureau system, Individuals, all state-owned land that has been approved to use the Jinan Railway Bureau system to enjoy legal land use rights in the province shall pay railway land occupation fees to the Jinan Railway Bureau. (B) Occupied railway land charges: 1. Station site. Stations and freight yards of special stations, one station and two stations are 0.35 yuan per square meter per month, stations and cargo yards in stations three and four are RMB 0.3 per square meter per month, and stations below the fourth station are RMB 0.15 per square meter per month. 2. Business land. The number of stations and locations in special railway stations, and other stations is 0.3–0.5 yuan per square meter per month; the stations and freight yards in the fourth-class stations are 0.3 yuan per square meter per month; the stations in the third and fourth stations are 0.20–0.4 yuan per square meter per month; Wait for the following stations and locations to be 0.15 yuan per square meter per month. 3. Residential land. The number of special stations, first and second stations and their locations are RMB 0.20 per square meter per month, and the other stations and locations are RMB 0.12 per square meter per month. 4. Special line land. The property rights of the special railway lines belong to the land occupied by the off-road units, which is 0.15 yuan per square meter per month. 5. Interval. The railway section is 0.08 yuan/m^2 monthly. In the provision of this fee standard, provisions have also been made on the specific circumstances of exemption from the occupation of railway land.

On December 28, 1999, the Liaoning Provincial Price Bureau issued the "Circular on Formulating Rental Standards and Related Issues for Non-Railway Units Using Railway Land to Pay for Use with Compensation" in order to strengthen the territory of Liaoning Province. Management of railway land to prevent the loss of revenue from state-owned land, according to the Provisional Regulations on the Transfer and Use of Urban State-owned Land Use Rights in the People's Republic of China and the State's regulations on the management of railway land, for non-railway units (collective enterprises including railways). The use of the railway land for paid use. The rent standard for the use of railway land borrowed (occupied) in Liaoning Province is: rent for the use of railway land as commercial service land, and rent rates shall be

[1] From January, 2009, the railway land-use charge will be changed from administrative charge to operation and service charge, and the standard remains unchanged.

determined according to the land class and the city's prosperous and non-prosperous areas respectively. The division of the downtown area and the non-prosperous area is determined by the railway land management department in accordance with the local land classification standard, but must be reported to the Provincial Price Bureau and the local price department for the record. For other types of land rents, the land is divided into industrial lands, land occupied by individuals above the third station (including yards), individual land occupied by stations below the fourth class (including yards), special railway lines, and highways (once) Sex collection), etc., respectively using the monthly 0.2 yuan per square meter, 0.1, 0.05, 0.20 yuan and other standards to collect rent.

On June 24, 2002, the "Review on Adjusting the Temporary Occupied Railway Use Fees in Hubei Province" issued by the Hubei Provincial Price Bureau held that the railway is the major artery of the national economy and railway land is the most basic production material for the railway transportation industry. In order to prevent some units and individuals from arbitrarily occupying railway land at the railway station area and along the railway line, stop the digging of ponds, open canals, and planting land at the foot of the roadbed and other serious threats to the national railway traffic and their own personal safety, according to the Land Administration of the People's Republic of China. The relevant provisions of the Law and the State's management of railway land use, in view of the special circumstances of temporary use of land for railways different from land transfer, and lease, have agreed to collect fees for temporary use of railway land in Hubei Province.

On August 22nd, 2012, the Sichuan Provincial Development and Reform Commission and the Ministry of Finance issued the Notice on Transfer of Railway Land Use Fees into Operational Service Charges and Standards. It is believed that with the national and railway departments' management of the allocation of land charges, the requirements are getting higher and higher, and the fees for railway land use are gradually being strengthened and improved. In order to effectively regulate railway land use behavior and further improve railway land use fees, it was decided to convert Chengdu Railway Bureau's railway land use fees into operating charges. The scope for collecting rents for railway land use is that units and individuals outside the Chengdu Railway Bureau system in Sichuan Province that use state-owned land that is legally authorized to use by the Chengdu Railway Bureau are all subject to the rental scope for railway land use rights. The charging standards for land use rent have been stipulated in detail according to different categories: 1. The monthly rent of land for the use of land in the six areas of Chengdu Wuhou, Jinjiang, Jinniu, Qingyang, Chenghua and Gaoxin and its urban built-up areas, operating cargo yards is RMB 3; the monthly rent for other business land per square meter Standard 5 yuan; state-owned large and medium-sized industrial enterprises in the railway freight yard, self-use temporary housing, supporting agricultural materials storage yards and other land per square meter monthly rent standard 1 yuan. 2. The monthly rent of land for railway vehicles in urban and state-level medium-sized cities within the city-constructed area, the land for operating goods yards is RMB 1 per square meter; the monthly rent for other business land is RMB 1.5 per square meter; the enterprises are located at railways The monthly rent standard of RMB

0.80 per square meter for the self-use cargo yard, self-use temporary housing, and supporting agricultural material storage yards. 3. The monthly rent of the railway stations within the county-level small cities and their urban built-up areas is 1 yuan per square meter of land used for operating freight yards, 1.2 yuan per month for other business land, and the company's own freight yard for railways. The monthly rent standard of RMB 0.40 per square meter for land for temporary use of self-use temporary housing and support for agricultural materials and stockyards. 4. Railway stations within towns below the county level and land within urban built-up areas, monthly rent standard of 0.6 yuan per square meter of land used for operating freight yards, and 1.0 yuan per month per square meter for other business land; enterprises set up for railway use The monthly rent standard for land used for goods yards, self-use temporary housing, and supporting agricultural materials yards is RMB 0.4 per square meter. 5. Line section land (excluding the above-mentioned built-up areas), regardless of use monthly rent standard per square meter 0.4 yuan; special line land, monthly rent standard per square meter 0.5–0.8 yuan. 6. Orphanages, welfare homes, nursing homes, maternal and child health stations, epidemic prevention stations, and disabled companies are exempted from land rents.

In addition to the above, most local railways in the country occupy railway land for neighboring units or individuals, and the two parties negotiate to collect the fee for use.

11.2.2 High-Speed Railway and Agricultural Land Use in the Region

High-speed rail can save land resources so as to protect the cultivated land as much as possible while increasing the supply of traffic. The high-speed railway adopts the "Bridge to Road" construction method. Compared with roads, ordinary railways, and air transport, high-speed railways can save certain land resources and protect cultivated land under the same capacity.

The width of common railway subgrade (including drainage ditch) is more than 40 m. The width of the bridge for high-speed railway bridges is only 18 m, with a reduction of more than 55%. With a single project analogy, the railway bridge project saves about 3 hm^2 per kilometre compared with the roadbed project. The Beijing–Tianjin–Hebei high-speed railway has a total length of 120 km, and the length of the bridge accounts for 87% of the total length of the line. This effectively solves the problem of crossroads affecting speed and safety, and saves more than 300 hm^2 of land. Take the Beijing–Shanghai high-speed railway as an example, its total length is 1318 kilometers, and the total length of the bridge accounts for 80% of the entire line. Compared with ordinary railways, it saves about 3000 hm^2 of land. Compared with roads, the area of land occupied by unit traffic is only 1/10 of that of roads. The six-lane expressway in China has a base width of 35 m and a four-lane 28 m, while the Beijing–Shanghai high-speed railway has a base width of 18 m, which is 1/2 of the expressway.

Compared with civil aviation, high-speed railways can be laid in areas where land values are relatively low and residents are relatively sparse. Some road sections can also use the existing railway land without additional occupancy of urban land. Building an airport requires a lot of land. At the same time, the land surrounding the airport will lose its own value due to the impact of strong noise. Airport land has higher requirements on the area and flatness. In general, the airport is far from the city. Highways and other ancillary facilities from the city to the airport need to be built, and the high-speed railway can directly travel between cities, which is convenient and relatively reduces the area. Similarly, taking the Beijing–Shanghai high-speed railway as an example, the land occupied by railways and other facilities along the railway lines covers an area of 4998.8 hm^2. As of the end of June 2012, the Beijing–Shanghai high-speed railway opened a total of approximately 87.72 million passengers during the opening year, with an average of 10,000 people, which covers an area of 0.55 hm^2. The total civil airport and special facilities occupy an area of about 50,000 hm^2. In 2011, the total number of passengers transported by China's civil aviation reached 292 million people, and the average annual passenger transport volume accounted for 1.71 hm^2, which was 3.1 times that of the high-speed railway dedicated to passenger transport. The high-speed railway covers less than 1/3 of the average level of civil aviation.

In general, compared with ordinary railways, highways, and civil aviation, high-speed rail can largely save land resources on the basis of ensuring transportation capacity.

11.2.3 High-Speed Railway and Regional Commercial and Industrial Land Use

The high-speed rail has eased land use pressure in other parts of the city. In most cities in China, economic imbalances have always existed. As far as the entire city is concerned, the economic development in the urban areas and county areas is uneven. In urban areas, the gap between the old urban areas and the new urban areas is different. Performance is the gap between the degree of development and utilization of land and efficiency. In the long run, the imbalance in the use of land resources and the mismatch in distribution will become contradictions in economic and social development.

Because high-speed railways have the characteristics of rapid and large-capacity transportation, higher requirements are placed on the distribution functions of the high-speed railway stations. At the same time, a large number of people will drive the improvement of commercial service facilities in the region. All cities along the route will use the high-speed railway station area as the main external transportation and commercial key area in the future. Near the high-speed railway station, high-speed railway stations, subway stations and other integrated transportation hubs are the core. Within a certain service radius, they radiate outwards in a ring-shaped structure and are divided into core areas, expansion areas, and impact areas; for the entire transportation hub area, In terms of the core area, the largest scope of influence and the most prominent role are. The general area of the core area is limited, with high value-added industries such as commerce, office, and finance as the mainstay, giving full play to the maximum benefits of the land. The operation of high-speed railways has strengthened the polarization of the regional economy. The areas surrounding the high-speed rail stations rely on their convenient traffic conditions and have huge advantages in the development of the regional economy. Therefore, the development intensity of the site areas is generally higher than that of ordinary urban areas. Local governments actively promote the construction of high-speed railways, which will not only help change the urban structure, but also add a considerable amount of financial revenue through land transfer for infrastructure construction, enrich and improve urban functions, and improve the level of public facilities such as living, studying and entertainment.

Cities along the high-speed rail line take advantage of high-speed rail construction to increase land planning and investment in the area where the site is located. It is hoped that the high-speed rail will drive the economic and social development of the area. For example, the Beijing–Shanghai high-speed railway Nanjing South Station is a typical urban–rural junction with weak development foundation and low economic aggregate. With the Beijing–Shanghai high-speed railway Nanjing South Station completed and put into operation, the region has ushered in unprecedented opportunities for development. The government put forward the requirement of "grasping the strategic opportunity to build a railway hub." In recent years, the high-speed rail station city has emerged a surge of "building a station". For example, Wuxi East Railway Station is located in Wuxi New City, which is planned by Wuxi City, and the new city is located in the geometric center of Xishan District. The planned area is 128 square kilometers, and the high-speed railway business area covers an area of 45 square kilometers. The construction of the HSR station complied with the plans of the entire city of East Extension in Wuxi. Changzhou invested to build a new high-speed rail city with a core area of 1.6 square kilometers. Although the area is small, it provides strong support for Changzhou's development goals of "one body and two wings". Suzhou has also planned to develop itself into a high-speed rail new city from a high starting point, which will be an important part of the overall plan for a city with one city and four cores.

Traffic and land use have complex interactions. In the early years of research, scholars concluded that the development and changes in transportation can lead to urban spatial accessibility, land rent curves, land use, and urban form changes. In recent years, scholars have reached a basic consensus on the study of the mutual-feedback relationship between transportation and land use, that is, the evolution of cities can be regarded as the evolution of land use and transportation integration.

11.3 High Speed Railway and Low Carbon Energy Strategy

Low carbonization is an important feature of China's "12th Five-Year" and "13th Five-Year" energy development. Adjusting the energy structure is an important strategy for energy development in the next decade. Its main points are: 1. Optimize the development of the coal industry. 2. Develop natural gas vigorously. 3. Accelerate the construction of large-scale hydropower and nuclear power. 4. Actively develop renewable clean energy. By 2020, China's water conservancy capacity will reach 300 million kilowatts or more, and 300 million tons of standard coal will be saved annually, reducing carbon dioxide emissions by about 750 million tons. By 2020, the installed capacity of nuclear power should reach 70 million kilowatts or more, save 18,000 tons of standard coal annually, and reduce carbon dioxide emissions by about 450 million tons. By 2020, the installed capacity of wind power generation will reach 100 million kilowatts, and the emission reduction of carbon dioxide will be about 160 million tons; the installed capacity of solar power generation will be more than 20 million kilowatts, and 10 million standard coal will be saved each year, and about 20 million tons of carbon dioxide will be reduced. Meanwhile, we will vigorously develop the use of biomass and other renewable energy sources, including the use of straw to produce biomass energy. In the optimization of the development of the coal industry, the proportion of clean coal used has increased by 10%. In the development of natural gas (including coalbed methane and shale gas), in 2030, domestically produced natural gas will be able to reach 300 billion square meters, plus imports up to 4000–500 billion square meters, which will account for 10% of primary energy and become one of the highlights and pillars of China's green energy development strategy. By 2020, the energy consumption per unit of GDP will be reduced by 50% compared with 2005, and will be reduced by 20% in 2015. By 2020, renewable energy consumption will reach 15% of total energy consumption, and 2015 will reach 13%.

The railway sector is one of the basic industrial sectors of China's national economy, and it is also the link between logistics and passenger flow in social and economic activities. It occupies an important place in the transportation sector. Like most countries in the world, the railway sector is also a major energy consumer in China. Among the five modes of transport, civil aviation transport units have the highest energy consumption, which is ten to hundred times that of other modes of transport; followed by road transport; again, railways, pipelines and water transport. From the point of view of energy efficiency alone, more adoption of water transport, pipelines, and rail transportation will undoubtedly have a significant effect on the rapid growth of energy use mitigation. However, the development of the transportation industry is aimed at ensuring and promoting the development of the national economy. With the progress of the times, the demand for transportation is shifted to efficiency and convenience. In these respects, aviation and road transport have certain irreplaceable advantages. Therefore, when determining the division of labor of various modes of transport, the transport system should fully consider the respective technical and economic characteristics of various modes of transport, and provide appropriate preferential policies for energy-saving and environmentally friendly modes of transportation such as railways while meeting the needs of socio-economic development.

In addition to the rational division of labor in various modes of transport, together with the promotion of China's future energy-saving and emission-reduction targets, various transport modes have their own potential for energy-saving and emission reduction. Energy-saving work in the modernization of the railway should follow the country's overall energy strategy and break down energy-saving and emission-reduction targets into various industries and sectors. From the perspective of changes in the energy intensity of railway modernization methods during the "10th Five-Year Plan" and "Eleventh Five-Year Plan" period, railway transportation has gradually increased the ratio of electric locomotives to diesel locomotives through the elimination of steam locomotives in recent years, and the energy intensity of the entire system has declined; With electrification and heavy-duty, there is still potential for further decline in the energy consumption of railway freight units; however, with the increase in high-speed passenger transportation and comfort (for example, installation of air-conditioning and other facilities), the energy consumption of passenger transport units is unlikely to decline.

According to expert analysis, the world's oil and natural gas reserves are statically available for only 40 years and 65 years respectively. Facing the problems of depletion of fossil energy resources, environmental pollution, and climate warming, improving energy efficiency, adjusting energy consumption structure, and developing new energy sources have become one of the most pressing and important issues in the world today. Compared with other modes of transport, railways have a clear comparative advantage in energy conservation. According to the statistics, China's railway uses 9% of the energy consumption of the national transportation industry, completing 32% of the industry's total transportation volume. First, the energy-saving effect of high-speed railways is even more pronounced. The Chinese Railway "Harmony" EMU adopts a low resistance design of the car body and the pantograph, which effectively reduces the air resistance of the train. By adopting lightweight body technology, the weight is lighter than the average railway passenger car by more than 30%, which significantly reduces the train's weight. Significantly reduce the weight of the train, a significant reduction in energy consumption; the use of train regenerative braking technology to improve energy efficiency. EMU trains use electric power to reduce the use of oil and natural gas. Electric energy belongs to clean energy and has little pollution to the environment. According to the technical experts of China South Locomotive Group, when the high-speed EMUs operate at a speed of 300 km per hour, the energy consumption per 100 km per person is only 3.64 kWh, which is equivalent to 1/12 of the passenger plane, 1/8 of the sedan, and 1/1 of the medium-sized bus. 3, The average per capita energy consumption for a trip is about 48 kWh. Assuming that high-speed EMU trains are fully loaded, the average train power consumption for a single high-speed train will be about 48,000 degrees, equivalent to 19.4 tons of standard coal. For example, according to the calculation of the annual turnover of the Beijing–Shanghai high-speed railway, a rough estimate of the Beijing–Shanghai high-speed railway will consume 1.138

billion kWh of electricity for a year, which is equivalent to 460,000 tons of standard coal. According to the 2011 Railway Statistical Bulletin, the national railway energy consumption is equivalent to 17,725,000 tons of standard coal, which shows that the energy consumption of high-speed railways accounts for a small proportion of the total railway consumption. Compared with aircraft, although the per capita energy consumption of passenger airplanes is 12 times that of high-speed railway trains, and the high-speed rail transport is about 10 times that of airplanes, so the energy consumption per kilometer is calculated for high-speed rail and passenger planes. The energy consumption ratio is 1:1.5. EMU trains rely on electric power, and electric energy can be obtained through clean energy such as wind power, nuclear power, and solar energy. Compared to aircraft and automobiles, the dependence on coal and petroleum is reduced.

Secondly, apart from using electric locomotives, high-speed railways can implement the "replacement with electricity" project. Its new station design has realized energy saving and environmental protection due to the adoption of new technologies. For example, the Beijing South Railway Station and the Tianjin Station that have been completed and put into use have all designed a large-area glass dome, and have also been treated with light transmission on all floors to fully utilize natural light illumination. Beijing South Railway Station adopts a thermoelectric cooling triple supply and sewage source heat pump technology, which can achieve energy cascade utilization. The annual power generated by the system can meet 49% of the station's electricity load. Beijing South Railway Station has also adopted solar photovoltaic power generation technology to make full use of solar energy.

In view of the relatively prominent problems in China's energy use and environmental protection, it has become a limiting factor in economic and social development. In order to fundamentally change this situation, the country regards sustainable development as a major strategy. Saving land, energy conservation and emission reduction have become the hard targets for assessing the economic development of various industries in various regions. Railways, especially high-speed rails, have obvious advantages in saving land, saving energy, and protecting the environment. Judging from the social burden of traffic, according to the results of the analysis of 17 European countries commissioned by the International Railway Federation, experts from relevant agencies in Germany and Switzerland, 92% of the external costs (accidents, noise, air pollution, congestion, etc.) caused by traffic. From road traffic, only 1.7% came from rail transit. 77% of external expenses come from passenger transport, while private cars account for 60%, road freight accounts for 21%, motorcycles account for 8%, aerospace passenger transportation accounts for 4.5%, and air cargo accounts for 1.5%. Research by the Organization for Economic Co-operation and Development (OCED) shows that among the carbon monoxide emissions that contribute to the global warming effect, the portion derived from traffic (mainly roads and aviation) already exceeds 1/5, and the proportion is still rising. Therefore, from the point of view of human protection of the living environment, the high-speed railway is more conducive to the sustainable development of the society than the other two modes of transportation, and the society must be in an appropriate manner (such as the internalization of external costs, "who pollutes, who pays"). Direct passenger transport demand to high-speed railways.

In response to the current problem of excessively high resource and environmental costs caused by economic growth, the Party's 17th National Congress of the Communist Party of China (NPC) report clearly set forth the goal of building an ecological civilization for the first time in the new requirements for the goal of building a well-off society in an all-round way. It is of utmost importance to put forward the idea of building an ecological civilization, whether it is to achieve people-centered, all-round, coordinated and sustainable development, or to improve the ecological environment, the quality of people's lives, and achieve the goal of building a well-off society in an all-round way.

The current economic growth model has made China's economic development meet the limit of growth. Mainly reflected in: First, serious environmental pollution. The second is economic growth at the expense of the quality of the ecological environment. According to estimates by the World Bank, China's annual share of air and water pollution alone accounts for more than 8% of GDP. Third, China's natural resources support system has been unable to sustain the economic growth under the current model. The increasingly prominent energy shortages and environmental degradation have become common problems faced by all countries in the world. The transportation industry is a large consumer of energy and one of the major sources of environmental pollution. Transportation is mainly fuel-efficient. Adding noise has become the main source of air pollution. From a worldwide perspective, the transportation system consumes about 1/3 of the world's energy. At present, European road traffic consumes more energy than industry, which accounts for 80% of the total energy consumption, and this figure is still rising. Therefore, the construction of a resource-conserving and environment-friendly society must be placed in a prominent position in the industrialization and modernization development strategy, and priority should be given to the development of energy-saving, light-discharge, and less-area rail transportation methods. This is a common practice of developed countries.

To achieve sustainable development requires the railway to give full play to its advantages. Since the beginning of the twenty-first century, the developed countries, led by the European Union and Japan, have paid close attention to the relationship between transport and environmental development from the perspective of guaranteeing humans to achieve sustainable development, and have internalized external costs of various modes of transport reflected in the government's policies and planning. In 2001, the European Union announced the European White Paper on Transport Policies for 2010, which uses internalization of external costs as a means to ensure fair competition among various modes of transport and expand investment in railway infrastructure. Its core content includes the improvement of railways in various countries, technology and operational interoperability, construction of high-speed railways, large capacity freight corridors, etc. At the same time, as far as possible, investment in externally less costly modes of transportation has been increased, and railway infrastructure is the focus. In July 2005, the European Union announced the "Trans-European Transport Network (TEN-T) Priority Project Development Plan" to determine that by 2020 the network will cover highways

89500 km; railways 94000 km, and railways will have nearly 5000 km more than highways. Among the 30 projects that need to be prioritized for ensuring input, there are 21 railway projects and railway-related projects, including 6 high-speed railway projects, and the rest are ordinary railway projects, heavy-haul railways, multimodal transport, and Galileo satellite communications. This is the first time since the establishment of the EU that the railway project will be placed first in the entire transportation project. The status of the railway in sustainable development is evident.

According to statistics, in 2006 China's average per capita arable land was only 0.0933 hectares (1.4 mu), which was less than 40% of the world average. The central government has adopted the most stringent land policy and determined that it should not exceed the 120 million hectares (1.8 billion mu) red line. The total energy consumption in China has increased year by year and has become a major driver of high energy consumption. From 2002 to 2006, the country's total energy consumption increased from 1.52 billion tons of standard coal to 2.46 billion tons of standard coal, and the total oil consumption increased from 248 million tons to 350 million tons. Among them, the oil consumption of the transportation industry increased from 61.57 million tons to 110 million tons, an increase of nearly 1 time. Estimated by international calibre, China's energy consumption in the transportation industry accounts for about 10% in the country's total consumption. Of particular concern is that due to the use of oil and gas as the main source of transportation, almost all gasoline, 60% diesel, and 80% kerosene are consumed by various modes of transport. This energy consumption model is difficult to remain.

The transport economy theory holds that transportation is an important part of national economic activities. It brings great benefits to the economy and also poses an increasingly serious threat to the ecological environment. The only way out is to achieve sustainable transport. In the twenty-first century, environmental problems have become more and more severe. In the transportation industry, comprehensive analysis has been conducted from the aspects of energy consumption per unit transport volume, occupancy of environmental resources, protection of environmental quality, adaptation to the natural environment, and operational safety. The advantage of the railway is most obvious. Therefore, after experiencing a tortuous road, European developed countries have re-examined and adjusted their transport policies and gradually shifted their focus back to the railways. An important part of their strategy is to plan and develop high-speed railways.

The history of world railway development proves that high-speed railway is an inevitable trend of economic and social development. Since the construction of the world's first railway in the United Kingdom in 1825, railways have become the backbone of transportation in various countries for a long historical period due to the advantages of transportation speed and transportation energy. Since the 1950s, rapid development of highways and air transport has placed the railways at a disadvantage position in terms of speed. Long-distance passenger transport has been crowded out by air transport, and short-distance passenger transport has been replaced by motor transport. After entering the 1970s, due to the energy crisis, environmental deterioration, traffic safety and other issues, people re-recognized the value of the railway. In particular, high-speed railways have adapted to the new needs of modern social economic development due to a series of technological advantages such as high speed, high transport capacity, low energy consumption and light pollution. On October 1, 1964, the Tokyo-Osaka Shinkansen Tokyo-Osaka high-speed railway was operated at 210 km/h, and France built the first high-speed railway (TGV SE line) in 1981, achieving a remarkable economy. Benefits like social benefits are that many countries and regions such as Europe, the Americas, and Asia have built, reconstructed, or planned the construction of high-speed railways. The development of high-speed railways has become a common trend in the development of the world's railways.

High-speed railways use on-board power to drive the wheels, which is an environmentally friendly and energy-saving fast transportation method. Therefore, its development prospects have always been optimistic. In order to improve the railway technology, traditional railways that used to rely on non-renewable energy have been converted to high-speed railways that rely on renewable energy to achieve sustainable development of resources; and because of the change in energy consumption, railways that used to emit polluting tail gas have been converted to low emissions due to the change of energy consumption, the railways that have discharged polluted tail gas in the past have been transformed into low-emission high-speed railways to achieve an environment-friendly development.

First of all, China's high-speed railways should further increase the level of human capital investment, technology research and development and technological innovation, so as to better research and explore new energy sources that are beneficial to the development of high-speed railways, thereby alleviating energy crisis and exhaust emissions, and improving energy efficiency. When the input of human capital and material elements is constant, a greater level of output is obtained. For example, if the power is put into a certain condition, the driving speed is increased; in high-cold areas, high-speed railways are normally driven.

Second, China should continue to support the role of high-speed railways in the sustainable development of the economy. China's development strategy is to take advantage of its latecomer, learn from the successful experiences of foreign countries, and combine China's national conditions to form China's core competitiveness. Therefore, in terms of energy security and environmental pollution, China must also take advantage of its latecomer, avoid the road of first pollution, adhere to the core idea of rejuvenating the country through science and technology, adhere to energy security strategies, develop green economy, and support the development of high-speed railways. It is an inevitable choice for the path of sustainable development.

With the progress of technology in the future, the increase in the efficiency of energy use, the research and development of energy-saving and environmental-friendly materials, and the environmental advantages of high-speed rail will be further exerted. However, with the current level of technology, the increase in speed will inevitably lead to increased energy consumption and pollution, and high-speed railways are no exception. Although the discharge of pollutants from high-speed railways is much less than that of roads and aircrafts, it does not mean that they are simply "zero emissions". For example, the high-speed ballastless track per kilometer consumes 3510 tons of cement and 93 tons of rebar more than conventional railways. Cement and steel are high-energy-consuming and high-emission industries. According to the average level of China's existing production technology, 0.8 tons of carbon dioxide and 1.8 tons of carbon dioxide are emitted for each ton of cement and 1 ton of steel. Therefore, the 1 km ballastless track emits 2975.4 tons of carbon dioxide more than traditional railways. However, seeing that carbon emissions are disposable, they can be allocated to the life cycle of ballastless tracks. At present, the design life of ballastless tracks in China is as long as 100 years, and the construction of 1 km of ballastless tracks equals 29.75 tons of carbon dioxide emissions per year. In terms of the energy conservation and environmental protection of high-speed railways, the construction of high-speed railways and the consumption of electrical energy must be put into full consideration.

11.4 High-Speed Railway and Technology Transfer Strategy

11.4.1 The Export History of China's High-Speed Rail Technology

On April 18, 2007, the national railways implemented the sixth-largest speed increase and new train operation chart. The speed of the speeding section is 200–250 km, which has reached the highest value of the speed increase of the existing railway line in the world. At the same time, the "Harmony" EMU was born. On February 26, 2008, the Ministry of Railways and the Ministry of Science and Technology signed a plan to jointly develop a new generation of high-speed trains operating at a speed of 380 km per hour. On August 1, 2008, China's first high-speed railway and Beijing–Tianjin inter-city railway with complete independent intellectual property rights and world-class standards was put into operation.

In October 2009, when Russian Prime Minister Vladimir Putin visited China to attend the meeting of the heads of government of the Shanghai Cooperation Organization member states, China and Russia signed a memorandum on the development of high-speed railway between China and Russia, and China will help Russia build a high-speed railway. In November 2009, the Chinese Ministry of Railways signed a memorandum of understanding with General Electric of the United States. The two sides pledged to seek cooperation in the high-speed railway projects with a speed of over 350 km per hour in the United States. On July 13, 2010, Chinese companies participated in the bidding for the first high-speed railway in Brazil that was officially constructed. From July 12 to July 15, Argentine President Fernandes signed a number of railway science and technology export contracts worth US$10 billion with China during the visit to China. This is the largest transaction for exporting railway technology in recent years (Kang 2010).[2]

On July 25, 2014, China's first high-speed rail project that was constructed overseas was opened to traffic in Turkey.[3] The second phase of the high-speed railway linking the Turkish capital of Ankara and Istanbul is the first electrified high-speed railway project contracted by Chinese companies overseas. This high-speed rail has a total length of 533 km and a design speed of 250 km per hour. The first phase of the project was built by Spain and Turkey, and the second phase by China and Turkey. The high-speed train runs for three and a half hours, which is more than half the time of conventional railways and kilometers.

From the initial core technology, it all depends on imports. Up to now, it has a number of patents and world-leading technologies. The Chinese railway represented by high-speed rail has achieved remarkable development in the world in the past decade. The former Ministry of Railways set up a coordination group for overseas cooperation projects such as China-U.S., China-Russia, China-Saudi Arabia, China-Myanmar, Sino-Burmese, China-Kyrgyzstan, China-India, etc. to organize the development of overseas railway projects contracting and equipment export markets for different markets. As of the end of 2010, railway projects undertaken by Chinese railway companies in overseas countries have covered more than 50 countries and regions in the world, with a contractual value of US$26 billion. Railway technology and equipment have been exported to more than 30 countries in Asia, Africa, Oceania, and the Americas. In addition, the locomotives and passenger cars produced by CSR have already assumed 90% of passenger traffic and 70% of cargo volume in Turkmenistan, and they have assumed more than 40% of Uzbekistan's passenger traffic and more than 30% of the cargo volume.

11.4.2 The Opportunity and Challenge of High-Speed Rail Technology Transfer in China

With the implementation of Premier Li Keqiang's "high-speed rail diplomacy" strategy, the "going out" of China's high-speed rail has now become a hot topic in China and even the world. At the same time, as resources and environmental constraints become increasingly strict on a global scale, the concept of sustainable development centered on "low carbon and conservation" has become a universal consensus of the international community. All countries in the world have adopted "low carbon and economy" as the fundamental starting point for economic development. Against this background, the development of the global high-speed railway

[2] Development of China's high-speed railway [EB/OL]. China Youth Daily. 2010–08-18. http://money.163.com/10/0818/08/6EBSMLO500253B0H.html.

[3] China Communications and Transportation Association, *China Communications and Transportation Association Newsletter*, 2014.8, No. 3, P. 21.

has ushered in its third wave, whether it is the traditional high-speed railway development powerhouses of Japan, France, Germany, Spain, or the "emerging members" of the United States, Russia, Australia, or other high-speed railways. The high-speed railway reflects a strong interest and proposes a long-term vision for the construction and development of a large-scale high-speed railway. The rapid development of high-speed railways in the world has brought rare opportunities for the development of China's high-speed railways to "go global", and has also provided a broad space for the development of China's high-speed railways to "go global". "China High Speed Rail" follows the technical guideline of "advanced, mature, economical, applicable, and reliable", and aims at the world's most advanced high-speed railway technology from the initial stage of development, through original innovation, integrated innovation and introduction, digestion and absorption, and re-innovation. The complete set of high-speed railway technology integrating design and construction, equipment manufacturing, vehicle control, system integration and operation management has formed a high-speed railway technology system with independent intellectual property rights and the world's advanced level. At present, Chinese enterprises have undertaken railway construction projects in more than 50 countries and regions. Railway equipment and spare parts have also spread to more than 50 countries and regions including Asia, Africa, Oceania and the Americas, including the United States, Russia, Brazil, Saudi Arabia and Turkey. Several dozen countries, including Poland, Venezuela, India, Myanmar, Cambodia, Laos, and Thailand, all hope to strengthen their exchanges and cooperation with China in the high-speed rail and general railway fields. Judging from the current state of development of China's high-speed railways, "China High Speed Rail", which has significant design advantages, integration advantages, technological advantages, construction advantages, and cost advantages, is fully capable of grasping this "going out of the country and going to the world". Unprecedented opportunity to establish and build high-quality, advanced "China High Speed Rail" brand in the world. In addition, after more than 30 years of rapid development through reform and opening up, China's economic strength and overall national strength have been continuously enhanced. It not only has more than US$2 trillion in foreign exchange reserves, but also provides strong financial support for China's high-speed rail expedition overseas, and it is also communicating with foreign countries. Cooperation and cooperation have accumulated rich and valuable experience and laid a good foundation for China's high-speed rails to "go global". In addition, China's high-speed trains will go to the international market with the relatively mature domestic cooperation model of "railroad + finance", which will provide more stable support and guarantee for China's high-speed rails to fully "go global".

At the same time, the issue of international competition faced by China's high-speed rail "going out" is unavoidable. For many years, the world's high-speed railway market has been occupied by companies in a few countries, such as Japan, Germany, and France. The "going out" development of high-speed rail in China will inevitably lead to the adjustment of the inherent pattern of the world's high-speed rail export market. However, in terms of long-term development, with the further expansion of the international high-speed railway scale, the future high-speed rail market "opportunities are more than challenges and cooperation is more than competition", and the "going out" of high-speed rails in China will further enhance the overall technical level of the world's high-speed rails. Through the deepening of international exchanges and cooperation, the "win–win" of the international community can be achieved. The issue of intellectual property rights is the most sensitive issue for China's high-speed rails to "go global". China is not the original country of high-speed railway technology. Many technologies are developed based on the introduction of foreign advanced technology, through original innovation, integrated innovation, and introduction, digestion and absorption, and re-innovation, and have completely independent intellectual property rights. Although the current level of technology has been higher than originality, it has not been recognized by the international community in some respects. Therefore, the development of China's high-speed railway has not been finished, we should look to the long-term, and continue to develop new technologies, new processes and new products of high-speed railways with better technical performance and widely recognized by the international community. At the same time, under appropriate conditions, foreign companies that have relevant intellectual property rights and patented technologies can jointly cooperate to participate in international projects. The highest realm of China's high-speed rail "going out" should be the output of Chinese standards. At present, the standard system of China's high-speed railways is not well-known in the international community. It lacks a systematic and complete foreign language translation. It affects the depth and overall level of China's high-speed rail "going out" to a certain extent, and it also affects the Chinese high-speed rail brand effect. In the future, China should accelerate the internationalization of high-speed rail standards in China and increase the international visibility and recognition of China's high-speed rail standards.

In short, in the wave of new rounds of international competition and cooperation, the "China High Speed Rail" with significant integration advantages, technological advantages, and cost advantages should follow the trend, strengthen cooperation, and fully "go out" while "bringing in". The "going out" of high-speed rails in China is no longer the traditional low-end level of labor exports, primary product output, semi-finished product output, and the output of "OEM" finished products, but rather the full cooperation in a truly high-end field.

11.4.3 Factors and Suggestions for a Technology Transfer Strategy for China's High-Speed Rail

11.4.3.1 Significant Factors to be Considered for Technology Transfer

In general, China's high-speed railways adopt the "going out" strategy and may consider choosing a BOT approach. In accordance with the BOT division of labor, domestic high-speed railway companies will appear as partners (funding parties).

Therefore, we must prepare early, do our homework, understand the situation of the project in a comprehensive, meticulous, and accurate manner. We can put it into practice only after professional evaluation and consultation. Before and after "going out", consider the following factors:

Policy environment and national conditions. First of all, we should understand the attitude of the target countries for the development of the high-speed railway industry and related policies and national conditions. From the perspective of high-speed railways with low carbon, environmental protection, and rapidity, and the huge impact brought about by China's successful construction and operation of high-speed railways, many countries in the world will actively develop railways, especially high-speed railways, in the near future. Therefore, we should currently consider the policies of the relevant countries that have signed high-speed railway cooperation agreements with our countries, in particular whether there are preferential policies for the development of high-speed railways and whether the policies have been implemented, for example, whether there are similar high-speed railways in our country. Long-term development planning, etc.; at the same time, through the Internet, national foreign affairs departments, and foreign exchange agencies, etc., to understand the system, economic development level, language, population, ethnicity, religion, customs, geographical location, topography and geology of the target country, the basic characteristics of climate characteristics, infrastructure levels, and the status of high-speed railway development.

Legal System. After the "going out" of China's high-speed railways, it will directly face an unfamiliar environment. In the course of implementation, cooperation projects will inevitably cause disputes for various reasons. The effective way to resolve disputes, reduce losses, and maximize profits is to fully rely on the target country's judicial system. The cooperating party (funding party) must have a deeper understanding of the target country's judicial system, legal system, and litigation procedures. It should mainly know whether the country is a party to the relevant international convention (treaty), what kind of legal system it is, and whether it has relevant laws and regulations and its content.

Assessment evaluation. Including whether the project can be used in the early stage of cooperation and the use of what kind of cooperation evaluation, as well as the project implementation process of risk assessment and project progress to a certain stage of summary and evaluation. Through scientific and rigorous assessment and evaluation, and propose solutions or emergency plans to ensure the smooth operation of cooperative projects under controllable goals.

Technical standard. The most important thing for China's high-speed railways to implement the strategy of "going out" is to export our technical standards. Only by unifying with our standards can we have the initiative to make full use of our engineering and technological advantages and to utilize the production capacity of our equipment manufacturing industry. It is necessary to understand the status of the technical standards system of the target cooperating countries, and strive to move closer to our technical standards in project cooperation.

Material equipment. Compared with many countries, China's vast land and resources, large quantities of sand, stone, cement, and steel that are needed for the construction of high-speed railways can be easily obtained. The high-speed rail transportation equipment and spare parts are relatively complete in the operational phase; and many national economies need to develop high-speed railways. The development is relatively backward, the land area and related resources are limited. Therefore, it is necessary to conduct a detailed investigation of the supply of materials and equipment for the countries to be cooperated and its neighboring countries, and to use them as close as possible, or in countries (regions) with relatively large number of cooperation projects, and domestic Related companies jointly establish their own material equipment production plant (plant) or transfer reserve warehouse to achieve localized production, and try to avoid long-distance transportation from the domestic some low value-added materials and equipment.

Insurance. The use of BOT cooperation, the project cycle is as low as 4–5 years (simple construction), as many as 20–30 years, such a longtime span, natural disasters and social security, engineering accidents and other natural and man-made disasters are difficult to avoid, to reduce Losses and transfer risks must be in accordance with international practice to participate in related insurance, such as engineering insurance, personal accident insurance and so on. This requires a prior investigation of the development of the insurance industry in the cooperating countries and the international insurance situation, participating in the corresponding insurance, including the insurance premiums into the project cost, and recalculating the profitability of the project.

Management level. During the cooperation period, the high-speed railway project must go through two major stages of project construction and operation and maintenance, especially in the early stages of construction. It is one or more of the roles of the relevant Chinese company as a construction management, survey and design, construction, consulting (supervisory) unit during project construction. Since we are far away from our country, we cannot use all domestic management personnel. Therefore, we should fully understand the project management level (management methods, number of management personnel, etc.) of the project cooperating countries, make full use of local human resources, and use their strengths to carry out project management. In this way, it can offset some of the international biases in the export of labor services in China. It can also help the smooth implementation of the project through the social relations of the local people and create a good social and interpersonal environment for the operation phase.

11.4.3.2 Suggestions

As the operation department of the high-speed railway of the Chinese government, the railway company should continue to carry forward the first style when we introduced international high-speed railway technology, adhere to the strategy of "going out" of the entire high-speed railway, play the leading role of the government, unify the organization, and make unified plans. The unification principle guides and coordinates all high-speed railway companies to hold a group to focus their operations, leading the entire high-speed railway industry to go abroad and go global, and ensure the smooth implementation of China's high-speed railway "going out" strategy.

Whether it is a railway company or a high-speed railway company, we must make sufficient efforts before going out and conduct investigations in a careful and careful manner to find out the relevant situations and learn from the overseas and successful cases of related industries and enterprises. According to the specific situation of the relevant target country, treat differently and select the specific cooperation method suitable for the project to negotiate the high-speed railway project.

All in all, it is critical that high-speed railways and the enterprises, research institutes, universities and colleges concerned take active action and look ahead to speed up the training and reserve of talents, technology, equipment and other necessary conditions for exporting Chinese HSR technology, and form a high-speed railway project cooperation model with Chinese characteristics, which is professional and standardized, making it the trendsetter of high-speed railway development in the world.

Reference

Kang Ping. New Choice of China High-speed Railway Era [EB/OL]. *China Youth Daily*. August 18th, 2010. Retrieved on zqb.cyol.com/content/2010–08/18/content_3379916.htm

Appendix

Chapter 3

See Table 1.

Table 1 Emissions of major air pollutants of the two indicators

Air pollutants (AIR)	(CHN)	(USA)	(OECD)
(Emissions of sulphur oxides (kg/cap)	17.3	49.4	27.8
Emissions per $1000 of GDP (kg/1000 USD GDP)	2.9	1.4	1.1
Annual emissions of nitrogen oxides (kg/cap))	9.3	63.9	34.3
Emissions per $1000 of GDP (kg/1000 USD GDP)	1.7	1.8	1.4
Annual Emissions of carbon dioxide (t./cap))	3.6	19.7	11.1
Emissions per $1000 of GDP (kg/l000 USD GDP)	0.61	0.53	0.44

OECD: Environmental Performance Reviews China [R]. (1999–2004)

X. Lin, *High-Speed Railways and New Structure of Socio-economic Development in China*, Research Series on the Chinese Dream and China's Development Path,
https://doi.org/10.1007/978-981-19-6387-2

See Table 2.

Table 2 Reference coefficient of standard coal converted from various energy sources

Energy resources	Net calorific power	Conversion coefficient of standard coal
Raw coal	20,934 kJ/kg	0.7143 kg standard coal/kg
Washed coal	26,377 kJ/kg	0.9000 kg standard coal/kg
Other coal washing	8374 kJ/kg	0.2850 kg standard coal/kg
Coke	28,470 kJ/kg	0.9714 kg standard coal/kg
Crude oil	41,868 kJ/kg	1.4286 kg standard coal/kg
Fuel oil	41,868 kJ/kg	1.4286 kg standard coal/kg
Petrol	43,124 kJ/kg	1.4714 kg standard coal/kg
Paraffin	43,124 kJ/kg	1.4714 kg standard coal/kg
Diesel	42,705 kJ/kg	1.4571 kg standard coal/kg
Liquefied petroleum gas (LPG)	47,472 kJ/kg	1.7143 kg standard coal/kg
Refinery dry gas	46,055 kJ/kg	1.5714 kg standard coal/kg
Natural gas	35,588 kJ/m^3	12.143 Tons/10,000 m^3
Coke-oven gas	16,746 kJ/m^3	5.714 –6.143 Tons/10,000 m^3
Other gas		3.5701 Tons/10,000 m^3
Heat power		0.03412 Ton/million kilojoules
Electric power		1.229 Ton/10,000 kWh

Source http://nyj.ndrc.gov.cn/

Chapter 4

The regression analysis results of SPSS software are as follows:

The variables entered/removed

Model	The input variable	The variable removed	Method
1	Virtual variable D, GDP per capita, total population[a]		Input

[a]All requested variables have been entered

Model summary

Model	R	R square	Modified R square	Estimated standard error
1	0.999[a]	0.999	0.998	60.61528

[a]Predictors: (constant), dummy variable D, GDP per capita, total population

Anova[b]

Model		Sum of squares	df	Mean square	F	Sig
1	Regressed value	3.345E7	3	1.115E7	3034.998	0.000[a]
	Residual error	47,764.751	13	3674.212		
	Total	3.350E7	16			

[a]Predictors: (constant), dummy variable D, GDP per capita, total population. [b]Dependent variable: Overall passenger turnover

Coefficient[a]

Model		Unstandardized coefficients		Standard coefficient	t	Sig
		β	Standard error	Trial		
1	(Constant)	– 10,552.119	1489.773		– 7.083	0.000
	Total population	1.328	0.166	0.462	7.990	0.000
	GDP per capital	0.003	0.000	0.502	9.589	0.000
	Dummy variable D	131.978	58.936	0.047	2.239	0.043

[a]Dependent variable: Overall passenger turnover

MIX
Papier aus verantwortungsvollen Quellen
Paper from responsible sources
FSC® C105338

If you have any concerns about our products,
you can contact us on
ProductSafety@springernature.com

In case Publisher is established outside the EU,
the EU authorized representative is:
Springer Nature Customer Service Center GmbH
Europaplatz 3, 69115 Heidelberg, Germany

Printed by Libri Plureos GmbH
in Hamburg, Germany